Lecture Notes in Ar[t]   4265

Edited by J. G. Carbonell an

Subseries of Lecture Notes in Computer Science

Ljupčo Todorovski   Nada Lavrač
Klaus P. Jantke (Eds.)

Discovery Science

9th International Conference, DS 2006
Barcelona, Spain, October 7-10, 2006
Proceedings

 Springer

Series Editors

Jaime G. Carbonell, Carnegie Mellon University, Pittsburgh, PA, USA
Jörg Siekmann, University of Saarland, Saarbrücken, Germany

Volume Editors

Ljupčo Todorovski
Nada Lavrač
Jožef Stefan Institute
Department of Knowledge Technologies
Jamova 39, 1000 Ljubljana, Slovenia
E-mail:{ljupco.todorovski, nada.lavrac}@ijs.si

Klaus P. Jantke
Technical University Ilmenau
Institut für Medien- und Kommunikationswissenschaft
PF 10 05 65, 98684 Ilmenau, Germany
E-mail: klaus-peter.jantke@tu-ilmenau.de

Library of Congress Control Number: 2006933944

CR Subject Classification (1998): I.2, H.2.8, H.3, J.1, J.2

LNCS Sublibrary: SL 7 – Artificial Intelligence

ISSN   0302-9743
ISBN-10   3-540-46491-3 Springer Berlin Heidelberg New York
ISBN-13   978-3-540-46491-4 Springer Berlin Heidelberg New York

Springer is a part of Springer Science+Business Media

springer.com

© Springer-Verlag Berlin Heidelberg 2006
Printed in Germany

Typesetting: Camera-ready by author, data conversion by Scientific Publishing Services, Chennai, India
Printed on acid-free paper   SPIN: 11893318   06/3142   5 4 3 2 1 0

Preface

The 9th International Conference on Discovery Science (DS 2006) was held in Barcelona, Spain, on 7–10 October 2006. The conference was collocated with the 17th International Conference on Algorithmic Learning Theory (ALT 2006). The two conferences shared the invited talks.

This LNAI volume, containing the proceedings of the 9th International Conference on Discovery Science, is structured in three parts. The first part contains the papers/abstracts of the invited talks, the second part contains the accepted long papers, and the third part the accepted regular (short) papers. Out of 87 submitted papers, 23 were accepted for publication as long papers, and 18 as regular papers. All the submitted papers were reviewed by two or three referees. In addition to the presentations of accepted papers, the DS 2006 conference program consisted of three invited talks, two tutorials, the collocated ALT 2006 conference and the Pascal Dialogues workshop.

We wish to express our gratitude to

- the authors of submitted papers,
- the program committee and other referees for their thorough and timely paper evaluation,
- DS 2006 invited speakers Carole Goble and Padhraic Smyth, as well as Andrew Ng as joint DS 2006 and ALT 2006 invited speaker,
- invited tutorial speakers Luis Torgo and Michael May,
- the local organization committee chaired by Ricard Gavaldà,
- DS 2006 conference chair Klaus P. Jantke,
- the DS steering committee, chaired by Hiroshi Motoda,
- Andrei Voronkov for the development of EasyChair which provided excellent support in the paper submission, evaluation and proceedings production process,
- Alfred Hofmann of Springer for the co-operation in publishing the proceedings,
- the ALT 2006 PC chairs Phil Long and Frank Stephan, as well as Thomas Zeugman and José L. Balcázar, for the cooperation and coordination of the two conferences, and finally
- we gratefully acknowledge the financial support of the Universitat Politècnica de Catalunya, Idescat — the Statistical Institute of Catalonia (for providing support to tutorial speakers), the PASCAL Network of Excellence (for supporting the Pascal Dialogues), the Spanish Ministry of Science and Education, the Slovenian Ministry of Higher Education, Science, and Technology, and Yahoo! Research for sponsoring the Carl Smith Student Award.

We hope that the week in Barcelona in early October 2006 was a fruitful, challenging and enjoyable scientific and social event.

August 2006   Nada Lavrač and Ljupčo Todorovski

Conference Organization

Conference Chair

Klaus Jantke   Technical University of Ilmenau, Germany

Steering Committee Chair

Hiroshi Motoda   AFOSR/AOARD & Osaka University, Japan

Program Committee Chairs

Nada Lavrač   Jožef Stefan Institute, Slovenia
Ljupčo Todorovski   Jožef Stefan Institute, Slovenia

Program Committee

José Luis Balcázar	Universitat Politècnica de Catalunya, Spain
Michael Berthold	University of Konstanz, Germany
Elisa Bertino	Purdue University, USA
Vincent Corruble	Pierre and Marie Curie University, France
Andreas Dress	Shanghai Institutes for Biological Sciences, China
Sašo Džeroski	Jožef Stefan Institute, Slovenia
Tapio Elomaa	Tampere University of Technology, Finland
João Gama	University of Porto, Portugal
Dragan Gamberger	Rudjer Bošković Institute, Croatia
Gunter Grieser	Technical University of Darmstadt, Germany
Fabrice Guillet	University of Nantes, France
Mohand-Saïd Hacid	Claude Bernard University Lyon 1, France
Udo Hahn	Jena University, Germany
Tu Bao Ho	Japan Advanced Institute of Science and Technology, Japan
Achim Hoffmann	University of New South Wales, Australia
Szymon Jaroszewicz	Szczecin University of Technology, Poland
Kristian Kersting	University of Freiburg, Germany
Ross King	University of Wales, UK
Kevin Korb	Monash University, Australia
Ramamohanarao Kotagiri	University of Melbourne, Australia
Stefan Kramer	Technical University of Munich, Germany
Nicolas Lachiche	Louis Pasteur University, France
Aleksandar Lazarević	University of Minnesota, USA
Jinyan Li	Institute for Infocomm Research, Singapore

Ashesh Mahidadia	University of New South Wales, Australia
Michael May	Fraunhofer Institute for Autonomous Intelligent Systems, Germany
Dunja Mladenić	Jožef Stefan Institute, Slovenia
Igor Mozetič	Jožef Stefan Institute, Slovenia
Ion Muslea	Language Weaver, USA
Lourdes Peña Castillo	University of Toronto, Canada
Bernhard Pfahringer	University of Waikato, New Zealand
Jan Rauch	University of Economics, Prague, Czech Republic
Domenico Saccà	University of Calabria, Italy
Rudy Setiono	National University of Singapore, Singapore
Einoshin Suzuki	Kyushu University, Japan
Masayuki Takeda	Kyushu Univeristy, Japan
Kai Ming Ting	Monash University, Australia
Alfonso Valencia	National Centre for Biotechnology, Spain
Takashi Washio	Osaka University, Japan
Gerhard Widmer	Johannes Kepler University, Austria
Akihiro Yamamoto	Kyoto University, Japan
Mohammed Zaki	Rensselaer Polytechnic Institute, USA
Filip Železný	Czech Technical University, Czech Republic
Djamel A. Zighed	Lumière University Lyon 2, France

Local Organization

Ricard Gavaldà (Chair)	Universitat Politècnica de Catalunya, Spain
José Luis Balcázar	Universitat Politècnica de Catalunya, Spain
Albert Bifet	Universitat Politècnica de Catalunya, Spain
Gemma Casas	Universitat Politècnica de Catalunya, Spain
Jorge Castro	Universitat Politècnica de Catalunya, Spain
Jesús Cerquides	Universitat de Barcelona, Spain
Pedro Delicado	Universitat Politècnica de Catalunya, Spain
Gábor Lugosi	Universitat Pompeu Fabra, Spain
Victor Dalmau	Universitat Pompeu Fabra, Spain

External Reviewers

Table of Contents

I Invited Papers

II Long Papers

III Regular Papers

e-Science and the Semantic Web: A Symbiotic Relationship

Carole Goble[1], Oscar Corcho[1], Pinar Alper[1], and David De Roure[2]

[1] School of Computer Science, University of Manchester,
Manchester M13 9PL, UK
{carole, ocorcho, penpecip}@cs.man.ac.uk
[2] School of Electronics and Computer Science, University of Southampton,
Southampton SO17 1BJ UK
dder@ecs.soton.ac.uk

Abstract. e-Science is scientific investigation performed through distributed global collaborations between scientists and their resources, and the computing infrastructure that enables this. Scientific progress increasingly depends on pooling know-how and results; making connections between ideas, people, and data; and finding and reusing knowledge and resources generated by others in perhaps unintended ways. It is about harvesting and harnessing the "collective intelligence" of the scientific community. The Semantic Web is an extension of the current Web in which information is given well-defined meaning to facilitate sharing and reuse, better enabling computers and people to work in cooperation. Applying the Semantic Web paradigm to e-Science has the potential to bring significant benefits to scientific discovery. We identify the benefits of lightweight and heavyweight approaches, based on our experiences in the Life Sciences.

1 Introduction

The term e-Science is normally used to describe computationally intensive science that is carried out collaboratively in highly distributed network environments [1]. Typically, a feature of such collaborative scientific enterprises is that they require access to very large data collections, very large scale computing resources and high performance visualisation back to the individual user scientists. Brain neuroscientists remotely control and collect data from the world's largest and most powerful transmission electron microscope in Japan (telescience.ucsd.edu). Astronomers steer telescopes from their offices, collect the data using remote archive repositories, and process it by exploiting the availability of machines of other institutions. The International Virtual Observatory Alliance (www.ivoa.net) makes available vast digital sky archives to all astronomers not just a lucky few. ^{my}Grid-Taverna (www.mygrid.org.uk) allows biologists to assemble personalised exploratory *in silico* experiments that interoperate between remotely and locally available applications, code-bases and databases to identify new genes [2]. The Human Genome effort is an

N. Lavrač, L. Todorovski, and K.P. Jantke (Eds.): DS 2006, LNAI 4265, pp. 1–12, 2006.

example of e-Science – over 500 datasets and tools are available on the web for bioinformaticians to piece together our understanding of life [3, 4].

The e-Science infrastructure supports and enhances the scientific process by enabling scientists to generate, analyse, share and discuss their insights, experiments and results in a more effective manner, particularly in the context of the deluge of data resulting from new experimental practices [5, 6].

Scientific progress increasingly depends on pooling resources, know-how and results; making connections between ideas, people, and data; and finding and interpreting knowledge generated by others, in ways that may not have been anticipated when it was created. It is about harvesting and harnessing the "collective intelligence" of the scientific community. It has as much to do with intelligent information management as with sharing scarce resources like large scale computing power or expensive instrumentation. It is about making connections between decoupled resources and people in the broadest context of diverse scientific activity, outside the bounds of localised experiments and closed projects, and enabling scientific endeavour "in the wild".

The Semantic Web is defined as an extension of the current Web in which information is given well-defined meaning, better enabling computers and people to work in cooperation [7]. Applying Semantic Web to Science [8] has attracted great interest particularly in the Life Sciences [9-13] which has been proposed as a "nursery" for incubating the required technological developments [14]. We take a perspective in which the Semantic Web is seen foremost as an infrastructure for gathering and exploiting collective intelligence; i.e. the capacity of human communities to evolve towards higher order complexity and integration through collaboration and innovation.

Section 2 introduces the Semantic Web and distinguishes between lightweight and heavyweight approaches. In Sections 3 to 5 we present three aspects of the Semantic Web – annotation, integration and inference – and sketch how the methods, techniques and tools used for each of them could provide benefits to scientists. In Section 6 we reflect on the symbiosis between e-Science and the Semantic Web.

2 The Semantic Web

Annotation is the process of associating metadata with an object. Metadata, defined as structured data about an object that supports functions associated with it, can be generated for any entity for which contextual data can be recorded [15]. We can annotate any object, be it a document, dataset, publication, codes, notebooks, and so on, within the scientific process – even a person or scientific instrument. Metadata can be expressed in a wide range of languages (from natural to formal ones) and with a wide range of vocabularies (from simple ones, based on a set of agreed keywords, to complex ones, with agreed taxonomies and formal axioms). It can be available in different formats: electronically or even physically (written down in the margins of a textbook). It can be created and maintained using different types of tools (from text editors to metadata generation tools), either manually or automatically.

The foundational pillars of the Semantic Web are the tagging of entities with machine-processable metadata which asserts facts using terms, and the associated languages which define these terms and their relationships [16]. These languages extend existing markup languages like HTML or XML in order to represent knowledge – they enable *Semantic Annotation*. The publishing and sharing of the annotations and languages is crucial to realising the benefits of this approach.

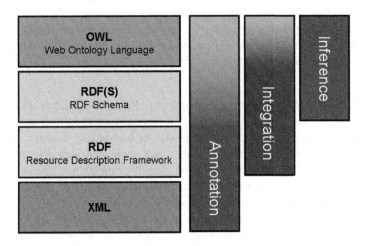

Fig. 1. Markup languages and the Semantic Web of Annotation, Integration and Inference

Figure 1 shows the mark-up languages that have been developed in recent years. Each language is based on the ones that are under it (i.e., OWL is based on RDF Schema, RDF and XML). The right-hand part of the figure shows that this stack of languages can be used for different purposes, distinguishing three functions that the Semantic Web can provide:

-    **Annotation.** The Semantic Web can be used to provide machine-processable descriptions of resources in any of the languages in the stack, asserted using RDF statements. Anything can be annotated, and annotations can be shared.
-    **Data Integration.** The Semantic Web can be used as a means of integrating information from diverse sources, connecting through shared terms (in RDF(S) or OWL) and shared instances (in RDF) and preserving context and provenance.
-    **Inference.** The Semantic Web can be used as a powerful tool to infer new knowledge or detect inconsistencies in the knowledge already described in it, provided in RDF Schema and OWL. Each of these languages has different expressiveness, and different reasoning mechanisms that can be applied to its descriptions, so the inferences that can be achieved are different in each case.

These three different but complementary views of the Semantic Web are discussed in the following sections in the context of e-Science.

3 The Semantic Web as the Annotation Web

3.1 Annotation in e-Science

The annotations associated with Web resources form their own overlaid and intertwingled Web – the Annotation Web. This is potentially a powerful tool which enables e-Science to be carried out collaboratively in highly distributed network environments. Not only does it provide a means for describing the resources being dealt with by e-Scientists (new findings, experiment results and provenance, etc.), but it also allows content and people to be connected, hence allowing the harvesting and harnessing of the collective intelligence of the scientific community. Let us see some examples:

- The ^{my}Grid-Taverna workflow environment [17] is a significant example of a new platform for scientific discovery [2]. Services are described according to a service ontology that specifies the model to be used to describe the service inputs and outputs. Following this approach, annotated services can easily be reused for the design and execution of scientific workflows. The annotation of services is done collectively using annotation tools like Pedro [18] and maintained by a curator in order to ensure their quality.
- The CombeChem project [19] (www.combechem.org) focuses on the notion of "publication at source", capturing comprehensive annotations in order to facilitate interpretation and reuse of results. Using RDF to interlink information in multiple datastores, both internally and externally, CombeChem has established a *Semantic DataGrid* containing tens of millions of RDF triples [20]. The annotation commences in the laboratory [21]. The project also captured scientific discourse as part of the provenance record, through provision of tools to annotate meetings [22].
- The Friend-Of-A-Friend (FOAF) initiative (www.foaf.org) is being adopted for science and scientific publications. Scientific FOAF (www.urbigene.com/foaf/) is an example of how so-called "Web 2.0 technologies" [15] can be used to relate content and people in order to improve e-Science. For each article in the NCBI Pubmed bibliographic database (www.ncbi.nlm.nih.gov/entrez/), users are asked if a specific person is unambiguously one of the authors and whether they know any of the co-authors (which may not be necessarily the case when there are a large number of collaborators). Authors' interests are defined using the MeSH (Medical Subject Headings) terms of their papers. All this information is used to generate the FOAF profile of scientists.
- Recording the provenance of scientific results is important in facilitating interpretation and reuse of data, as well as for regulatory purposes. Semantic Web technologies are used to provide a flexible and extensible record of the experimental process in the *in silico* experimentation of ^{my}Grid [23] and commencing in the laboratory in CombeChem [24].

3.2 Lightweight and Heavyweight Annotation

We can make a distinction between what we call *lightweight* and *heavyweight* annotation, by considering the vocabularies used to create metadata. Both forms of annotation are relevant to scientific tasks such as those above but they have significantly different characteristics. Our classification is analogous to that of lightweight and heavyweight ontologies, introduced by Studer [25].

Lightweight annotation

We define lightweight annotation as the process of associating metadata with a resource, where the metadata does not necessarily refer to an existing ontology but consists of tags defined by the person(s) in charge of the annotation, which express the meaning of the information source by using terms instead of logical expressions. Furthermore, if the annotation is based on existing ontologies, it normally identifies instances of the ontology concepts, but not the relationships between those instances.

For example, should we have a Web document that describes a gene and the processes in which the gene is involved, a lightweight annotation would consist of tags assigned to different parts of the document. These tags may be: "Gene", "BRCA2", "Breast Cancer", "disease", etc. The CombeChem and SciFOAF examples above also illustrate lightweight annotation.

The advantage of this type of annotation is that it eases the collection of the most important terms and relations of a community, and can be a good starting point to achieve agreement in a specific area. Lightweight annotation has appeared in the context of the Web 2.0 initiative, which refers to a second generation of tools and services on the Web that lets people collaborate and share information online [26]. While lightweight annotation does not focus on relating annotations to existing common vocabularies, vocabularies can be created out of them in what we call *folksonomies* [27].

Heavyweight annotation

We define heavyweight annotation as the process of associating metadata with a resource, where the metadata refers to an existing ontology that is implemented in a formal language (e.g., RDF Schema, OWL, etc.); and the annotation does not only identify instances of ontology concepts but also relationships between those instances, which are compliant with the underlying ontology.

For example, should we have a Web document that describes a gene and the processes in which the gene is involved, a heavyweight annotation of this document would comprise metadata based on an ontology about genes, such as the Gene Ontology (www.geneontology.org). The metadata consists of instances of concepts that are defined in the Gene Ontology, plus instances of relations between those concepts, including genes, the processes where they are involved, etc. The myGrid-Taverna service discovery framework uses heavyweight annotation [5].

Heavyweight annotations normally give precise information about the contents of a document. This information is not necessarily exhaustive and is normally costly, since it is usually authored by domain experts. The ontologies in which those annotations are based are normally shared in a community.

3.3 Tools for Annotation

Tools for creating more lightweight annotations are based on the Web 2.0 philosophy and emerging tools. SemanticMediaWiki allows adding semantic information to a Web document while creating it; the semantic information added consists of typed links and typed attributes (wiki.ontoworld.org/index.php/Semantic_MediaWiki). del.icio.us (http://del.icio.us/) is a social tagging tool that allows annotating Web resources by adding tags that are not specifically connected to an existing ontology, but that can be used to derive folksonomies. Flickr (www.flickr.com) is a social tagging tool that allows annotating images in a similar way to del.icio.us.

Tools for creating heavyweight annotations (ontology-based annotators) are primarily designed to allow inserting and maintaining ontology-based mark-ups in Web documents [28] [29]. First conceived as tools that could be used to alleviate the burden of including ontology-based annotations manually into Web resources, many annotators have evolved into more complete environments that use Information Extraction, Machine Learning and Natural Language techniques to propose semi-automatic annotations for any type of Web resources. In general, the more automated the annotation process is, the lower the quality of the annotations obtained.

Other tools are aimed at annotating data available in data sets by establishing mappings between the data set model and a set of terms coming from ontologies [30]. This annotation improves data discovery and integration, among others.

4   The Semantic Web as the Integration Web

Scientific discovery involves bringing together information from diverse sources, many of which may be available in the form of databases rather than Web pages. The set of these Web resources, also known as the "Deep Web" or ""Data Web", represents more than 80% of the total amount of data available on the Web [31].

Many of these databases contain overlapping or complementary information about the same individuals. For instance, the same protein might appear in UniProt and PDB, or two proteins might share the same Gene Ontology code as part of their annotation. One of the main challenges for the *Integration Web* is how to align the data that is available in heterogeneous distributed databases so that the appropriate connections are made among the pieces of information described in different disconnected databases.

Semantic technologies can be applied to this vital task of integrating information from multiple sources [13] [32]. They can describe the content of a set of databases according to a shared model, or according to a set of semantic models that can be aligned using ontology alignment or merging techniques (www. ontologymatching. org). Using these approaches we can leverage the information available in the Deep Web to information whose meaning is clearly described according to a common understood and/or agreed model of the domain.

In e-Science the information available in documents (e.g., journal publications) represents only a small percentage of the total information available. According to [33], "traditional biological journals will become just one part of various biological data resources as the scientific knowledge in published papers is stored and used more like a database". One of the examples that support this vision is that of reading a description of an active site of biological molecule in a paper and being able to access immediately the atomic coordinates specifically for that active site, and then using a tool to explore the intricate set of hydrogen-bonding interactions described in the paper. A similar illustration is provided by the "Crystal EPrints" interface in CombeChem [20].

There are many examples in the literature that show how heterogeneous data sources can be integrated, supporting users by providing a common view of them and shortening the time needed to find and relate information. Here are two approaches: a data web and a wikipedia.

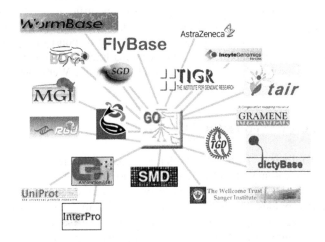

Fig. 2. Resources cross-linked through using the Gene Ontology

A **Data Web** approach encourages databases to export their data self-described in RDF, breaking down the barriers of different schemas. If the same instance – a protein, say – appears in many databases we can integrate the various RDF descriptions. The data entries are commonly annotated within the databases using controlled terms such as GO ids from the Gene Ontology (Fig 2). These "within record" annotations are exposed on the Data Web, and we can use shared terms to link between different database entries, and between the database records and other resources, such as documents, that are annotated using the Gene Ontology [34]. YeastHub is an RDF data warehouse that integrates different types of yeast genome data provided by different resources in different formats [35]. BioDASH takes advantage of UniProt's RDF export facility and the BioPAX ontology to associate disease, compounds, drug progression stages, molecular biology, and pathway knowledge for a team of users. CombeChem used RDF to annotate entities but also to hold chemical data in RDF format [36], recognising the flexibility of this approach. The Collaboratory for Multi-scale Chemical Science (CMCS) [37] has developed an open source toolkit addressing provenance tracking and lightweight federation of data and application resources into cross-scale information flows, and provides another case study in the chemical data arena.

A **Wikipedia** is an example of how Web 2.0 technologies can be used to support information collection in e-Science. The Gene Wikipedia is proposed as a comprehensive knowledge collection about genes, gathering information from multiple data sources: GenBank, UniProt, Bind, Kegg, etc (www.bioinformatics. org/ genewiki/wiki/). It is based on a common set of ontologies: NCBI taxonomy, Gene Symbol, Disease/Phenotype ontology, Protein Interaction databases, GenBank, Pathway and Gene Ontology allowing searches on the information based on these resources. The first wiki pages, and the curation of the available data, is based on the use of text mining tools, ensuring data quality and assurance.

The whole notion of "self-describing experiments" is taking hold as part of the W3C's Health Care and Life Sciences Interest Group's exploration of RDF, and initiatives such as King's EXPO ontology of scientific experiments [38] and collection standards expressed in RDF [39]. Calls to the scientific publishers to annotate papers at publication with ontologies [40] and responses by publishers such as Nature to embrace collaborative tagging systems like Connotea (www. connotea. org) hold out the promise that we could build Integration Webs between data and publications.

5 The Semantic Web as the Inference Web

BioPax (the Biological Pathways Exchange) makes inferences over biological pathway data; an automated OWL reasoner is used to find overlapping and non-overlapping molecular interactions between two pathway datasets [41]. The ^{my}Grid-Taverna project provides the ability to search over scientist-centric descriptions in a subject specific way, using taxonomy information in the associated ontology [42]. These are both examples of inference: metadata that uses terms from OWL ontologies can be reasoned over using logic-based decision procedures, which means that new relationships can be inferred between statements that had not been explicitly made before.

There are several languages in the Semantic Web, with different expressivity and complexity. Depending on the formal language selected, different types of reasoning functions will be available. If we use RDF Schema to describe our ontologies (which means that we will use RDF to describe our annotations), we will be able to execute queries over the models created, using RDF query languages like SPARQL. We can do this to retrieve annotations, as well as during the data integration process once the data source to be queried has been identified and selected. The query engines that process these languages are able to perform basic taxonomic reasoning and consistency checking regarding the domain and range of properties. A more expressive language like OWL, based on the description logic formalism, enables us to perform more complex reasoning processes. For instance, we will be able to detect inconsistencies in the vocabulary that we are creating during the modelling process, such as finding missing terms in the Gene Ontology [43], and we will be able to derive concept taxonomies automatically from the descriptions that we have provided. Furthermore, we will be able to infer the concepts to which an individual belongs given the information that we have from it, or we will be able to detect whether an individual description is inconsistent due to the constraints expressed in the ontology, for example finding new properties of proteins leading to possible drug discovery targets [39]. The complexity of modelling in OWL, however, should not be underestimated [44, 45].

Other logical formalisms, such as rules, allow inferring conditional relationships between individuals that were not possible with the previous formalisms – for example correspondences between data are not necessarily obvious to detect, requiring specific rules [46].

6 Symbiosis

The practice of science and e-Science can be enhanced by the technologies of the Semantic Web, enabling a dramatic reduction in the time needed to create new results and consequently the so-called "time to insight" and discovery [47]. Semantic Web-based applications promise help in: the development of controlled vocabularies, flexible metadata modelling, intelligent searching and document discovery, social and knowledge networking, advanced content syndication and publishing, data integration, aggregation and cross linking, application interoperability and knowledge mining [48]. A heavyweight approach, ("pain today") can be combined with a lightweight approach ("pain tomorrow") to facilitate community participation. Sometimes a little semantics goes a long way [49].

Significantly we have taken a holistic view of the scientific process. Rather than looking at specific, self-contained projects, we observe that scientific discovery occurs "in the wild" using diverse resources. The Semantic Web permits this decoupling between the people, the content of information sources, the metadata about the content, and the time when this content and metadata is created. The person who creates the data is not necessarily the one who digests it. Furthermore, data and metadata can be used for objectives that were not expected when they were created, and at any point in time since its creation.

The natural match between Semantic Web and Life Sciences has been recognised by W3C in their focus on Life Sciences [50]. The Semantic Web benefits from e-Science, since its emerging technologies are being tested on a large-scale thanks to their use in e-Science applications [14] and the piloting of Semantic Webs for Life Science such as Sealife (www.biotec.tu-dresden.de/sealife/) and SWAN [51]. Realising the benefits of bringing Semantic Web and Science together depends on establishing awareness and understanding across those communities [52].

In fact the Semantic Web technologies also have a role inside the e-Science infrastructure. The Semantic Grid [53] is characterised as an open system in which people, services and computational resources (all owned by different stakeholders) are brought together flexibly and dynamically. While Semantic Web technologies can be used *on* the Grid, the Semantic Grid emphasises applying the Semantic Web technologies *within* the e-Science infrastructure – the machine processable descriptions then enable the services and resources to be brought together automatically [54].

We suggest that there is in fact a deeper mutual benefit for e-Science and the Semantic Web. The Semantic Web will thrive in an environment where the annotations – heavyweight and lightweight – are produced and consumed through the routine use of the infrastructure. To reach that point there needs to be a mechanism for bootstrapping the semantic infrastructure, and this will occur where there is incentive and a low cost to entry. We have described the incentives which lead to new scientific discovery. The low entry cost emphasises the need for tools for e-Scientists and also highlights the value of the lightweight tagging approach to move the endeavour forward. The Web flourished through ease of use and incentive to publish as well as to consume content, and e-Science appears to provide similar circumstances for the Semantic Web.

Acknowledgements

The authors express their thanks to their many colleagues in the e-Science programme and beyond. We acknowledge the support of the EPSRC through the ^myGrid (GR/R67743/01, EP/C536444/1) and CombeChem e-Science projects (GR/R67729/01, EP/C008863/1). This work is also supported by the EU FP6 OntoGrid project (STREP 511513) funded under Grid-based Systems for solving complex problems, and by the Marie Curie fellowship RSSGRID (FP6-2002-Mobility-5-006668).

References

1. Hey, T. and A.E. Trefethen, *Cyberinfrastructure for e-Science.* Science, 2005. **308**(5723): p. 817-821.
2. Stevens, R., et al. *Exploring Williams-Beuren Syndrome Using myGrid.* in *12th International Conference on Intelligent Systems in Molecular Biology.* 2004. Glasgow, UK: Bioinformatics
3. NAR1, *Web Server issue.* Nucleic Acids Research 2006. **34**.
4. NAR2, *Database Issue* Nucleic Acids Research 2006. **34**.
5. Hey, A.J.G. and A.E. Trefethen, *The Data Deluge: An e-Science Perspective*, in *Grid Computing - Making the Global Infrastructure a Reality*, F. Berman, G.C. Fox, and A.J.G. Hey, Editors. 2003, Wiley and Sons. p. 809-824.
6. Szalay, A. and J. Gray, *2020 Computing: Science in an exponential world.* Nature, 2006. **440**: p. 413-414.
7. Berners-Lee, T., J. Hendler, and O. Lassila, *The Semantic Web.* Scientific American, 2001. **284**(5): p. 34-43.
8. Hendler, J., *Science and the Semantic Web.* Science 2003. **299**: p. 520-521.
9. Baker, C.J.O. and K.-H. Cheung, eds. *Semantic Web: Revolutionizing Knowledge Discovery in the Life Sciences.* 2006, In Press.
10. Schroeder, M. and E. Neumann, *Special Issue on Semantic Web in Life Sciences.* Journal of Web Semantics, 2006. **4**(3).
11. Stevens, R., O. Bodenreider, and Y.A. Lussier. *Semantic Webs for Life Sciences Session Introduction.* in *Pacific Symposium on Biocomputing.* 2006.
12. Kazic, T. *Putting Semantics into the Semantic Web: How Well Can It Capture Biology?* in *Pacific Symposium on Biocomputing.* 2006.
13. Neumann, E., *A Life Science Semantic Web: Are We There Yet?* Sci. STKE, 2005(283).
14. Goble, C. *Using the Semantic Web for e-Science: inspiration, incubation, irritation.* in *4th International Semantic Web Conference (ISWC2005).* 2005. Galway, Ireland: Springer.
15. Greenberg, J., *Metadata and the World Wide Web.* The encyclopedia of library and information science, 2002. **72**: p. 244-261.

16. Antoniou, G. and F. van Harmelen, *A Semantic Web Primer*. 2004: MIT Press

17. Oinn, T., et al., *Taverna: A tool for the composition and enactment of bioinformatics workflows*. Bioinformatics Journal, 2004. **20**(17): p. 3045-3054.

18. Garwood, K., et al. *Pedro ontology services: A framework for rapid ontology markup*. in *2nd European Semantic Web Conference*. 2005. Heraklion, Greece: Springer-Verlag.

19. Frey, J.G., D. De Roure, and L.A. Carr. *Publication At Source: Scientific Communication from a Publication Web to a Data Grid*. in *Euroweb 2002 Conference, The Web and the GRID: from e-science to e-business*. 2002. Oxford, UK: BCS.

20. Taylor, K., et al. *A Semantic Datagrid for Combinatorial Chemistry*. in *6th IEEE/ACM International Workshop on Grid Computing*. 2005. Seattle.

21. Hughes, G., et al., *The Semantic Smart Laboratory: A system for supporting the chemical e-Scientist*. Organic & Biomolecular Chemistry., 2004. **2**(22): p. 3284-3293.

22. Bachler, M., et al. *Collaborative Tools in the Semantic Grid*. in *GGF11 - The Eleventh Global Grid Forum*. 2004. Honolulu, Hawaii, USA: Global Grid Forum.

23. Zhao, J., et al. *Using Semantic Web Technologies for Representing e-Science Provenance*. in *3rd International Semantic Web Conference ISWC2004*. 2004. Hiroshima, Japan: Springer.

24. Frey, J., et al. *CombeChem: A Case Study in Provenance and Annotation using the Semantic Web*. in *International Provenance and Annotation Workshop (IPAW'06)*. 2006. Chicago, USA: Springer.

25. Corcho, O., M. Fernández-López, and A. Gómez-Pérez, *Methodologies, tools andlanguages for building ontologies. Where is their meeting point?* Data & Knowledge Engineering, 2003. **46**(1): p. 41-64.

26. O'Reilly, T. *What Is Web 2.0*. 2005 [cited July 2006]; Available from: www. oreillynet. com/go/web2.

27. Gruber, T. *Ontology of Folksonomies: A Mash-up of Apples and Oranges*. 2005 [cited July 2006]; Available from: http://tomgruber.org/writing/ontology-of-folksonomy.htm.

28. Corcho, O., *Ontology-based document annotation: trends and open research problems*. International Journal of Metadata, Semantics and Ontologies, 2006. **1**(1): p. 47-57.

29. Uren, V., et al., *Semantic Annotation for Knowledge Management: Requirements and a Survery of the State of the Art*. Journal of Web Semantics, 2006. **4**(1).

30. Bizer, C. *D2R MAP – A DB to RDF Mapping Language*. in *12th International World Wide Web Conference (WWW2003)*. 2003. Budapest, Hungary.

31. Bergman, M.K., *The deep Web: surfacing hidden information*. The Journal of Electronic Publishing, 2001. **7**(1).

32. Stephens, S., et al., *Aggregation of bioinformatics data using Semantic Web technology*. Journal of Web Semantics, 2006. **4**(3).

33. Bourne, P., *Will a Biological Database Be Different from a Biological Journal?* PLoS Comput Biol, 2005. **1**(3): p. e34.

34. Bechhofer, S., R. Stevens, and P. Lord. *Ontology Driven Dynamic Linking of Biology Resources*. in *Pacific Symposium on Biocomputing*. 2005. Hawaii.

35. Cheung, K.H., et al., *YeastHub: a semantic web use case for integrating data in the life sciences domain*. Bioinformatics 2005. **21**(1: i85-i96).

36. Taylor, K.R., et al., *Bringing Chemical Data onto the Semantic Web*. J. Chem. Inf. Model., 2006. **46**: p. 939-952

37. Myers, J.D., et al. *A Collaborative Informatics Infrastructure for Multi-scale Science*. in *Second International Workshop on Challenges of Large Applications in Distributed Environments*. 2004. Honolulu, Hawaii.

38. Newscientist.com news service and *Translator lets computers "understand" experiments*. 2006 [cited July 2006]; Available from: http://www.newscientist.com/article/dn9288-translator-lets-computers-understand-experiments-.html.

39. Wang, X., R. Gorlitsky, and J.S. Almeida, *From XML to RDF: How Semantic Web Technologies Will Change the Design of 'Omic' Standards.* Nature Biotechnology, 2005. **23**(9): p. 1099-103.
40. Blake, J., *Bio-ontologies—fast and furious.* Nature Biotechnology 2004. **22**: p. 773-774.
41. Cheung, K.-H., et al., *A semantic web approach to biological pathway data reasoning and integration.* Journal of Web Semantics, 2006. **4**(3).
42. Lord, P., et al. *Feta: A light-weight architecture for user oriented semantic service discovery.* in *2nd European Semantic Web Conference.* 2005. Heraklion, Greece: Springer-Verlag.
43. Wroe, C.J., et al. *Methodology To Migrate The Gene Ontology To A Description Logic Environment Using DAML+OIL.* in *8th Pacific Symposium on Biocomputing (PSB).* 2003. Hawaii.
44. Stevens, R., et al., *Managing OWL's Limitations in Modelling Biomedical Knowledge.* Intl Journal of Human Computer Systems, Accepted for publication.
45. Zhang, S., O. Bodenreider, and C. Golbreich. *Experience in Reasoning with the Foundational Model of Anatomy in OWL DL.* in *Pacific Symposium on Biocomputing.* 2006.
46. Neumann, E.K. and D. Quan. *Biodash: A Semantic Web Dashboard for Drug Development.* in *Pacific Symposium on Biocomputing.* 2006.
47. Gates, W. *Opening Keynote.* in *SuperComputing 2005 (SC05).* 2005. Seattle, Washington.
48. Goble, C., R. Stevens, and S. Bechhofer, *The Semantic Web and Knowledge Grids.* Drug Discovery Today: Technologies, 2005. **2**(3): p. 193-302.
49. Wolstencroft, K., et al. *A little semantics goes a long way in Biology.* in *4th International Semantic Web Conference (ISWC2005).* 2005. Galway, Ireland.
50. W3C, *Workshop on Semantic Web for Life Sciences.* 2004: Cambridge, Massachusetts USA.
51. Gao, Y., et al., *SWAN: A distributed knowledge infrastructure for Alzheimer disease research.* Journal of Web Semantics, 2006. **4**(3).
52. Goble, C.A. and D. De Roure. *The Semantic Grid: Building Bridges and Busting Myths.* in *16th European Conference in Artificial Intelligence (ECAI 2004).* 2004. Valencia, Spain.
53. De Roure, D., N.R. Jennings, and N.R. Shadbolt, *The Semantic Grid: Past, Present, and Future.* Proceedings of the IEEE, 2005. **93**(3): p. 669-681.
54. Goble, C.A., et al., *Enhancing Services and Applications with Knowledge and Semantics,* in *The Grid 2: Blueprint for a New Computing Infrastructure,* I. Foster and C. Kesselman, Editors. 2004, Morgan-Kaufmann. p. 431-458.

Data-Driven Discovery
Using Probabilistic Hidden Variable Models

Padhraic Smyth

Department of Computer Science, University of California, Irvine, USA
smyth@ics.uci.edu

Abstract. Generative probabilistic models have proven to be a very useful framework for machine learning from scientific data. Key ideas that underlie the generative approach include (a) representing complex stochastic phenomena using the structured language of graphical models, (b) using latent (hidden) variables to make inferences about unobserved phenomena, and (c) leveraging Bayesian ideas for learning and prediction. This talk will begin with a brief review of learning from data with hidden variables and then discuss some exciting recent work in this area that has direct application to a broad range of scientific problems. A number of different scientific data sets will be used as examples to illustrate the application of these ideas in probabilistic learning, such as time-course microarray expression data, functional magnetic resonance imaging (fMRI) data of the human brain, text documents from the biomedical literature, and sets of cyclone trajectories.

N. Lavrač, L. Todorovski, and K.P. Jantke (Eds.): DS 2006, LNAI 4265, p. 13, 2006.

Reinforcement Learning and Apprenticeship Learning for Robotic Control*

Andrew Ng**

Computer Science Department, Stanford University, Stanford, U.S.A.
`ang@cs.stanford.edu`

Abstract. Many control problems, such as autonomous helicopter flight, legged robot locomotion, and autonomous driving are difficult because (i) It is hard to write down, in closed form, a formal specification of the control task (for example, what is the cost function for "driving well"?), (ii) It is difficult to learn good models of the robot's dynamics, and (iii) It is expensive to find closed-loop controllers for high dimensional, highly stochastic domains. Using apprenticeship learning—in which we learn from a human demonstration of a task—as a unifying theme, I will present formal results showing how many control problems can be efficiently addressed given access to a demonstration. In presenting these ideas, I will also draw from a number of case studies, including applications in autonomous helicopter flight, quadruped obstacle negotiation, snake robot locomotion, and high-speed off-road navigation.

Finally, I will also describe the application of these ideas to the STAIR (STanford AI Robot) project, which has the long term goal of integrating methods from all major areas of AI—including spoken dialog/NLP, manipulation, vision, navigation, and planning—to build a general-purpose, "intelligent" home/office robotic assistant.

---

* The full version of this paper is published in the Proceedings of the 17th International Conference on Algorithmic Learning Theory Lecture Notes in Artificial Intelligence Vol. 4264.

** Joint work with Pieter Abbeel, Adam Coates, Ashutosh Saxena, Jeremy Kolter, Honglak Lee, Yirong Shen, Justin Driemeyer, Justin Kearns, and Chioma Osondu.

The Solution of Semi-Infinite Linear Programs Using Boosting-Like Methods*

Gunnar Rätsch

Friedrich Miescher Laboratory of the Max Planck Society
Spemannstr. 39, 72076 Tübingen, Germany
Gunnar.Raetsch@tuebingen.mpg.de

Abstract. We consider methods for the solution of large linear optimization problems, in particular so-called Semi-Infinite Linear Programs (SILPs) that have a finite number of variables but infinitely many linear constraints. We illustrate that such optimization problems frequently appear in machine learning and discuss several examples including maximum margin boosting, multiple kernel learning and structure learning. In the second part we review methods for solving SILPs. Here, we are particularly interested in methods related to boosting. We review recent theoretical results concerning the convergence of these algorithms and conclude this work with a discussion of empirical results comparing these algorithms.

---

* The full version of this paper is published in the Proceedings of the 17th International Conference on Algorithmic Learning Theory Lecture Notes in Artificial Intelligence Vol. 4264.

Spectral Norm in Learning Theory: Some Selected Topics*

Hans Ulrich Simon**

Fakultät für Mathematik, Ruhr-Universität Bochum, 44780 Bochum, Germany
`simon@lmi.rub.de`

Abstract. In this paper, we review some known results that relate the statistical query complexity of a concept class to the spectral norm of its correlation matrix. Since spectral norms are widely used in various other areas, we are then able to put statistical query complexity in a broader context. We briefly describe some non-trivial connections to (seemingly) different topics in learning theory, complexity theory, and cryptography. A connection to the so-called Hidden Number Problem, which plays an important role for proving bit-security of cryptographic functions, will be discussed in somewhat more detail.

---

* The full version of this paper is published in the Proceedings of the 17th International Conference on Algorithmic Learning Theory Lecture Notes in Artificial Intelligence Vol. 4264.

** This work was supported in part by the IST Programme of the European Community, under the PASCAL Network of Excellence, IST-2002-506778. This publication only reflects the authors' views.

N. Lavrač, L. Todorovski, and K.P. Jantke (Eds.): DS 2006, LNAI 4265, p. 16, 2006.

Classification of Changing Regions Based on Temporal Context in Local Spatial Association

Jae-Seong Ahn[1], Yang-Won Lee[2,*], and Key-Ho Park[1]

[1] Department of Geography, College of Social Sciences, Seoul National University
[2] Center for Spatial Information Science, University of Tokyo
Cw-503 IIS, 4-6-1 Komaba, Meguro-ku, Tokyo 153-8505, Japan
Tel.: +81-3-5452-6417; Fax: +81-3-5452-6414
jwlee@iis.u-tokyo.ac.jp

Abstract. We propose a method of modeling regional changes in local spatial association and classifying the changing regions based on the similarity of time-series signature of local spatial association. For intuitive recognition of time-series local spatial association, we employ Moran scatterplot and extend it to QS-TiMoS (Quadrant Sequence on Time-series Moran Scatterplot) that allows for examining temporal context in local spatial association using a series of categorical variables. Based on the QS-TiMoS signature of nodes and edges, we develop the similarity measures for "state sequence" and "clustering transition" of time-series local spatial association. The similarity matrices generated from the similarity measures are then used for producing the classification maps of time-series local spatial association that present the history of changing regions in clusters. The feasibility of the proposed method is tested by a case study on the rate of land price fluctuation of 232 administrative units in Korea, 1995-2004.

1 Introduction

One of the most notable concepts in spatial data analysis is Tobler's first law of geography [17]: "everything is related to everything else, but near things are more related than distant things." Central to this law are the words "related" and "near" [12]. The relationship or association among geographic entities is modeled in spatial autocorrelation statistics to analyze the dependence relative to nearness [13, 8, 9]. The nearness or closeness among geographic entities is typically defined based on distance or contiguity relationship [3, 10, 4, 15]. In addition to the global measures of spatial association that present the overall degree of spatial dependence among regions in the whole area, local measures of spatial association provide the information as to how much each region and its neighbors are spatially associated.

Two things may be stressed with regard to the local spatial association: temporal and clustering aspects. First, as to the temporal aspect, since the relationship of a region and its neighbors changes by time, the local spatial association

---

* Corresponding author.

N. Lavrač, L. Todorovski, and K.P. Jantke (Eds.): DS 2006, LNAI 4265, pp. 17–28, 2006.

needs to be approached by time-series concept. Secondly, as to the clustering aspect, since near things are more related, the related things can have more possibility to be near located (or clustered), vice versa. Moreover, with the two aspects combined, the time-series of local spatial association may be captured in a clustered pattern. If we measure the similarity of a series of local spatial association accumulated by time, we can group together similar regions that have experienced similar changes in local spatial association. Thus, understanding local spatial association in temporal context and finding out common history of spatially associated regions can facilitate the analysis of spatial and temporal characteristics of changing regions.

The objective of this paper is to propose a method of modeling regional changes in local spatial association and classifying the changing regions based on the similarity of time-series signature of local spatial association. We deal with the regional changes in local spatial association as a time-ordered sequence in which region's state of local spatial association is continuously connected at certain intervals. For intuitive recognition of time-series local spatial association, we employ Moran scatterplot and extend it to QS-TiMoS (Quadrant Sequence on Time-series Moran Scatterplot) that allows for examining temporal context in local spatial association using a series of categorical variables. Based on the QS-TiMoS signature of nodes and edges, we develop the similarity measures for "state sequence" and "clustering transition" of the time-series of local spatial association. The similarity matrices generated from the similarity measures are then used for producing the classification maps of time-series local spatial association that present the history of changing regions in clusters. Through a case study on the rate of land price fluctuation of 232 administrative units in Korea, 1995-2004, the feasibility of the proposed method is tested for the similarity analysis of sequential changes in local spatial association and the clustered classification of changing regions.

2   Related Work

A number of techniques for measuring spatial association have been developed to examine the nature and extent of spatial dependence. Global measures of spatial association (or autocorrelation) such as Moran's I [13], Geary's C [8], and G statistic of Getis & Ord [9] derive a single value for the whole area. Since these global value may not be universally applicable throughout the whole area, local measures of spatial association alternatively examine the spatial dependence in subsets focusing on the variation within the study area [6, 7, 5]. As local indicators of spatial association (LISA), local Moran's I, local Geary's C [1], and local G and G^* of Ord & Getis [14] produce a local value for each subset region.

In particular, Local Moran's I can be extended to Moran scatterplot [2] that assesses local instability of spatial association in a study area. As in Figure 1, the horizontal axis denotes Z-score of each subset region, and the vertical axis denotes spatially lagged Z-score, namely, mean of the neighbors' Z-score of corresponding region. The slope of regression line corresponds to the global

Moran's *I*. Letting the intersection of horizontal and vertical means be an origin, four quadrants of Moran scatterplot are interpreted as in Table 1. The Moran Scatterplot is thought to be a very useful tool for revealing several aspects of local spatial association. First, as the spatial-lag pairs become concentrated in quadrant I/III or quadrant II/IV, the overall level of spatial association in the distribution strengthens. Secondly, outliers in terms of spatial-lag pairs that deviate from the overall trend (high or low outliers) can be easily identified. Thirdly, clusters of local spatial association (hot or cold spots) can also be identified in the scatterplot [16].

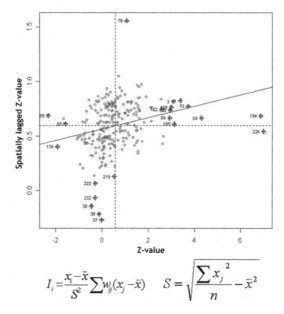

$$ I_i = \frac{x_i - \bar{x}}{S^2} \sum w_{ij}(x_j - \bar{x}) \qquad S = \sqrt{\frac{\sum x_j^{\,2}}{n} - \bar{x}^2} $$

Fig. 1. Moran scatterplot for assessing local pattern of spatial association

Table 1. Interpretation of Moran scatterplot

State	Quadrant	Interpretation
HH	I	Cluster - "I'm high and my neighbors are high."
LH	II	Outlier - "I'm a low outlier among high neighbors."
LL	III	Cluster - "I'm low and my neighbors are low."
HL	IV	Outlier - "I'm a high outlier among low neighbors."

If considering temporal changes in Moran scatterplot, a series of state {LL → LL → LH → LH → HH → HH}, for instance, can be interpreted as "I was low like my neighbors (LL). While I was still low, my neighbors became high (LH). Then, I became high like my neighbors (HH)." In order to examine such changes, Rey [16] has proposed a space-time transition measure using Moran scatterplot and Markov chain. This method provides a refined analysis on the probability of a region's changes along with its neighbors' changes. Five transition types are defined in the analysis. Type 0 denotes no move of both a region and its neighbors. Type I denotes a relative move of only the region, such as HH → LH, HL → LL, LH → HH, and LL → HL. Type II denotes a relative move of only the neighbors, such as HH → HL, HL → HH, LH → LL, and LL → LH. Type III denotes a move of both a region and its neighbors to a different side: Type III-A includes joint upward move (LL → HH) and joint downward move (LL → HH) while Type III-B includes diagonal switch such as HL → LH and LH → HL. The transition types by time-span are incorporated into a discontinuous Markov chain; hence, each transition is taken as an individual observation. For example, $\{t_0 \to t_1 \to t_2 \to \cdots \to t_7 \to t_8 \to t_9\}$ includes nine individual observations for the time-span $\{t_n \to t_{n+1}\}$, and each observation participates in calculating the probability of regional changes by transition type, regardless of time order.

3   Proposed Method

Since the space-time transition measure mentioned above is based on discontinuous Markov chain (that is, $\{t_0 \to t_1 \to t_2\}$ is taken as two separate data $\{t_0 \to t_1\}$ and $\{t_1 \to t_2\}$), the probability of regional changes by transition type may not sufficiently reflect the sequential aspect of regional changes. As an alternative to this, we deal with the regional changes in local spatial association as a time-ordered sequence in which a region's state of local spatial association is continuously connected at certain intervals. Although regional characteristics are changing seamless in real life, statistical data has no choice but to be gathered at certain intervals like day, month, or year. Therefore, in this paper we assume the interval basis discrete data represents the sequential state of an object, likewise the time-series local spatial association. Local Moran's I derived from the data appears on Moran scatterplot in the form of a Z-score, but for intuitive recognition of time-series local spatial association, we use quadrant category as a simplified representative of the Z-score. With the similarity analysis on the time-series local spatial association, the final result would be the classification maps for time-series local spatial association that present the history of changing regions in clusters. Figure 2 illustrates the workflow of the proposed method.

3.1   QS-TiMoS Signature

In order to formalize the time-series signature of local spatial association in terms of continuous regional changes, we employ Moran scatterplot and extend it to the time-series Moran scatterplot by the accumulation at certain intervals. If we connect n corresponding points of certain region on time-series Moran scatterplot, the connected line composed of n nodes and n-1 edges forms a time-series signature of local spatial association for the region. A node of a region's signature represents the state of local spatial association at certain time (black point in Figure 3(a)), and an edge of a region's signature represents the change of local spatial association at certain period (thick line in Figure 3(b)). For the notations of a node's state

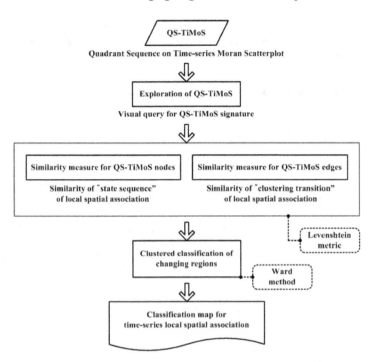

Fig. 2. Method for clustered classification of time-series local spatial association

of local spatial association, we use I/II/III/IV after four quadrants of Moran scatterplot. In addition, the edge connecting two nodes possesses the information as to "clustering transition," such as upward/downward clustering and declustering. We set up six notations for the "clustering transition" as in Table 2.

Table 2. Notations for "clustering transition" of local spatial association

	I (HH$_i$)	II (LH$_i$)	III (LL$_i$)	IV (HL$_i$)
I (HH$_i$)	O	DD	JD	UD
II (LH$_i$)	UC	O	DC	UD
III (LL$_i$)	JU	DD	O	UD
IV (HL$_i$)	UC	DD	DC	O

Upward clustering (UC) includes the transitions such as LH → HH and HL → HH. Downward clustering (DC) includes the transitions such as LH → LL and HL → LL. Upward declustering (UD) includes the transitions that a region (relatively) moves up while its neighbors (relatively) move down, such as HH → HL and LL → HL. Downward declustering (DD) includes the transitions that a region (relatively) moves down while its neighbors (relatively) move up, such as HH → LH and LL → LH. Joint upward clustering (JU) denotes LL → HH, and

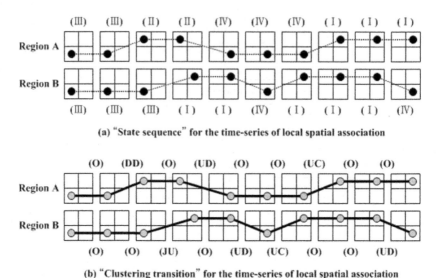

(a) "State sequence" for the time-series of local spatial association

(b) "Clustering transition" for the time-series of local spatial association

Fig. 3. QS-TiMoS signatures: "state sequence" and "clustering transition"

joint downward clustering (JD) denotes LL → HH transition. No change in the transition is represented as (O).

3.2   Similarity Measure for QS-TiMoS Signature

Each region has its own signature of QS-TiMoS. Some signatures may be similar to one another or may be not. The similarity measure for QS-TiMoS signature is a key to the classification of changing regions because the similarity of QS-TiMoS signature explains as to which regions have experienced similar history in local spatial association and thus can be grouped together.

In order to obtain the similarity matrix for region pairs from QS-TiMoS signature, we employ Levenshtein metric [11], the most widely used method for sequence comparison of categorical data. Levenshtein metric is defined as the minimum number of edit operation (insertion/deletion/substitution) needed to transform one sequence into the other, namely, a unit cost for the edit operations. The algorithm of Levenshtein metric for two sequences s and t is described in Table 3, producing the degree of dissimilarity (distance) of them. Using the Levenshtein metric, the similarity measures for the time-series signature of local spatial association indicates how far (or dissimilar) each region pair is in terms of "state sequence" (in the notations I/II/III/IV) and "clustering transition" (in the notations UC/DC/UD/DD/JU/JD).

3.3   Clustered Classification of Changing Regions

The clustered classification of changing regions in terms of time-series local spatial association is conducted using the similarity matrices for "state sequence"

Table 3. Algorithm of Levenshtein metric

Step	Description
1	Set n to be the length of s.
	Set m to be the length of t.
	Construct a matrix containing $0..m$ rows and $0..n$ columns.
2	Initialize the first row to $0..n$.
	Initialize the first column to $0..m$.
3	Examine each character of s (i from 1 to n).
4	Examine each character of t (j from 1 to m).
5	If $s[i]$ equals $t[j]$, the cost is 0.
6	Set cell $d[i,j]$ of the matrix equal to the minimum of:
	(a) The cell immediately above plus 1: $d[i\text{-}1,j] + 1$.
	(b) The cell immediately to the left plus 1: $d[i,j\text{-}1] + 1$.
	(c) The cell diagonally above and to the left plus the cost: $d[i\text{-}1,j\text{-}1] + \text{cost}$.
7	After the iteration steps (3, 4, 5, 6) are completed, the distance is found in cell $d[n,m]$.

and "clustering transition" in QS-TiMoS signature. For a cluster analysis, we employ Ward method [18] that minimizes the sum of squares of any two (hypothetical) clusters.

4   A Case Study

As a feasibility test of the proposed method, we conduct a case study using yearly basis data of land price fluctuation of 232 administrative units in Korea, 1995-2004. This experiment aims to examine the sequential changes of local spatial association in land price fluctuation and thus to classify the changing regions based on the similarity of QS-TiMoS signature. With this data, we first explore data distribution pattern, and then carry out a similarity analysis of time-series local spatial association using the similarity measures for QS-TiMoS nodes and edges. Through the similarity analysis, we present a clustered classification of changing regions in terms of "state sequence" and "clustering transition" of time-series local spatial association. Necessary functions for this case study are implemented using Microsoft Visual Studio .NET.

4.1   Data Exploration

As in the time-series parallel coordinate plot of Figure 4(a), the rate of land price fluctuation has experienced a sudden drop caused by IMF crisis around 1998 and a steep rise owing to deregulations and development plans around 2002. The local G^* statistic that considers the neighbors within certain distance including itself may be appropriate for visualizing the distinction of local spatial association among regions. As in Figure 4(b), the local G^* maps of a few years (1995, 1998, 2002, and 2004) show that upward regions and stagnant regions respectively are somewhat spatially clustered.

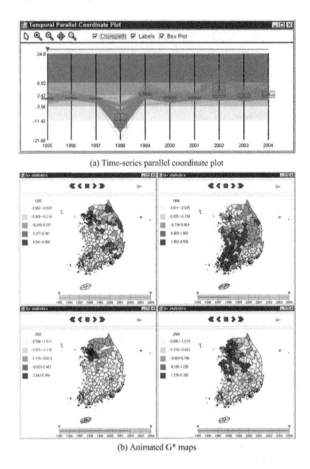

(a) Time-series parallel coordinate plot

(b) Animated G* maps

Fig. 4. Exploration of land price fluctuation in Korea, 1995-2004

4.2   State Sequence of Time-Series Local Spatial Association

In order to capture the change characteristics of local spatial association and the similarity among changing regions, we build a time-series Moran scatterplot by accumulating the local Moran's I with the reciprocal of centroid distance as a spatial proximity weight. From the 10 years' QS-TiMoS signature, a "state sequence" similarity matrix of region pairs is extracted for cluster analysis. Using Ward method, seven clusters that have experienced similar history in the state change of local spatial association are classified (Figure 5). The cluster 3 and 4 are thought to be cold spots that have stayed in quadrant III for almost ten years. These regions have not been involved in remarkable development plans during the period. The cluster 5 and 6 are thought to be hot spots that have stayed in quadrant I for almost ten years. Seoul, the capital of Korea is located in the cluster; hence, its neighboring regions may have been affected by the development plans for alleviating metropolis concentration during the period. In Figure 6,

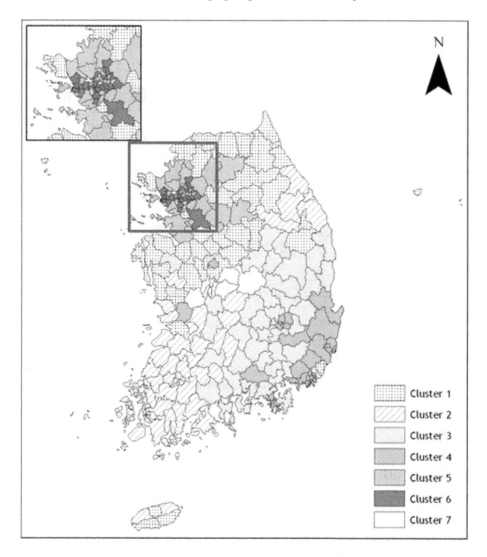

Fig. 5. Classification map for "state sequence" of time-series local spatial association

the number of each cluster's region participating in each quadrant on Moan scatterplot is provided in the form of a proportional grayscale representation for illustrating the change characteristics of local spatial association.

4.3 Clustering Transition of Time-Series Local Spatial Association

While the analysis of "state sequence" derived from QS-TiMoS signature is based on each region's state change in relation to its neighbors, the analysis of "clustering transition" is about the change in the tendency of clustering with neighbors,

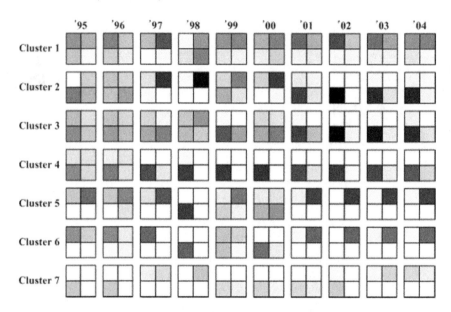

Fig. 6. Change characteristics of local spatial association for each cluster

such as upward/downward clustering and declustering. From the "clustering transition" of QS-TiMoS signature, five clusters are classified as in Figure 7: the cluster 4 and 5 are characterized by upward clustering particularly since 2001 while the cluster 3 is characterized by downward clustering for almost the whole years.

5 Concluding Remarks

In order to model regional changes in local spatial association and thus to classify the changing regions based on the similarity of time-series signature, we developed the similarity measures for "state sequence" and "clustering transition" of local spatial association by extending and integrating related methods. We deal with the regional changes in local spatial association as a time-ordered sequence in which region's state of local spatial association is continuously connected at certain intervals. Based on the QS-TiMoS signature incorporating both "sequential state" (e.g., hot/cold spot or high/low outlier) and "clustering transition" (e.g., upward/downward clustering or declustering), our method performs the similarity analysis using QS-TiMoS nodes and edges to produce the similarity matrices for clustered classification of changing regions. Through a case study on the rate of land price fluctuation of 232 administrative units in Korea, 1995-2004, we confirmed the feasibility of the proposed method that could find out the temporal context in spatial changes.

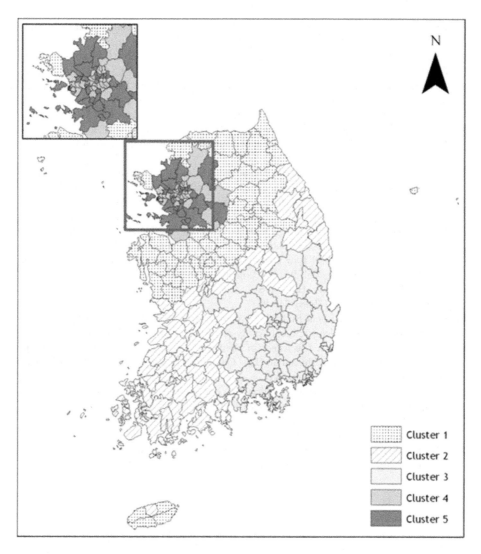

Fig. 7. Classification map for "clustering transition" of time-series local spatial association

References

1. Anselin, L.: Local Indicators of Spatial Association - LISA. Geographical Analysis, Vol. 27, No. 2 (1995) 93-115
2. Anselin, L.: The Moran Scatterplot as an ESDA Tool to Assess Local Instability in Spatial Association. In Fisher, M., Scholten, H.J., Unwin, D. (Eds.): Spatial Analytical Perspectives on GIS, Taylor & Francis, London (1996) 111-125
3. Bailey, T.C., Gatrell, A.C.: Interactive Spatial Data Analysis, Longman Scientific & Technical, Essex UK (1995)

4. Bera, R., Claramunt, C.: Topology-based Proximities in Spatial Systems. Journal of Geographical Systems, Vol. 5, No. 4 (2003) 353-379
5. Boots, B.: Developing Local Measures of Spatial Association for Categorical Data. Journal of Geographical Systems, Vol. 5, No. 2 (2003) 139-160
6. Fotheringham, A.S.: Trends in Quantitative Methods I: Stressing the Local. Progress in Human Geography, Vol. 21, No. 1 (1997) 88-96
7. Fotheringham, A.S., Brunsdon, C.: Local Forms of Spatial Analysis. Geographical Analysis, Vol. 31, No. 4 (1999) 340-358
8. Geary, R.C.: The Contiguity Ratio and Statistical Mapping. Incorporated Statistician, Vol. 5, No. 3 (1954) 115-145
9. Getis, A., Ord, J.K.: The Analysis of Spatial Association by Use of Distance Statistics. Geographical Analysis, Vol. 24, No. 3 (1992) 186-206
10. Lee, J., Wong, D.: Statistical Analysis with ArcView GIS, John Wiley & Sons, New York (2001)
11. Levenshtein, V.I.: Binary Codes Capable of Correcting Deletions, Insertions, and Reversals. Soviet Physics-Doklandy, Vol. 10, No. 8 (1966) 707-710
12. Miller, H.J.: Tobler's First Law and Spatial Analysis. Annals of the Association of American Geographers, Vol. 94, No. 2 (2004) 284-289
13. Moran, P.: The Interpretation of Statistical Maps. Journal of Royal Statistical Society, Vol. 10, No. 2 (1948) 243-251
14. Ord, J.K., Getis, A.: Local Spatial Autocorrelation Statistics: Distribution Issues and an Application. Geographical Analysis, Vol. 27, No. 4 (1995) 286-306
15. Park, K.-H.: A Study on the Effect of Spatial Proximity Weight Matrices on the Spatial Autocorrelation Measures: The Case of Seoul Administrative Units. Research of Seoul & Other Cities, Vol. 5, No. 3 (2004) 67-83
16. Rey, S.J.: Spatial Empirics for Regional Economic Growth and Convergence. Geographical Analysis, Vol. 33, No. 3 (2001) 195-214
17. Tobler, W.: A Computer Movie Simulating Urban Growth in the Detroit Region. Economic Geography, Vol. 46, No. 2 (1970) 234-240
18. Ward, J.: Hierarchical Grouping to Optimize an Objective Function. Journal of the American Statistical Association, Vol. 58, No. 301 (1963) 236-244

Kalman Filters and Adaptive Windows for Learning in Data Streams*

Albert Bifet and Ricard Gavaldà

Universitat Politècnica de Catalunya, Barcelona, Spain
{abifet, gavalda}@lsi.upc.edu

Abstract. We study the combination of Kalman filter and a recently proposed algorithm for dynamically maintaining a sliding window, for learning from streams of examples. We integrate this idea into two well-known learning algorithms, the Naïve Bayes algorithm and the k-means clusterer. We show on synthetic data that the new algorithms do never worse, and in some cases much better, than the algorithms using only memoryless Kalman filters or sliding windows with no filtering.

1 Introduction

We deal with the problem of distribution and concept drift when learning from streams of incoming data. We study the combination of a classical estimation method in automatic control theory, the Kalman filter, with the also classical idea in machine learning of using a window of recently seen data items for learning. Many of the previous works in the machine learning area use windows of a fixed length. We use instead an algorithm that we proposed recently [2] for adaptively changing the size of the window in reaction to changes observed in the data.

In automatic control theory, many modern complex systems may be classed as estimation systems, combining several sources of (often redundant) data in order to arrive at an estimate of some unknown parameters. Among such systems are terrestrial or space navigators for estimating such parameters as position, velocity, and altitude, and radar systems for estimating position and velocity.

One of the most widely used estimation algorithms is the Kalman filter, an algorithm that generates estimates of variables of the system being controlled by processing available sensor measurements.

Kalman filtering and related estimation algorithms have proved tremendously useful in a large variety of settings. Automatic machine learning is but one of them; see [12,6] among many others. There is however an important difference in the control theory and machine learning settings: In automatic control, we assume that system parameters are known or easily detectable; these parameters

---

* Partially supported by the 6th Framework Program of EU through the integrated project DELIS (#001907), by the EU PASCALNetwork of Excellence, IST-2002-506778, and by the DGICYT MOISES-BAR project, TIN2005-08832-C03-03. Home pages: http://www.lsi.upc.edu/~{abifet, gavalda}".

N. Lavrač, L. Todorovski, and K.P. Jantke (Eds.): DS 2006, LNAI 4265, pp. 29–40, 2006.

are physical properties of devices, and therefore fixed. In contrast, in most machine learning situations the distribution that generates the examples is totally unknown, and there is no obvious way to measure its statistics, other than estimating them from the data. In addition, these statistics may vary impredictably over time, either continuously at a slow rate, or abruptly from time to time.

Besides occasional uses of filtering, most previous work on learning and time change has used variations of the sliding-window idea: at every moment, one window (or more) is kept containing the most recently read examples, and only those examples are considered relevant for learning. A critical point is the choice of a window size, and several strategies have been proposed [4].

We have recently proposed a different strategy [2]: keeping a window whose length is adjusted dynamically to reflect changes in the data. When change seems to be occurring, as indicated by some statistical test, the window is shrunk to keep only data items that still seem to be valid. When data seems to be stationary, the window is enlarged to work on more data and reduce variance. Our algorithm, called ADWIN for Adaptive Windowing, can be implemented with a low amount of memory and time per data item, and experimentally outperforms every window of a fixed size S when the time scale of change is either much smaller or much larger than S. In a way, ADWIN adjusts its window size to "best" one for the data it is seeing.

1.1 Proposal and Results of This Paper

In this paper, we combine ADWIN and Kalman filter and compare experimentally the performance of the resulting algorithm, K-ADWIN, with other estimator algorithms. The intuition why this combination should be better than ADWIN alone or the Kalman filter alone is as follows.

The Kalman filter is a memoryless algorithm, and it can benefit from having a memory aside. In particular, running a Kalman filter requires knowledge of at least two parameters of the system, named *state covariance* and *measurement covariance*, that should be estimated a priori. These are generally difficult to measure in the context of learning from a data stream, and in addition they can vary over time. The window that ADWIN maintains adaptively is guaranteed to contain up-to-date examples from which the current value of these covariances can be estimated and used in the Kalman filter.

On the other hand, ADWIN is somewhat slow in detecting a gradual change, because it gives the same weight to all examples in the window – it is what we will call a *linear* estimator. If there is a slow gradual change, the most recent examples should be given larger weight. This is precisely what the Kalman filter does in its estimation.

As in [2], we test K-ADWIN on two well-known learning algorithms where it is easy to observe the effect of distribution drift: the Naïve Bayes classifier and the k-means clusterer. We also perform experiments that directly compare the ability of different estimators to track the average value of a stream of real numbers that varies over time. We use synthetic data in order to control precisely the type and amount of distribution drift. The main conclusions are:

- In all three types of experiments (tracking, Naïve Bayes, and k-means), K-ADWIN either gives best results or is very close in performance to the best of the estimators we try. And each of the other estimators is clearly out-performed by K-ADWIN in at least some of the experiments. In other words, no estimator ever does much better than K-ADWIN, and each of the others is outperformed by K-ADWIN in at least one context.
- More precisely, in the tracking problem K-ADWIN and ADWIN automatically do about as well as the Kalman filter with the best set of fixed covariance parameters (which, in general, can only be determined after a good number of experiments). And these three do far better than any fixed-size window.
- In the Naïve Bayes experiments, K-ADWIN does somewhat better than ADWIN and far better than any memoryless Kalman filter. This is, then, a situation where having a memory clearly helps.
- For k-means, again K-ADWIN performs about as well as the best (and difficult to find) Kalman filter, and they both do much better than fixed-size windows.

The paper is structured as follows: In Section 2 we describe a general framework for discussing estimator algorithms, as made of three modules: Memory, Estimator, and Change detector. In Section 3 we describe the Kalman filter (an example of Estimator) and the CUSUM test (an example of Change Detector). In Section 4 we review the ADWIN algorithm. and its theoretical guarantees of performance. In Section 5 we describe how ADWIN and the Kalman filter can be combined, and describe the tracking experiments. In Sections 6 and 7 we describe the experiments with Naïve Bayes and k-means, respectively.

NOTE: Many proofs, experimental results, and discussions are omitted in this version due to the page limit. A full version is available on-line from the authors' homepages.

2   Time Change Detectors and Predictors: A General Framework

Most approaches for predicting and detecting change in streams of data can be discussed in the general framework: The system consists of three modules: a Memory module, an Estimator Module, and a Change Detector or Alarm Generator module. These three modules interact as shown in Figure 1, which is analogous to Figure 8 in [11].

In general, the input to this algorithm is a sequence $x_1, x_2, \ldots, x_t, \ldots$ of data items whose distribution varies over time in an unknown way. The outputs of the algorithm are, at each time step

- an estimation of some important parameters of the input distribution, and
- a signal alarm indicating that distribution change has recently occurred.

In this paper we consider a specific, but very frequent case, of this setting: that in which all the x_t are real values. The desired estimation is usually the

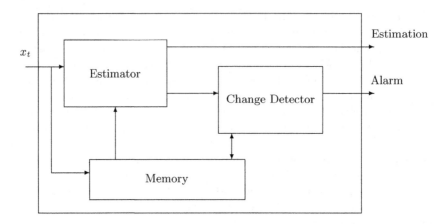

Fig. 1. General Framework

expected value of the current x_t, and less often another distribution statistics such as the variance. The only assumption on the distribution is that each x_t is drawn independently from each other.

Memory is the component where the algorithm stores all the sample data or summary that considers relevant at current time, that is, that presumably shows the current data distribution.

The Estimator component estimates the desired statistics on the input data, which may change over time. The algorithm may or may not use the data contained in the Memory. The simplest Estimator algorithm for the expected is the *linear estimator,* which simply returns the average of the data items contained in the Memory. Other examples of run-time efficient estimators are Auto-Regressive, Auto Regressive Moving Average, and Kalman filters.

The change detector component outputs an alarm signal when it detects change in the input data distribution. It uses the output of the Estimator, and may or may not in addition use the contents of Memory.

We distinguish four classes of predictors, according to the presence or absence of the Memory and Change detector modules.

– *Type I: Estimator only.* The simplest one is modelled by

$$\hat{x}_k = (1 - \alpha)\hat{x}_{k-1} + \alpha \cdot x_k.$$

The linear estimator corresponds to using $\alpha = 1/N$ where N is the width of a virtual window containing the last N elements we want to consider. Otherwise, we can give more weight to the last elements with an appropriate constant value of α. The Kalman filter tries to optimize the estimation using a non-constant α (the K value) which varies at each discrete time interval.
– *Type II: Estimator with Change Detector.* An example is the Kalman Filter together with a CUSUM test change detector algorithm, see for example [6].
– *Type III: Estimator with Memory.* We add Memory to improve the results of the Estimator. For example, one can build an Adaptive Kalman Filter

that uses the data in Memory to compute adequate values for the process variance Q and the measure variance R. In particular, one can use the sum of the last elements stored into a memory window to model the Q parameter and the difference of the last two elements to estimate parameter R.

- Type IV: Estimator with Memory and Change Detector. This is the most complete type. Two examples of this type, from the literature, are:
  - A Kalman filter with a CUSUM test and fixed-length window memory, as proposed in [11]. Only the Kalman filter has access to the memory.
  - A linear Estimator over fixed-length windows that flushes when change is detected [7], and a change detector that compares the running windows with a reference window.

In [2], we proposed another Type IV algorithm called ADWIN, which is a new Memory+Change detector module together with a linear estimator. We showed it performs advantageously with respect to strategies that maintain a fixed-length window and linear estimators. The goal of this paper is to propose an even better Type IV algorithm by combining the best of both worlds: a Kalman filter as a "good" Estimation module and ADWIN as "good" Memory+Change detector modules.

3   The Kalman Filter and the CUSUM Test

One of the most widely used Estimation algorithms is the Kalman filter. We give here a description of its essentials; see [13] for a complete introduction.

The Kalman filter addresses the general problem of trying to estimate the state $x \in \Re^n$ of a discrete-time controlled process that is governed by the linear stochastic difference equation

$$x_k = Ax_{k-1} + Bu_k + w_{k-1}$$

with a measurement $z \in \Re^m$ that is

$$Z_k = Hx_k + v_k.$$

The random variables w_k and v_k represent the process and measurement noise (respectively). They are assumed to be independent (of each other), white, and with normal probability distributions

$$p(w) \sim N(0, Q) \qquad p(v) \sim N(0, R).$$

In essence, the main function of the Kalman filter is to estimate the state vector using system sensors and measurement data corrupted by noise.

In our case we consider a sequence of real values $z_1, z_2, \ldots, z_t, \ldots$ as the measurement data. The difference equation of our discrete-time controlled process is simpler, with $A = 1, H = 1, B = 0$. So the equations are simplified to:

$$K_k = P_{k-1}/(P_{k-1}+R), \quad X_k = X_{k-1}+K_k(z_k - X_{k-1}), \quad P_k = P_k(1-K_k)+Q.$$

The performance of the Kalman filter depends on the accuracy of the a-priori assumptions:

- linearity of the difference stochastic equation
- estimation of covariances Q and R, assumed to be fixed, known, and follow normal distributions with zero mean.

When applying the Kalman filter to data streams that vary arbitrarily over time, both assumptions are problematic. The linearity assumption for sure, but also the assumption that parameters Q and R are fixed and known – in fact, estimating them from the data is itself a complex estimation problem.

The cumulative sum (CUSUM algorithm),(see [6,10,11]) is a change detection algorithm that gives an alarm when the mean of the input data is significantly different from zero.

4   The ADWIN Algorithm

In this section we review ADWIN, an algorithm for estimating, detecting change, and dynamically adjusting the length of a data window. For details see [2].

The inputs to the algorithm are a confidence value $\delta \in (0, 1)$ and a (possibly infinite) sequence of real values x_1, x_2, x_3, ..., x_t, ... The value of x_t is available only at time t. Each x_t is generated according to some distribution D_t, independently for every t. We denote with μ_t the (unknown) expected value of x_t when it is drawn according to D_t.

Algorithm ADWIN uses a sliding window W with the most recently read x_i. Let n denote the length of W, $\hat{\mu}_W$ the (known) average of the elements in W, and μ_W the (unknown) average of μ_t for $t \in W$.

The idea behind ADWIN is simple: whenever two "large enough" subwindows of W exhibit "distinct enough" averages, we conclude that the corresponding expected values are different, and drop the older portion of the window. Formally, "large enough" and "distinct enough" translate into the computation of a cut value ϵ_c (which depends on δ, the length of the subwindows, and the averages of their contents). The computation of ϵ_c (not discussed here) is crucial to the performance of the algorithm. In particular, the main technical result in [2] about the performance of ADWIN is:

Theorem 1. *[2] At every time step we have:*

1. *(Few false positives guarantee) If μ_t remains constant within W, the probability that ADWIN shrinks the window at this step is at most δ.*
2. *(Few false negatives guarantee) If for any partition W in two parts $W_0 W_1$ (where W_1 contains the most recent items) we have $|\mu_{W_0} - \mu_{W_1}| > \epsilon$, and if*

$$\epsilon \geq 4 \cdot \sqrt{\frac{3 \max\{\mu_{W_0}, \mu_{W_1}\}}{\min\{n_0, n_1\}} \ln \frac{4n}{\delta}}$$

then with probability $1 - \delta$ ADWIN shrinks W to W_1, or shorter.

Thus, ADWIN is in a sense conservative: it is designed to sustain the null hypothesis "no change" as long as possible. This leads to a very good false positive rate (essentially 0 in all experiments) but a somewhat slow reaction time to slow, gradual changes. Abrupt changes, on the other hand, are detected very quickly.

This first version of ADWIN is computationally expensive, because it checks exhaustively all "large enough" subwindows of the current window for possible cuts. Furthermore, the contents of the window is kept explicitly, with the corresponding memory cost as the window grows. To reduce these costs we presented in [2] a new version, ADWIN2 that uses ideas developed in data stream algorithmics [1,8,3] to find a good cutpoint quickly. We summarize the behavior of this policy in the following theorem.

Theorem 2. The ADWIN2 algorithm maintains a data structure with the following properties:

- It can provide the exact counts of 1's for all subwindows whose lengths are of the form $\lfloor (1 + \epsilon)^i \rfloor$, for some design parameter ϵ, in $O(1)$ time per point.
- It uses $O(\frac{1}{\epsilon} \log W)$ memory words (assuming a memory word can contain numbers up to W).
- The arrival of a new element can be processed in $O(1)$ amortized time and $O(\log W)$ worst-case time.

ADWIN2 tries $O(\log W)$ cutpoints, so the processing time per example is $O(\log W)$ (amortized) and $O(\log^2 W)$ (worst-case). The choice of the internal parameter ϵ controls the amount of memory used and the desired density of checkpoints. It does *not* reflect any assumption about the time scale of change. Since points are checked at a geometric rate anyway, this policy is essentially scale-invariant.

In the sequel, whenever we say ADWIN we really mean its efficient implementation, ADWIN2.

5   K-ADWIN = ADWIN + Kalman Filtering

ADWIN is basically a linear Estimator with Change Detector that makes an efficient use of Memory. It seems a natural idea to improve its performance by replacing the linear estimator by an adaptive Kalman filter, where the parameters Q and R of the Kalman filter are computed using the information in ADWIN's memory.

We have set $R = W^2/50$ and $Q = 200/W$, where W is the length of the window maintained by ADWIN. While we cannot rigorously prove that these are the optimal choices, we have informal arguments that these are about the "right" forms for R and Q, on the basis of the theoretical guarantees of ADWIN.

Let us sketch the argument for Q. Theorem 1, part (2) gives a value ϵ for the maximum change that may have occurred within the window maintained by ADWIN. This means that the process variance within that window is at most ϵ^2, so we want to set $Q = \epsilon^2$. In the formula for ϵ, consider the case in which $n_0 = n_1 = W/2$, then we have

$$\epsilon \geq 4 \cdot \sqrt{\frac{3(\mu_{W_0} + \epsilon)}{W/2}} \cdot \ln \frac{4W}{\delta}$$

Isolating from this equation and distinguishing the extreme cases in which $\mu_{W_0} \gg \epsilon$ or $\mu_{W_0} \ll \epsilon$, it can be shown that $Q = \epsilon^2$ has a form that varies between c/W and d/W^2. Here, c and d are constant for constant values of δ, and $c = 200$ is a reasonable estimation. This justifies our choice of $Q = 200/W$. A similar, slightly more involved argument, can be made to justify that reasonable values of R are in the range W^2/c to W^3/d, for somewhat large constants c and d.

When there is no change, ADWIN window's length increases, so R increases too and K decreases, reducing the significance of the most recent data arrived. Otherwise, if there is change, ADWIN window's length reduces, so does R, and K increases, which means giving more importance to the last data arrived.

5.1   Experimental Validation of K-ADWIN

We compare the behaviours of the following types of estimators:

- *Type I*: Kalman filter with different but fixed values of Q and R. The values $Q = 1$, $R = 1000$ seemed to obtain the best results with fixed parameters.
- *Type I*: Exponential filters with $\alpha = 0.1, 0.25, 0.5$. This filter is similar to Kalman's with $K = \alpha, R = (1 - \alpha)P/\alpha$.
- *Type II*: Kalman filter with a CUSUM test Change Detector algorithm. We tried initially the parameters $v = 0.005$ and $h = 0.5$ as in [6], but we changed to $h = 5$ which systematically gave better results.
- *Type III*: Adaptive Kalman filter with R as the difference of $x_t - x_{t-1}$ and Q as the sum of the last 100 values obtained in the Kalman filter. We use a fixed window of 100 elements.
- *Types III and IV*: Linear Estimators over fixed-length windows, without and with flushing when changing w.r.t. a reference window is detected. Details are as in [2].
- *Type IV*: ADWIN and K-ADWIN. K-ADWIN uses a Kalman filter with $R = W^2/50$ and $Q = 200/W$, where W is the length of the ADWIN window.

We build a framework with a stream of synthetic data consisting of some triangular wavelets, of different periods, some square wavelets, also of different periods, and a staircase wavelet of different values. We generate 10^6 points and feed all them to all of the estimators tested. We calculate the mean L_1 distances from the prediction of each estimator to the original distribution that generates the data stream. Finally, we compare these measures for the different estimators.

Unfortunately, due to the space limitation, we cannot report the detailed results of these experiments. They can be found in the full version of the paper. A summary of the results is as follows: The results for K-ADWIN, ADWIN, the Adaptive Kalman filter, and the best fixed-parameter Kalman filter are the best ones in most cases. They are all very close to each other and they outwin each other in various ways, always by a small margin. They all do about as well as the best fixed-size window, and in most cases they win by a large amount. The exception are wavelets of very long periods, in which a very large fixed-size window wins. This is to be expected: when change is extremely rare, it is best to use a large window. Adaptivity necessarily introduces a small penalty, which is a waste in this particular case.

6   Example 1: Naïve Bayes Predictor

We test all the different estimators in a classical Naïve Bayes predictor. To compute the class of each new instance $I = (x_1 = v_1, \ldots, x_k = v_k)$ that arrives, we use an estimator for every probability $\Pr[x_i = v_j \wedge C = c]$ and $\Pr[C = c]$.

The experiments with synthetic data use a changing concept based on a rotating hyperplane explained in [5]. A hyperplane in d-dimensional space is the set of points x that satisfy

$$\sum_{i=1}^{d} w_i x_i \geq w_0$$

where x_i, is the ith coordinate of x. Examples for which $\sum_{i=1}^{d} w_i x_i \geq w_0$ are labeled positive, and examples for which $\sum_{i=1}^{d} w_i x_i < w_0$ are labeled negative. Hyperplanes are useful for simulating time-changing concepts because we can change the orientation and position of the hyperplane in a smooth manner by changing the relative size of the weights.

We test our algorithms on a classical Naïve Bayes predictor. We use 2 classes, 8 attributes, and 2 values per attribute. The different weights w_i of the hyperplane vary over time, at different moments and different speeds for different attributes i. All w_i start at 0.5 and we restrict to two w_i's varying at the same time, to a maximum value of 0.75 and a minimum of 0.25.

To test the performance of our Naïve Bayes predictor we do the following: At every time t we build a static Naïve Bayes model M_t using a data set of 1000 points generated from the distribution at time t. Model M_t is taken as a "baseline" of how well a Naïve Bayes model can do on this distribution. Then we generate 2000 fresh points, and compute the error rate of both this static model M_t and the different sliding-window models built from the t points seen so far. The ratio of these error rates is averaged over all the run.

Table 1 shows accuracy results. The "%Static" column shows the accuracy of the statically built model – the same for all rows, except for small variance. The "%Dynamic" column is the accuracy of the dynamically built model, using the estimator in the row. The last column shows the quotient of columns 1 and 2, i.e., the relative accuracy of the estimator-based model Naïve Bayes model with respect that of the statically computed one. Again, in each column (a test), we show in boldface the result for K-ADWIN and for the best result.

The results can be summarized as follows: K-ADWIN outperforms plain ADWIN by a small margin, and they both do much better than all the memoryless Kalman filters. Thus, having a memory clearly helps in this case. Strangely enough, the winner is the longest fixed-length window, which achieves 98.73% of the static performance compared to K-ADWIN's 97.77%. We have no clear explanation of this fact, but believe it is an artifact of our benchmark: the way in which we vary the attributes' distributions might imply that simply taking the

Table 1. Naïve Bayes benchmark

	Width	%Static	%Dynamic	% Dynamic/Static
ADWIN		83,36%	80,30%	96,33%
Kalman $Q = 1, R = 1000$		83,22%	71,13%	85,48%
Kalman $Q = 1, R = 1$		83,21%	56,91%	68,39%
Kalman $Q = .25, R = .25$		83,26%	56,91%	68,35%
Exponential Estimator $\alpha = .1$		83,33%	64,19%	77,03%
Exponential Estimator $\alpha = .5$		83,32%	57,30%	68,77%
Exponential Estimator $\alpha = .25$		83,26%	59,68%	71,68%
Adaptive Kalman		83,24%	76,21%	91,56%
CUSUM Kalman		83,30%	50,65%	60,81%
K-ADWIN		**83,24%**	**81,39%**	**97,77%**
Fixed-sized Window	32	83,28%	67,64%	81,22%
Fixed-sized Window	128	83,30%	75,40%	90,52%
Fixed-sized Window	512	83,28%	80,47%	96,62%
Fixed-sized Window	2048	83,24%	**82,19%**	**98,73%**
Fixed-sized flushing Window	32	83,28%	67,65%	81,23%
Fixed-sized flushing Window	128	83,29%	75,57%	90,73%
Fixed-sized flushing Window	512	83,26%	80,46%	96,64%
Fixed-sized flushing Window	2048	83,25%	82,04%	98,55%

average of an attribute's value over a large window has best predictive power. More experiments with other change schedules should confirm or refute this idea.

7   Example 2: k-Means Clustering

We adapt in essence the incremental version from [9]. In that version, every new example is added to the cluster with nearest centroid, and every r steps a recomputation phase occurs, which recomputes both the assignment of points to clusters and the centroids. To balance accuracy and computation time, r is chosen in [9] to be the square root of the number of points seen so far.

We change this algorithm to deal with time change in the following way. We create an estimator E_{ij} for every attribute centroid i and every attribute j. The algorithm still interleaves phases in which centroids are just incrementally modified with incoming points and phases where global recomputation of centroids takes place. Recomputation phases occur for two reasons. First, when any of the E_{ij} shrinks its window (for ADWIN-type estimators), we take this as a signal that the position of centroid i may have changed and recompute. In the case of estimators that use windows of a fixed size s, we recompute whenever this window becomes full. Second, when the average point distance to closest centroid has changed more than an ϵ factor, where ϵ is user-specified. This is taken as an indication that a certain number of points might move to another cluster if recomputation took place now.

The synthetic data used in our experiments consist of a sample of 10^5 points generated from a k-gaussian distribution with some fixed variance σ^2, and

Table 2. k-means sum of distances to centroids, with $k = 5$, 10^5 samples and change's velocity of 10^{-3}

	Width	$\sigma = 0.15$		$\sigma = 0.3$		$\sigma = 0.6$	
		Static	Dynamic	Static	Dynamic	Static	Dynamic
ADWIN		9,72	21,54	19,41	28,58	38,83	46,48
Kalman $Q = 1, R = 1000$		9,72	19,72	19,41	27,92	38,83	**46,02**
Kalman $Q = 1, R = 100$		9,71	17,60	19,41	**27,18**	38,77	46,16
Kalman $Q = .25, R = .25$		9,71	22,63	19,39	30,21	38,79	49,88
Exponential Estimator $\alpha = .1$		9,71	21,89	19,43	27,28	38,82	46,98
Exponential Estimator $\alpha = .5$		9,72	20,58	19,41	29,32	38,81	46,47
Exponential Estimator $\alpha = .25$		9,72	17,69	19,42	27,66	38,82	46,18
Adaptive Kalman		9,72	18,98	19,41	31,16	38,82	51,96
CUSUM Kalman		9,72	18,29	19,41	33,82	38,85	50,38
K-ADWIN		**9,72**	**17,30**	**19,40**	**28,34**	**38,79**	**47,45**
Fixed-sized Window	32	9,72	25,70	19,40	39,84	38,81	57,58
Fixed-sized Window	128	9,72	36,42	19,40	49,70	38,81	68,59
Fixed-sized Window	512	9,72	38,75	19,40	52,35	38,81	71,32
Fixed-sized Window	2048	9,72	39,64	19,40	53,28	38,81	73,10
Fixed-sized Window	8192	9,72	43,39	19,40	55,66	38,81	76,90
Fixed-sized Window	32768	9,72	53,82	19,40	64,34	38,81	88,17
Fixed-sized flushing Window	32	9,72	35,62	19,40	47,34	38,81	65,37
Fixed-sized flushing Window	128	9,72	40,42	19,40	52,03	38,81	70,47
Fixed-sized flushing Window	512	9,72	39,12	19,40	53,05	38,81	72,81
Fixed-sized flushing Window	2048	9,72	40,99	19,40	56,82	38,81	75,35
Fixed-sized flushing Window	8192	9,72	45,48	19,40	60,23	38,81	91,49
Fixed-sized flushing Window	32768	9,72	73,17	19,40	84,55	38,81	110,77

centered in our k moving centroids. Each centroid moves according to a constant velocity. We try different velocities v and values of σ in different experiments.

Table 2 shows the results of computing the distance from 100 random points to their centroids. Again, in each column (a test), we show in boldface the result for K-ADWIN and for the best result.

The results can be summarized as follows: The winners are the best fixed-parameter Kalman filter and, for small variance, K-ADWIN. ADWIN follows closely in all cases. These three do much better than any fixed-size window strategy, and somewhat better than Kalman filters with suboptimal fixed-size parameters.

8   Conclusions and Future Work

The experiments on synthetic data give strong indications that the combination of Kalman filtering and a system that dynamically manages a window of examples has good potential for learning from data streams. More precisely, it seems to give better results than either memoryless Kalman Filtering or sliding windows with linear estimators. Furthermore, it tunes itself to the data stream at hand, with no need for the user to hardwire or precompute parameters that describe how the data stream changes over time.

Future work goes in two directions: on the one hand, these ideas should be tested on real-world, not only synthetic data. This is a notoriously difficult evaluation problem, since it is generally difficult to assess the real drift present in a real-world data set, hence compare meaningfully the performance of different strategies. On the other hand, other learning algorithms should be tried; clear candidates are algorithms for induction of decision trees.

References

1. B. Babcock, S. Babu, M. Datar, R. Motwani, and J. Widom. Models and issues in data stream systems. In *Proc. 21st ACM Symposium on Principles of Database Systems*, 2002.
2. A. Bifet and R. Gavaldà. Learning from time-changing data with adaptive windowing. Technical report, Universitat Politècnica de Catalunya, 2006. Available from www.lsi.upc.edu/~abifet.
3. M. Datar, A. Gionis, P. Indyk, and R. Motwani. Maintaining stream statistics over sliding windows. *SIAM Journal on Computing*, 14(1):27–45, 2002.
4. J. Gama, P. Medas, G. Castillo, and P. Rodrigues. Learning with drift detection. In *SBIA Brazilian Symposium on Artificial Intelligence*, pages 286–295, 2004.
5. G. Hulten, L. Spencer, and P. Domingos. Mining time-changing data streams. In *7th ACM SIGKDD Intl. Conf. on Knowledge Discovery and Data Mining*, pages 97–106, San Francisco, CA, 2001. ACM Press.
6. K. Jacobsson, N. Möller, K.-H. Johansson, and H. Hjalmarsson. Some modeling and estimation issues in control of heterogeneous networks. In *16th Intl. Symposium on Mathematical Theory of Networks and Systems (MTNS2004)*, 2004.
7. D. Kifer, S. Ben-David, and J. Gehrke. Detecting change in data streams. In *Proc. 30th VLDB Conf., Toronto, Canada*, 2004.
8. S. Muthukrishnan. Data streams: Algorithms and applications. In *Proc. 14th Annual ACM-SIAM Symposium on Discrete Algorithms*, 2003.
9. C. Ordonez. Clustering binary data streams with k-means. In *ACM SIGMOD Workshop on Research Issues on Data Mining and Knowledge Discovery*, 2003.
10. E. S. Page. Continuous inspection schemes. *Biometrika*, 41(1/2):100–115, 1954.
11. T. Schön, A. Eidehall, and F. Gustafsson. Lane departure detection for improved road geometry estimation. Technical Report LiTH-ISY-R-2714, Dept. of Electrical Engineering, Linköping University, SE-581 83 Linköping, Sweden, Dec 2005.
12. M. Severo and J. Gama. Change detection with Kalman Filter applied to apnoeas disorder. In *2nd. Intl. Workshop on Knowledge Discovery from Data Streams*, Porto (Portugal), 2005.
13. G. Welch and G. Bishop. An introduction to the Kalman Filter. Technical report, University of North Carolina at Chapel Hill, Chapel Hill, NC, USA, 1995.

Scientific Discovery: A View from the Trenches

Catherine Blake and Meredith Rendall

University of North Carolina at Chapel Hill, School of Information and Library Science,
214A Manning Hall, Chapel Hill, NC, USA 27599
{cablake, mbr}@email.unc.edu

Abstract. One of the primary goals in discovery science is to understand the human scientific reasoning processes. Despite sporadic success of automated discovery systems, few studies have systematically explored the socio-technical environments in which a discovery tool will ultimately be embedded. Modeling day-to-day activities of experienced scientists as they develop and verify hypotheses provides both a glimpse into the human cognitive processes surrounding discovery and a deeper understanding of the characteristics that are required for a discovery system to be successful. In this paper, we describe a study of experienced faculty in chemistry and chemical engineering as they engage in what Kuhn would call "normal" science, focusing in particular on how these scientists characterize discovery, how they arrive at their research question, and the processes they use to transform an initial idea into a subsequent publication. We discuss gaps between current definitions used in discovery science, and examples of system design improvements that would better support the information environment and activities in normal science.

Keywords: Socio-technical, information behaviors, knowledge discovery.

1 Introduction

As scientists, we often find the magic that surrounds a scientific discovery captivating. How could we forget Kekulé's dream of a snake eating its tail that revealed to him the elusive benzene ring structure; or the contamination in Fleming's lab that lead to the profoundly important discovery of penicillin? Kuhn would refer to these landmark discoveries as "scientific revolutions"[7]. Although revolutionary discoveries are powerful ways to attract new-comers to a field, or to keep ourselves motivated, scientists spend much of their time on day-to-day activities or what Kuhn would call "normal" science.

The underlying premise of our research is that capturing the day-to-day activities used by scientists as they develop and verify hypotheses will provide both a glimpse into the human cognitive processes surrounding discovery and into the complex socio-technical environments in which successful discovery tools will eventually be embedded. Simon, the unequivocal father of discovery science, and his colleagues, stated that "Discovery systems which solve tasks cooperatively with a domain expert are likely to have an important role, because in any nontrivial domain, it will be virtually impossible to provide the system with a complete theory which is anyway

N. Lavrač, L. Todorovski, and K.P. Jantke (Eds.): DS 2006, LNAI 4265, pp. 41–52, 2006.

constantly evolving" [11]. The need for additional user involvement in discovery systems is echoed by Langley who predicted that "as developers realize the need to provide explicit support for human intervention, we will see even more productive systems and even more impressive discoveries" [8].

Our position is that in addition to the difficulty in encoding prior knowledge and the need to design more interactive computational discovery systems, we need to understand the broader context in which science takes place to ensure the habitual adoption of discovery systems.

If adoption is one of the criteria of a successful discovery system, then it is critical that that we understand the work environment in which the discovery system will eventually be embedded. In this paper, we explore the processes used by experienced faculty in chemistry and chemical engineering. We focus in particular on how scientists define discovery, how they arrive at their research question and the processes used to transform research questions into published manuscripts. To frame our conversations, we interviewed 21 experienced scientists, using the critical incident technique[3] that employed two papers where each scientist was principal investigator and one paper they considered seminal to the field. Although we concur with Simon, et al., that a single publication captures only a highly circumscribed problem [11]; a published manuscript seems a natural level of analysis because we refer to work at this level through citations and because institutions measure the productivity of a scientist in terms of the number and quality of publications for promotion. Langley for example used publication as the "main criterion for success" of computer-aided scientific discoveries [8].

2   Research Design

Kuhn (1996) described discovery as "an inherently complex task". Although much of the previous research in discovery science has emphasized automated discovery, we seek to develop a model of the reasoning processes that surrounds a scientist's day-to-day activities. In this study, we focus on the processes used to develop a research question (hypothesis development) and to transition from that question to a final published manuscript (hypothesis verification), activities that consume much of a scientist's time. Our strategy is to conduct and record a series of interviews with experts in chemistry and chemical engineering. Chemistry has been a fruitful area for automated discovery systems, such as in MECHEM [14] and FAHRENHEIT [16].

2.1   Recruitment

The interviews reported in this paper were the first step in a two-year Text Mining Chemistry Literature project funded by the NSF in conjunction with the Center for Environmentally Responsible Solvents and Processes (CERSP). The primary investigators for the Center were strong supporters of the proposed interviews and sent the initial invitation for participation. Of the 25 scientists in the Center 21 members participated, resulting in a response rate of 84%. Inclusion in the interview process was determined solely by whether the participant was currently or had recently been involved in chemistry research. To ensure that subjects were familiar with the scientific discovery process, we required that they had previously published at least three articles and written one successful grant. Semi-structured interviews were conducted, recorded, transcribed verbatim, and analyzed during the Spring and Summer 2006.

2.2 Methodology

After approval by the Institutional Review Board of the University of North Carolina at Chapel Hill, we conducted the interviews in each scientist's office. We confirmed their title, role, and affiliations and then asked targeted questions regarding their definition of a discovery, the factors that limited the adoption and deployment of a new discovery, the processes they used to come up with a research question and verify the hypothesis. We also asked questions about the role scientific literature plays during their discovery processes and about their information use behaviors with respect to information overload, but the latter two themes are beyond the scope of this paper.

Data was collected using the critical incident technique [3] based on three articles or research grants: two the scientist wrote (one recently completed project and one project of which they were particularly proud), and a third they considered seminal to the field. For those scientists who were unable to provide these articles, we recommended frequently cited articles that we identified from a corpus of 103000 full text chemistry articles.

In addition to our interview questions, we asked scientists to reflect on the process they used for each of the two papers they had written. We provided the following activity cards: reading, thinking, online searching, books, journals, experimenting, analyzing, writing, discussing, and organizing; we asked the scientists to organize the cards in a way that reflected the scientific process used in the first paper (we also provided repeat cards and invited scientists to add steps). We then asked them to review the cards to ensure that the process reflected their second and subsequent research processes. We asked they simultaneously describe the process. We recorded and subsequently transcribed the entire interview and used grounded theory[5] to identify reoccurring themes.

3 Results

The average duration of the recorded interviews was 52 minutes. Of the 21 participants, half provided articles, and we suggested articles for the remaining 11 participants. None of the scientists provided a research grant. Figure 1 contains a summary of the participants including their ID letter that we will use in the remainder of this paper. Interviews were transcribed (including those that were not recorded) and analyzed using NVivo 7 [9]. Once collected the data was analyzed using a grounded theory approach [5].

ID	Interview (mins)	Title	Area of Research	Experience (yrs)
A	67	Director	Biochemical Engineering	32
B	55	Assistant Professor	Colloid Science and Engineering	10
C	51	Associate Professor	Polymer Design and Synthesis	12
D	51	Professor	Semiconductor Surface Chemistry	34
E	43	Professor	Polymer Chemistry	16
F	60	Professor	Nanoelectronics and Photonics	10
G	50	Professor	Electronic Materials Synthesis	26
H	59	Director	Polymer Design and Synthesis	35
I	39	Assistant Professor	Colloidal & Macromolecular Physics	14
J	61	Professor	Nanoelectronics and Photonics	36
K	58	Associate Professor	Bioorganic Chemistry	7
L	44	Professor	Rheology	13
M	41	Professor	Organometallic Chemistry	31
N	54	Professor	Polymer Theory	23
O	41	Professor	Electrochemistry	46
P	56	Professor	Synthetic Organometallic Chemistry	37
Q	5 *	Associate Professor	Surface and Interface Polymers	10
R	†	Associate Professor	Polymer Thin Films	20
S	53	Director	Polymer Synthesis	16
X	56	Professor	Neutron Scattering	35
Y	33	Professor	Chemical Reaction Engineering	40

*This time reflects a partial recording. † Declined permission to record the interview.

Fig. 1. Summary of participants

3.1 Definition of a Scientific Discovery

The evaluation of a discovery system is problematic without a definition of scientific discovery. One of the most cited discovery definitions is borrowed from Knowledge Discovery in Databases (KDD), which has been defined as "the nontrivial process of identifying valid, novel, potentially useful, and ultimately understandable patterns in data" [2]. Valdés-Pérez revised this statement to, "Discovery in science is the generation of novel, interesting, plausible and intelligible knowledge about the objects of study" [13]. In addition to asking directly, "What is your definition of discovery?", the seminal papers provided a framework for the discovery conversations. Nineteen of the 21 participants indicated the discovery characteristics, which fell into five key themes: novelty, building on existing ideas, a practical application, experimentation and theory, and simplicity.

3.1.1 Novelty
It was of little surprise that novelty was the most common theme surrounding discovery (11 out of 19 cases). Terminology comprised "not previously seen" (M);

"new insight" (G and M); "obviously novel and new and doesn't exist in the literature" (L); "finding something new and unexpected" (P); "learning something that hasn't really been well understood before (G)"; and "it [discovery] opens the door to exploration" (O).

It was interesting that in addition to new materials and substances, scientists provided examples of new transformations and processes. For example, scientist J said, "Novelty meaning novelty in what you are going to look at, how you are going to look at it, what you expect." Scientist R best characterized this difference between the actual output and the way that the problem had been characterized.

> You need to understand there are two different types of seminal papers. There are the some people working in certain fields, accumulating a lot of data with current models. They run experiments which are 80-90% of the time in agreement with the current models. 10-20% of the time they disagree with the model. At a certain point, these people reach a critical point when they can support a claim that the model is wrong.
>
> At this point, they put forth a new point of view/model, which raises the level of science. Perhaps this provides new language, ways of explaining what is going on, and perhaps develops a new/tangential line of science.
>
> Then there are other seminal papers that discuss a fundamental topic in a new light. For instances, hundreds of groups are working on the same topic, coming from the same angle, all trying to make it through the doorway first. But there are different approaches, and if you can find a different approach, you can find another door that unlocks the mystery everyone else is trying to solve simply because you took a different tack than any of the other groups working on the issue.

Combined with the expectation of novelty and the need for something unexpected, scientist P said, "A discovery is more important than something partially anticipated"(Q), whereas another scientist thought discoveries could be "planned or serendipitous".

3.1.2 Building on Existing Ideas

Ironically, the second most frequently occurring theme surrounding discovery was an improved understanding of an existing mechanism (7 out of 19 cases). Several scientists doubted that anything completely new existed. "Even supposedly the most creative people...I don't think things are cut from a whole cloth anymore. I think there aren't any more cloths without big holes anymore" (I); "Everything has precedent, in my opinion" (M); "from my standards, one has only less than ten completely new ideas in their lifetime, and so, most of the time you are sort of doing some modifications on a new idea or something" (N).

3.1.3 Practical Application

On par with the number of scientists who suggested building on previous ideas was a discussion for a practical application of the discovery (7 out of 19 cases), which is akin to the "useful" requirement proposed by Fayyad and Valdés-Pérez. However, in contrast to explaining what was known our data revealed a tension between differentiating discovery and practice and the need for a practical or commercial application. "I'd much rather work in the discovery mode: we discover something, get enough data to publish it, and think about how might this be commercialized, what would we need to make it commercializable, who might be interested in commercializing it. But that's kind of a secondary question. The primary question is 'what did we learn?' "What new insight can we bring to the field?" Rather than is this going to make my funding agent happy" (G).

Scientist M's experience when working for a funding agency captured the tension between practice and theory "Gee, I got tired of reading that this was the first this and the first that." And I was, dependent on my outlook how frank I am, but I said, "This is the first time in April that I've reviewed a paper for [removed to maintain anonymity]!" But who cares? So there are lots of firsts that are wearing the shirt of a different color or making a new isorun that nobody cares about. I don't put those in a discovery category. On the other hand, NSF likes to try to support research that is going to lead to discoveries that will be transformational and start new areas of research. Those are hard things to do and don't happen very often."

3.1.4 Experimentation and Theory

Scientists grappled with the duality of being both a theorist and an experimentalist. "… as an experimentalist, I treat analyzing data as 'let me try to decide whether or not what I have measured is real before I get too excited about it'. It goes on and on, analyzing in the context of discussing and thinking about what's occurring. I think when you try to do experiments, you have to disconnect yourself instead of getting all excited about results and thinking you have found something when you might not have" (I). This tension between developing theories that truly reflects the available evidence is a hallmark of a good scientist.

In one case, the scientist had not yet published his theory because they were "waiting for more proofs" (N). In another, the scientist reflected on the importance of experimentation to test their theoretical knowledge, "There's this added level of, almost, engineering that we build a system to see if we can mimic what nature does. If we can mimic what nature does then that tells us that we do understand it as well as we think that we do" (K).

Valdés-Pérez observed the theoretical motivations in chemistry when he stated that chemistry is, "more of theory-driven than many of the recent tasks addressed by data-driven discovery research" [15]. Our results suggest that scientists require discoveries to include both a theoretical foundation, and supporting experimental evidence.

3.1.5 Simplicity

Fayyad, et al,'s discovery definition required that the process be non-trivial. Stemming from the conversation of the importance of "a good description" was the need for simplicity which is typified by the following comments: "A discovery doesn't have to be something that is very hard. It could be something very simple that you can get to work. It doesn't have to be tedious work or years spent. Some people to see something simple might say it is way too simple, but as long as it is an elegant thing and hasn't been thought out already; I think that's fine. It's a discovery" (L), and "I was trying to make as simple as possible understanding of how this long molecules move sort of like spaghetti. ... So this was my attempt to make to simplify to the minimum possible description simple as possible description of this very complicated problem" (N).

3.2 Arriving at a Research Question

Discovery systems have been used to confirm or refute an existing hypotheses or to generate hypotheses. Our second goal is to explore how scientists arrive at their research questions. We conjecture that an improved understanding of this process will lead to improvements in automated discovery systems that assist in hypothesis generation.

One scientist noted that crystallizing the question and refining the problem were the fundamental elements of success in his projects, saying, "From my point of view, to get a good research question you have to define the problem properly. If you don't define the problem, you cannot do anything. Once you define the problem clear enough, you can, if the person is smart, you can find a solution. The difficulty of not being able to find a solution is just because you haven't crystallized and clearly posed the problem" (N).

All 21 scientists described how they arrived at their research question. Their sources fell into four key themes: discussion, previous projects, combining expertise and the literature.

3.2.1 Discussion

Discussions with colleagues and students were the most heavily cited source of idea generation (14 out of 21 cases). Scientists consulted with colleagues one-on-one and during Center presentations to clarify ideas in a paper or to investigate new avenues of research. Scientist I best captured the importance of discussions: "It's just like anything else, each time you start explaining something to somebody you realize how much you know and how well you understand it. To me that's discussing things, and actively discussing the data in general."

Several scientists acknowledged that the level of cooperation depended on the degree to which they were in competition or collaboration with another research group. The greatest cooperation occurred between different areas of science rather than within one genre, unless the scientists were collaborating, in which case there was a regular stream of information.

One scientist identified the importance of informal friendships, where he spent "40 days together over a two-year time period doing some DARPA work and a lot of bus rides, finding out what each other does. He does that field. I simply asked the question. What are the [removed to ensure anonymity] problems he has? He was able to spell them out and I said, we can fix that. He said no, people have been trying to fix that. I said, give us a shot, and we did. That's how it started" (S). Another scientist emphasized the importance of external conversations "Talking to people outside my field is something I would always do. I would never consider it a waste of my time to talk with people" (Q).

Conferences provided another opportunity for discussions. Attending presentations and participating in conversations were integral to the generation of new ideas and collaborations. One scientist noted that 'hearing about other stuff' (K) provided the source of their ideas, and incorporating ideas and processes into their research from related areas was beneficial. Discussions also arose from lectures, seminars, and interactions with visiting scholars.

3.2.2 Previous Projects

It is of little surprise that on-going projects also determined new research questions (13 out of 21 cases). Though our scientists acknowledged the evolution of ideas, many remained in the same general area as their graduate or post-doctoral work, stating: "I've been doing that since 1972" (D) and "This hard earned body of knowledge is what must determine future projects, whether mechanistic or experimental" (L).

Several of the discoveries identified by these scientists stemmed from experimental "mistakes", which one scientist challenged with, "Is it really wrong? No. It might be interesting" (E). Another stated that "There are times when an investigation finds itself off course and heading in an unanticipated direction, which may be for a variety of reasons including the original idea looking less and less promising to the unexpected outcome is very exciting and potentially a new area of science" (P).

One scientist called his group the 'follow your nose' group, meaning they based their next set of experiments on the data from that day's experiments, asking what had they learned, what were the obstacles, and what did that mean?" (E). Deciding to follow unexpected results or not depended on a matrix of situational elements, but the key elements seemed to be the purpose of the research and the perceived likelihood of success. If the purpose of the project was a mechanistic understanding of a system, then research groups were less likely to follow their noses. In some cases, the outcome of a project was the need to develop a new tool or technique. At least two of the scientists stated they had the only tool or technique in the world.

3.2.3 Combining Expertise

A re-occurring theme in arriving at a research question was collaborations with colleagues from different areas of chemistry. Scientist N reflected the international dimension of those collaborations on his visit to a lab in France "They had some very strange results they couldn't explain, and on the other side we were working on this different part of the same problem. So the two things clicked. That's how, from a discussion in France and our research at EKC, this idea came that this must be new." This theme relates closely to discussions in general (see section 3.2.1); however, the differing backgrounds and geographical locations might have implications to a discovery system that supports these behaviors.

3.2.4 Reading Literature

Chemists read literature more articles per year and spend more time reading compared with scientists in other fields[13]. It is no surprise then, that literature was included as a source for new research questions. The scientists cited discrepancies and errors in articles as a strong motivator, with one scientist stating that he had to conduct the research because he found the article "chemically offensive" (E).

3.3 The "Normal" Scientific Process

Our third goal was to model the process used by scientists to transform their initial research hypothesis into a published manuscript (hypothesis verification). We have created transition and co-occurrence models that reflect the process diagrams collected in this study; however due to space limitations in this paper, we provide recommendations that would close the gap between existing computational discovery system designs and the functionality required to support 'normal' science.

Figure 2 shows one of the process diagrams (selected at random). The repeat cycles for cards on left and on the bottom right of the image were typical of process diagrams collected in this study. The most frequent iteration occurred between experimentation and analysis. A discovery system that integrated activities would better support iteration than existing stand-alone designs. For example, a discovery system that integrated the data collected during experimentation with analysis tools could then identify patterns that agreed with or refuted the scientist's current experimental evidence.

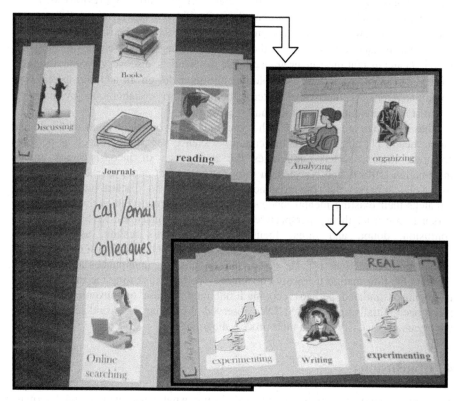

Fig. 2. An example of the process used to write a published manuscript

Consider again Figure 2, where the scientist's first iteration includes books, journals, online searching and reading. Perhaps the biggest gap between our observations of a scientist's day-to-day activities and existing computational discovery systems is the need to incorporate literature into the system. While discovery systems typically distinguish between structured and unstructured (aka data mining versus text mining), our observations suggest that data in both representations are critical to the process of verifying or refuting the validity of a new hypothesis. A discovery system that incorporated both could provide scientists with related articles and summaries of how previous work compares with current experimental findings. Such a system could also ease the transition from reading and analysis to writing by providing support for citation management.

Our data collected during this study suggests that scientist do not arrive at the discovery process with no a priori expectations. Instead, they start with a hypothesis projection, "the purely conjectural proliferation of a whole gamut of alternative explanatory hypotheses that are relatively plausible, a proliferation based on guesswork - though not 'mere' guesswork, but guesswork guided by a scientifically trained intuition."[8] To support this behavior, a discovery system should identify patterns that relate to a user's previous studies. The characterizations of discovery (see section 3.1.2) also reflect the need to relate current findings to previous work.

The process diagrams suggest that the scientific process is inherently a social endeavor, yet few discovery systems support interactions between scientists, or between scientists and students. Although mechanisms such as email (as shown in Figure 2) and instant messenger are available to scientists, discovery system designers have yet to incorporate these technologies into their design. In addition to real time communications, a discovery system could support collaboration by enabling scientists to share data, annotations, and manuscripts.

These findings are consistent with a previous study of scientists in medicine and public health [1] and with existing cognitive science perspectives, such as personal construct theory, which emphasizes the importance of inconsistencies among information artifacts and between an information artifact and a user's mental model [6]. Kelly suggests that inconsistencies force a user to discard information that threatens their existing mental models or formulate a tentative hypothesis. From the personal construct theory perspective, our study deemphasizes the early stages of confusion, doubt, and threat focusing on hypothesis testing, assessing, and reconstruing. Gardner's cognitive model suggests that synthesis plays a pivotal role in information interactions [4]. He states, "the organism ... manipulates and otherwise reorders the information it freshly encounters – perhaps distorting the information as it is being assimilated, perhaps recoding it into more familiar or convenient form once it has been initially apprehended." We observed re-ordering as several scientists used the organizing activity card in their process diagram.

4   Conclusions

This study explores activities conducted during what Kuhn would call "normal" science, including how scientists arrive at a research question (hypothesis development) and the processes used to transform a research question into a published manuscript (hypothesis verification). Such studies are critical if we are to design discovery systems that "solve tasks cooperatively with a domain expert" [10].

The most frequent themes surrounding the definition of discovery were novelty, building on existing ideas, practical application, conflict between experimentation and theory, and the need for simplicity. Such themes provide discovery systems designers with new criterion by which to measure the success of their automated discovery systems.

This study reveals the iterative nature of the scientific process, and the inherently social context in which normal science place. It is only by embedding computational methods of discovery into such an environment that we can ensure their habitual adoption.

Acknowledgments. Nancy Baker contributed to earlier discussions surrounding the activity cards and methodology used to collect the process diagrams from scientists. This material is based upon work supported in part by the STC Program of the National Science Foundation under Agreement No. CHE-9876674. Any opinions, findings, and conclusions or recommendations expressed in this material are those of the authors and do not necessarily reflect the views of the National Science Foundation.

References

[1] Blake, C. & Pratt, W. (In Press, 2006). Collaborative information synthesis I: A model of information behaviors of scientists in medicine and public health. To appear in *Journal of the American Society of Information Science and Technology*.

[2] Fayyad, U., Piatetsky-Shapiro, G., Smyth, P. and Uthurusamy, R. *Advances in Knowledge Discovery and Data Mining*. AAAI Press, 1996.

[3] Flanagan, J.C. The Critical Incident Technique. *Psychological Bulletin, 51* (4). 327-359, 1954.

[4] Gardner, H. *The mind's new science : a history of the cognitive revolution*. Basic Books, New York, 1985.

[5] Glaser, B.G. and Strauss, A.L. *The discovery of grounded theory. Strategies for qualitative research*. Chicago, Aldine Pub. Co, Chicago, 1967.

[6] Kelly, G.A. *A theory of personality: The psychology of personal constructs*. Norton, New York, 1963.

[7] Kuhn, T.S. *The Structure of Scientific Revolutions*. The University of Chicago Press, Chicago, 1996.

[8] Langley, P. The Computer-Aided Discovery of Scientific Knowledge. *Proceedings of the First International Conference on Discovery Science, 1* (4). 423-452, 1998.

[9] Rescher, N. (1978). Peirce's philosophy of science critical studies in his theory of induction and scientific method. Notre Dame, London: University of Notre Dame Press.

[10] QSR International Pty Ltd. www.qsrinternational.com/products/productoverview

[11] Simon, H.A., Valdés-Pérez, R.E. and Sleeman, D.H. Scientific discovery and simplicity of method. *Artificial Intelligence, 91* (2). 177-181, 1997.

[12] Tenopir, C., King, D. W., Boyce, P., Grayson, M., Zhang, Y., & Ebuen, M. (2003). Patterns of journal use by scientists through three evolutionary phases. D-Lib Magazine, 9, 1-15.

[13] Valdés-Pérez, R. Principles of human computer collaboration for knowledge discovery in science. *Artificial Intelligence, 107* (2). 335-346, 1999.

[14] Valdés-Pérez, R.E. Some Recent Human-Computer Discoveries in Science and What Accounts or Them. *AI Magazine*, 16 (3). 37-44, 1995.

[15] Zytkow, J.M., Combining many searches in the FAHRENHEIT discovery system. in *Proceedings of the 4th International workshop on machine learning*, (San Mateo, 1987),281-7.

[16] Zytkow, J.M. Integration of knowledge and method in real-world discovery. *ACM SIGART Bulletin*, 2 (4). 179-184, 1991.

Optimal Bayesian 2D-Discretization for Variable Ranking in Regression

Marc Boullé and Carine Hue

France Télécom R&D Lannion
Firstname.Name@orange-ft.com

Abstract. In supervised machine learning, variable ranking aims at sorting the input variables according to their relevance w.r.t. an output variable. In this paper, we propose a new relevance criterion for variable ranking in a regression problem with a large number of variables. This criterion comes from a discretization of both input and output variables, derived as an extension of a Bayesian non parametric discretization method for the classification case. For that, we introduce a family of discretization grid models and a prior distribution defined on this model space. For this prior, we then derive the exact Bayesian model selection criterion. The obtained most probable grid-partition of the data emphasizes the relation (or the absence of relation) between inputs and output and provides a ranking criterion for the input variables. Preliminary experiments both on synthetic and real data demonstrate the criterion capacity to select the most relevant variables and to improve a regression tree.

1 Introduction

In a data mining project, the data preparation step aims at providing a dataset for the modeling step [CCK+00]. Variable (or feature) selection consists in selecting a subset of the variables which is useful for a given problem. This selection process is an essential part of data preparation, which becomes critical in case of databases having large numbers of variables (order of thousands of). Indeed, the risk of overfitting the data quickly increases with the number of input variables, which is known as the curse of dimensionality. The objective of variable selection is three-fold: to improve the performance of predictors, to provide faster and more cost-effective predictors and to allow an easier interpretation of the prediction [GE03]. Variable selection methods generally use three ingredients [LG99]: a criterion to evaluate the relevance of a variable subset and compare variable subsets, a search algorithm to explore the space of all possible variable subsets and a stopping criterion. Variable selection is often linked to variable ranking which aims at sorting the variables according to their relevance. These two problems clearly differ as a subset of useful variables may exclude redundant but relevant variables. Conversely, the subset of the most relevant variables can be suboptimal among the subsets of equal size. Compared to variable selection, variable ranking is much more simple as it does not need any search algorithm

N. Lavrač, L. Todorovski, and K.P. Jantke (Eds.): DS 2006, LNAI 4265, pp. 53–64, 2006.
© Springer-Verlag Berlin Heidelberg 2006

but only the evaluation of the relevance criterion for each variable. For linear dependencies, the classical relevance criterion is the correlation coefficient or its square. To capture non linear dependencies, the mutual information is more appropriate but it needs estimates of the marginal and joint densities which are hard to obtain for continuous variables.

In this paper, we introduce a new relevance criterion for variable ranking in a regression problem with a large number of input variables. This criterion is based on the discretization of both the input and the output variables. Discretization has been widely studied in the case of supervised classification [Cat91] [Hol93] [DKS95] [LHTD02]. Our discretization method for regression extends our discretization method for the classification case to deal with numeric output variables. We apply a non parametric Bayesian approach to find the most probable discretization given the data. Owing to a precise definition of the space of discretization models and to a prior distribution on this model space, we derive a Bayes optimal evaluation criterion. We then use this criterion to evaluate each input variable and rank them. Besides a new variable ranking criterion, our method provides a robust discretization-based interpretation of the dependence between each input variable and the output variable and an estimator of the conditional densities for the considered regression problem.

The remainder of the document is organized as follows. Section 2 presents our discretization method for regression, with its criterion and optimization algorithm. In Section 3, we show preliminary experimental results both on synthetic and real data.

2   The MODL Discretization Method for Regression

We begin by recalling the principles of the Bayesian approach and the MDL approach [Ris78] for the model selection problem. We then present our approach (called MODL) which results in a Bayesian evaluation criterion of discretizations and the greedy heuristic used to find a near Bayes optimal discretization. We first present the principle of our discretization method for classification, and then extend it to the case of regression.

2.1   Bayesian Versus MDL Model Selection Techniques

In the Bayesian approach, the searched model is the one which maximizes the probability $p(Model|Data)$ of the model given the data. Using Bayes rule and since the probability is constant while varying the model, this is equivalent to maximizing:

$$p(Model)p(Data|Model) \tag{1}$$

Given a *prior* distribution of the models, the searched model can be obtained provided that the calculation of the probabilities $p(Model)$ and $p(Data|Model)$ is feasible. For classical parametric model families, these probabilities are generally intractable and the Bayesian Information Criterion (BIC) [Sch78] is a well-known penalized Bayesian selection model criterion. As detailed in the sequel, our approach, called MODL, conducts to an exact Bayesian model selection criterion.

To introduce the MDL approach, we can reuse the Bayes rule, replacing the probabilities by their negative logarithms. These negative logarithms of probabilities can be interpreted as Shannon code lengths, so that the problem of model selection becomes a coding problem. In the MDL approach, the problem of model selection is to find the model that minimizes:

$$DescriptionLength(Model) + DescriptionLength(Data|Model) \qquad (2)$$

The relationship between the Bayesian approach and the MDL approach has been examined by [VL00]. The Kolmogorov complexity of an object is the length of the shortest program encoding an effective description of this object. It is asymptotically equal to the negative log of a probability distribution called the *universal distribution*. Using these notions, the MDL approach turns into *ideal MDL*: it selects the model that minimizes the sum of the Kolmogorov complexity of the model and of the data given the model. It is asymptotically equivalent to the Bayesian approach with a universal prior for the model. The theoretical foundations of MDL allow focusing on the coding problem: it is not necessary to exhibit the prior distribution of the models. Unfortunately, the Kolmogorov complexity is not computable and can only be approximated.

To summarize, the Bayesian approach allows selecting the optimal model relative to the data, once a prior distribution of the models is fixed. The MDL approach does not need to define an explicit prior to find the optimal model, but the optimal description length can only be approximated and the approach is valid asymptotically.

2.2   The MODL Approach in Supervised Classification

The objective of a supervised discretization method is to induce a list of intervals which splits the numerical domain of a continuous input variable, while keeping the information relative to the output variable. A compromise must be found between information quality (homogeneous intervals in regard to the output variable) and statistical quality (sufficient sample size in every interval to ensure generalization). For instance, we present on left of Figure 1 the number of instances of each class of the Iris dataset w.r.t the sepal width variable. The problem is to find the split of the domain $[2.0, 4.4]$ in intervals which gives us optimal information about the repartition of the data between the three classes.

In the MODL approach [Bou06], the discretization is turned into a model selection problem. First, a space of discretization models is defined. The parameters of a specific discretization are the number of intervals, the bounds of the intervals and the output frequencies in each interval. Then, a prior distribution is proposed on this model space. This prior exploits the hierarchy of the parameters: the number of intervals is first chosen, then the bounds of the intervals and finally the output frequencies. The choice is uniform at each stage of the hierarchy. A Bayesian approach is applied to select the best discretization model,

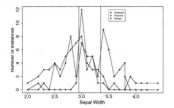

	]2.0,2.95[	[2.95, 3.35[	[3.35, 4.4[
Versicolor	34	15	1
Virginica	21	24	5
Setosa	2	18	30
Total	57	57	36

Fig. 1. MODL discretization of the Sepal Width variable for the classification of the Iris dataset in 3 classes

which is found by maximizing the probability $p(Model|Data)$ of the model given the data. Using the Bayes rule and since the probability $p(Data)$ is constant under varying the model, this is equivalent to maximize $p(Model)p(Data|Model)$. Let N be number of instances, J the number of output values, I the number of intervals for the input domain. $N_{i.}$ denotes the number of instances of input value in the interval i (total per column), $N_{.j}$ is the number of instances of class j (total per raw), and N_{ij} the number of instances of output value j in the interval i. In the context of supervised classification, the number of classes J and the number of instances per class $N_{.j}$ are supposed known. A discretization model is then defined by the parameter set $\left\{I, \{N_i\}_{1\leq i\leq I}, \{N_{ij}\}_{1\leq i\leq I, 1\leq j\leq J}\right\}$. We remark that the data partition obtained by applying such a discretization model is invariant by any monotonous variable transformation since it only depends on the variable ranks. Owing to the definition of the model space and its prior distribution, the Bayes formula is applicable to exactly calculate the prior probabilities of the models and the probability of the data given a model. Taking the negative log of the probabilities, this provides the evaluation criterion given in formula (3):

$$\log N + \log \binom{N+I-1}{I-1} + \sum_{i=1}^{I} \log \binom{N_i + J - 1}{J-1} + \sum_{i=1}^{I} \log \frac{N_i!}{N_{i,1}!N_{i,2}!\ldots N_{i,J}!}$$
(3)

The first term of the criterion stands for the choice of the number of intervals and the second term for the choice of the bounds of the intervals. The third term corresponds to the choice of the output distribution in each interval and the last term encodes the probability of the data given the model. The complete proof can be found in [Bou06].

Once the optimality of the evaluation criterion is established, the problem is to design a search algorithm in order to find a discretization model that minimizes the criterion. In [Bou06], a standard greedy bottom-up heuristic is used to find a good discretization. The method starts with initial single value intervals and then searches for the best merge between adjacent intervals. The best merge is performed if the MODL value of the discretization decreases after the merge and the process is reiterated until no further merge can decrease the criterion. In order to further improve the quality of the solution, the MODL algorithm performs post-optimizations based on hill-climbing search in the neighborhood of a discretization. The neighbors of a discretization are defined with combinations of interval splits and interval merges. Overall, the time complexity of the algorithm is $O(JNlog(N))$. The MODL discretization method for classification provides the most probable discretization given the data sample. Extensive comparative experiments report high quality performance. For the example given, the three obtained intervals are shown on left of Figure 1. The contingency table on the right gives us comprehensible rules such as "for a sepal width in $[2.0, 2.95]$, the probability of occurence of the Versicolor class is $34/57 = 0, 60$".

2.3   Extending the Approach to Regression

In order to illustrate the regression problem, we present in figure 2 the scatter-plot of the Petal Length and Sepal Length variables of the Iris dataset [Fis36]. The figure shows that Iris plants with petal length below 2 cm always have a sepal length below 6 cm. If we divide the sepal length values into two output intervals of values (below or beyond 6 cm), we can provide rules to describe the correlation between the input and the output variable. The regression problem is now turned into a classification problem. In this context, the objective of the MODL 2D-discretization method is to describe the distribution of the output intervals given the rank of the input value. This is extented to the regression case, where the issue is now to describe the rank of the output value given the rank of the input value. Discretizing both the input and output variable allows such a description, as shown in figure 3. The problem is still a model selection problem. Compared to the classification case, one additional parameter has to

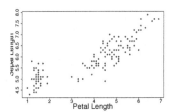

Fig. 2. Scatter-plot of the Petal Length and Sepal Length variables of the Iris dataset

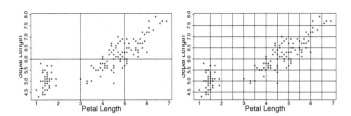

Fig. 3. Two discretization grids with 6 or 96 cells, describing the correlation between the Petal Length and Sepal Length variables of the Iris dataset

be optimized: the number of output intervals. A compromise has to be found between the quality of the correlation information and the generalization ability, on the basis of the grain level of the discretization grid. Let us now formalize this approach using a Bayesian model selection approach. A regression discretization model is defined by the parameter set $\left\{I,\ J,\ \{N_i\}_{1\leq i\leq I},\ \{N_{ij}\}_{1\leq i\leq I,\ 1\leq j\leq J}\right\}$. Unlike the supervised classification case, the number J of intervals in the output domain is now unknown but the number of instances $N_{.j}$ can be deduced by adding the N_{ij} for each interval. We adopt the following prior for the parameters of regression discretization models:

1. the numbers of intervals I and J are independent from each other, and uniformly distributed between 1 and N,
2. for a given number of input intervals I, every set of I interval bounds are equiprobable,
3. for a given input interval, every distribution of the instances on the output intervals are equiprobable,
4. the distributions of the output intervals on each input interval are independent from each other,
5. for a given output interval, every distribution of the rank of the output values are equiprobable.

The definition of the regression discretization model space and its prior distribution leads to the evaluation criterion given in formula (4) for a discretization model M:

$$c_{reg}(M) = 2\log(N) + \log\binom{N+I-1}{I-1} + \sum_{i=1}^{I}\log\binom{N_i+J-1}{J-1}$$
$$+ \sum_{i=1}^{I}\log\frac{N_i!}{N_{i,1}!N_{i,2}!\dots N_{i,J}!} + \sum_{j=1}^{J}\log N_j! \tag{4}$$

Compared with the classification case, there is an additional $log(N)$ term which encodes the choice of the number of output intervals, and a new last term (sum of $\log(N_j!)$) which encodes the distribution of the output ranks in each output interval. To give a first intuition, we can compute that for $I = J = 1$ the criterion value is $2\log(N) + \log(N!)$ (about 615 for $N = 150$) and for $I = J = N$ it gives $2\log(N) + \log\binom{2N-1}{N-1} + N\log(N)$ (about 224 for $N = 150$).

We adopt a simple heuristic to optimize this criterion. We start with an initial random model and alternate the optimization on the input and output variables. For a given output distribution with fixed J and N_j, we optimize the discretization of the input variable to determine the values of I, N_i and N_{ij}. Then, for this input discretization, we optimize the discretization of the output variable to determine new values of J, N_j and N_{ij}. The process is iterated until convergence, which usually takes between two and three steps in practice. The univariate discretization optimizations are performed using the MODL discretization algorithm. This process is repeated several times, starting from different random initial solutions. The best solution is returned by the algorithm. The evaluation criterion $c_{reg}(M)$ given in formula (4) is related to the probability that a regression discretization model M explains the output variable. We then propose to use it to build a relevance criterion for the input variables in a regression problem. The input variables can be sorted by decreasing probability of explaining the output variable. In order to provide a normalized indicator, we consider the following transformation of c_{reg}:

$$g(M) = 1 - \frac{c_{reg}(M)}{c_{reg}(M_\emptyset)},$$

where $M_\emptyset$ is the null model with only one interval for the input and output variables. This can be interpreted as a compression gain, as negative log of probabilities are no other than coding lengths [Sha48]. The compression gain $g(M)$ hold its values between 0 and 1, since the null model is always considered in our optimization algorithm. It has value 0 for the null model and is maximal when the best possible explanation of the output ranks conditionally to the input ranks is achieved.

Our method is non parametric both in the statistical and algorithmic sense : any statistical hypothesis needs to be done on the data distribution (like Gaussianity for instance) and, as the criterion is regularized, there is no parameter to tune before minimizing it. This strong point enables to consider large datasets.

3   Experimental Evaluation

In this section we first present the performance of the MODL 2D-discretization method on artificial datasets. Then, we apply it to rank the input variables of the Housing dataset from U.C. Irvine repository [DNM98] and show the interest of such a ranking criterion to improve regression tree performance.

3.1   Synthetic Data Experiments

We first test our method on a *noise pattern* dataset of size 100 where the input and output variables are independent and uniformly distributed on $[0; 1]$. As expected, the absence of relevant information in X to predict Y produces a 1 by 1 partition, i.e., a null compression gain.

Secondly, we test the ability of the MODL 2D-discretization to partition a noisy XOR pattern. Our dataset contains one hundred instances uniformly distributed in the square $[0; 0.5] \times [0; 0.5]$ and one other hundred in the square $[0.5; 1] \times [0.5; 1]$. Fifty instances have been added uniformely in the square $[0; 1] \times [0; 1]$. The optimal MODL partition with compression gain of 0.074 precisely detects the noisy XOR pattern as shown in Fig. 4. The associated contingency table gives the number of instances in each cellule of the partition. It enables to construct conditional density estimators as follows : from the first column we can say that if x is in $[0; 0.5]$, then y is in $0; 0.5]$ with probability $\frac{116}{116+13} = 0.90$.

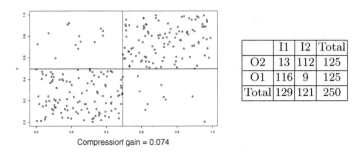

Compression gain = 0.074

	I1	I2	Total
O2	13	112	125
O1	116	9	125
Total	129	121	250

Fig. 4. Correlation diagram with optimal MODL grid for noisy XOR dataset

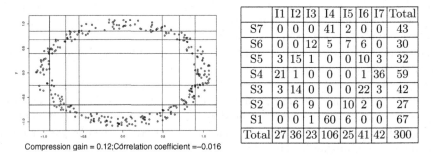

Compression gain = 0.12;Correlation coefficient =−0.016

	I1	I2	I3	I4	I5	I6	I7	Total
S7	0	0	0	41	2	0	0	43
S6	0	0	12	5	7	6	0	30
S5	3	15	1	0	0	10	3	32
S4	21	1	0	0	0	1	36	59
S3	3	14	0	0	0	22	3	42
S2	0	6	9	0	10	2	0	27
S1	0	0	1	60	6	0	0	67
Total	27	36	23	106	25	41	42	300

Fig. 5. Correlation diagram with optimal MODL grid for noisy circle synthetic data

The last synthetic experiment shows how the proposed method detects the presence of relevant information when two variables are not linearly correlated. We generate for this purpose a *circle* data set: three hundred instances have been generated on the circle of radius 1 with an additional noise such that their module is uniformely distributed in [0.9 ; 1.1]. As the empirical correlation is equal to −0.0169, any method based on the search of linear dependence fails. In contrast, the MODL 2D-discretization method underlines the relation between the two variables since the obtained compression gain is not zero. The optimal grid clearly identifies interesting regions as shown in Fig. 5.

3.2 Housing Data

In this section, we study the regression problem of the Housing MEDV variable which describes housing values in suburbs of Boston. The Housing dataset contains 506 instances, 13 numeric variables (including output variable MEDV) and 1 binary-valued variable which are described in Table 1.

We have split the Housing dataset in a 70% learning set and a 30% test set. Using the learning set, we have computed the optimal MODL 2D-discretizations for all of the twelve numeric variables. Considered as a relevance criterion, the associated compression gains are used to sort the variables according to the predictive information they contained w.r.t. the MEDV variable. To illustrate

Table 1. Description of the 13 variables of the Housing dataset

CRIM	per capita crime rate by town
ZN	proportion of residential land zoned for lots over 25,000 sq.ft.
INDUS	proportion of non-retail business acres per town
CHAS	Charles River dummy variable (= 1 if tract bounds river; 0 otherwise)
NOX	nitric oxides concentration (parts per 10 million)
RM	average number of rooms per dwelling
AGE	proportion of owner-occupied units built prior to 1940
DIS	weighted distances to five Boston employment centres
RAD	index of accessibility to radial highways
TAX	full-value property-tax rate per $10,000
PTRATIO	pupil-teacher ratio by town
B	$1000(Bk - 0.63)^2$ where Bk is the proportion of blacks
LSTAT	% lower status of the population
MEDV	Median value of owner-occupied homes in $1000's

Table 2. Sorted compression gains and empirical correlation coefficients for the 12 numerical variables of the Housing regression dataset

Input variable	Compression gain	Correlation coefficient
LSTAT	0.092	-0.748
RM	0.0617	0.715
NOX	0.0444	-0.414
CRIM	0.0397	-0.377
INDUS	0.0395	-0.462
PTRATIO	0.0365	-0.523
AGE	0.0346	-0.384
DIS	0.0280	0.239
TAX	0.0252	-0.435
RAD	0.017	-0.36
B	0.0115	0.3
ZN	0.0109	0.358

the interest of the MODL ranking criterion for variable selection, we then use this relevance criterion to improve CART regression trees [BFOS84]: we estimate such a tree with only the best MODL variable LSTAT, then with the two best variables and so on until the tree obtained with all the variables.

The sorted compression gains with the corresponding empirical correlation coefficients are shown in Table 2. For lack of space, the correlation diagram with the optimal MODL grids and the associated contingency table are shown in Fig 6 for the two better variables LSTAT and RM and in Fig 8 for the worst variable ZN. The PTRATIO variable seems also an interesting variable as its empirical correlation coefficient ranks it at the third position whereas the MODL criterion places it at the sixth (cf Fig 7). All these examples show the capacity of our MODL 2D-discretization algorithm to deal with complex datasets : the optimal grids present coarse grain when there is no predictive information or when the

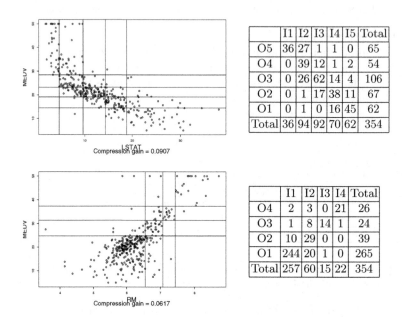

	I1	I2	I3	I4	I5	Total
O5	36	27	1	1	0	65
O4	0	39	12	1	2	54
O3	0	26	62	14	4	106
O2	0	1	17	38	11	67
O1	0	1	0	16	45	62
Total	36	94	92	70	62	354

	I1	I2	I3	I4	Total
O4	2	3	0	21	26
O3	1	8	14	1	24
O2	10	29	0	0	39
O1	244	20	1	0	265
Total	257	60	15	22	354

Fig. 6. Correlation diagram with optimal MODL grid for (LSTAT,MEDV) and (RM,MEDV) Housing variables

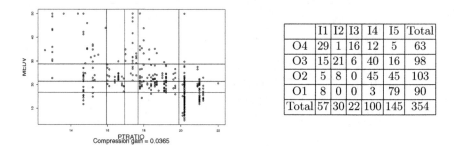

	I1	I2	I3	I4	I5	Total
O4	29	1	16	12	5	63
O3	15	21	6	40	16	98
O2	5	8	0	45	45	103
O1	8	0	0	3	79	90
Total	57	30	22	100	145	354

Fig. 7. Correlation diagram with optimal MODL grid for (PTRATIO,MEDV) Housing variables

data are too noisy and capture fine details as soon as there is enough instances. We then estimate the twelve trees with the 70% learning set. The first is estimated using only the best LSTAT variable, the second with the two best variables LSTAT and RM and so on until the twelfth tree estimated with all the variables. The obtained trees are used to predict the MEDV variable for both the learning and the test set. The resulting root mean squared errors are plotted in Fig. 9 for learning and test datasets. For both sets, we notice that:

- considering all the variables to estimate the regression tree is less efficient than considering only the more relevant ones according to the MODL criterion.

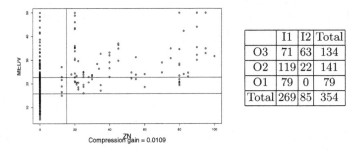

	I1	I2	Total
O3	71	63	134
O2	119	22	141
O1	79	0	79
Total	269	85	354

Fig. 8. Correlation diagram with optimal MODL grid for (ZN,MEDV) Housing variable

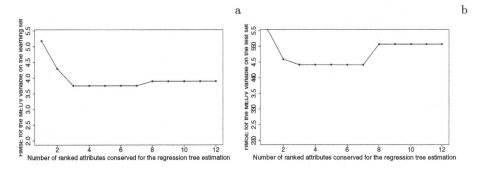

Fig. 9. Root Mean Squared Errors for the MEDV variable predicted from the regression trees estimated on a 70% learning dataset using an increasing number of ranked MODL variables (a) on the learning set (b) on the test set

- the optimal tree is obtained with the three best variables and degrades after the incorporation of the eighth.

We can then conclude that, for this dataset, choosing the third tree during training step conducts to the best choice for the test set.

4   Conclusion and Future Work

The MODL 2D-discretization method proposed in this paper is a Bayesian model selection method for discretization grid models. The exact MODL criterion obtained enables to find the most probable discretization-based explanation of the data. Using a heuristic iterative algorithm which alternatively performs the discretization of the input and of the output variable, the obtained partitions accurately show linear and non linear relation and their compression gain can be used as a relevance criterion for the input variable ranking problem. It seems a very promising method to efficiently detect the relevant variables in large datasets during the data preparation step of regression problems.

In a future work, we plan to pursue the validation of our approach on larger numerous datasets and to use it to build a multivariate naive Bayes regressor exploiting the MODL discretized grids.

References

[BFOS84] L. Breiman, J. H. Friedman, A. Olshen, and C. J. Stone. *Classification and Regression Trees*. Wadsworth, Belmont, 1984.

[Bou06] M. Boullé. Modl :a Bayes optimal discretization method for continuous attributes. *Machine Learning*, accepted for publication 2006.

[Cat91] J. Catlett. On changing continuous variables into ordered discrete variables. In *In Proceedings of the European Working Session on Learning*, pages 87–102, 1991.

[CCK+00] P. Chapman, J. Clinton, R. Kerber, T. Khabaza, T. Reinartz, C. Shearer, and R. Wirth. Crisp-dm 1.0: step-by-step data mining guide. *Applied Statistics Algorithms*, 2000.

[DKS95] J. Dougherty, R. Kohavi, and M. Sahami. Supervised and unsupervised discretization of continuous features. In *In Proceedings of the 12th International Conference on Machine Learning*, pages 194–202, 1995.

[DNM98] C.L. Blake D.J. Newman, S. Hettich and C.J. Merz. UCI repository of machine learning databases, 1998.

[Fis36] R.A. Fisher. The use of multiple measurements in taxonomic problems. *Annual Eugenics*, 7, 1936.

[GE03] I. Guyon and André Elissef. An introduction to variable and feature selection. *Journal of Machine Learning Research*, 2003.

[Hol93] R.C. Holte. Very simple classification rules perform well on most commonly used datasets. *Machine Learning*, 1993.

[LG99] Ph. Leray and P. Gallinari. Feature selection with neural networks. *Behaviormetrika*, 1999.

[LHTD02] H. Liu, F. Hussain, C.L. Tan, and M. Dash. Discretization: An enabling technique. *Data Mining and Knowledge Discovery*, 2002.

[Ris78] J. Rissanen. Modeling by shortest data description. *Automatica*, 1978.

[Sch78] G. Schwarz. Estimating the dimension of a model. *Annals of Statistics*, 1978.

[Sha48] C.E. Shannon. A mathematical theory of communication. *Bell systems technical journal*, 1948.

[VL00] P.M.B. Vitanyi and M. Li. Minimum description length induction, Bayesianism, and Kolmogorov complexity. *IEEE Trans. Inform. Theory*, 2000.

Text Data Clustering by Contextual Graphs

Krzysztof Ciesielski and Mieczysław A. Kłopotek

Institute of Computer Science, Polish Academy of Sciences,
ul. Ordona 21, 01-237 Warszawa, Poland
{kciesiel, klopotek}@ipipan.waw.pl

Abstract. In this paper, we focus on the class of graph-based clustering
models, such as growing neural gas or idiotypic nets for the purpose
of high-dimensional text data clustering. We present a novel approach,
which does not require operation on the complex overall graph of clusters,
but rather allows to shift majority of effort to context-sensitive, local subgraph and local sub-space processing. Savings of orders of magnitude in
processing time and memory can be achieved, while the quality of clusters
is improved, as presented experiments demonstrate.

1 Introduction

Visual presentation of Web mining results has been increasingly appreciated with
the advent of clustering in document processing. A prominent position among
the techniques of Visual Web-content mining is taken by the WebSOM (Self
Organizing Maps) of Kohonen and co-workers [12]. However, the overwhelm-
ing majority of the existing document clustering and classification approaches
rely on the assumption that the particular structure of the currently available
static document collection will not change in the future. This seems to be highly
unrealistic, because both the interests of the information consumer and of the
information producers change over time.

The WebSOM-like document map representation is regrettably time and space
consuming, and rises also questions of scaling and updating of document maps. In
our previous papers we have described some approaches we found useful in coping
with these challenges. Among others, techniques like Bayesian networks, growing
neural gas, SVD analysis and artificial immune systems have been studied [11].
We created a full-fledged search engine for collections of documents (up to a few
millions) capable of representing on-line replies to queries in graphical form on
a document map, based on the above-mentioned techniques, for exploration of
free text documents by creating a navigational 2-dimensional document map in
which geometrical vicinity would reflect conceptual closeness of the documents.

Hung et al. [9] demonstrated deficiencies of various approaches to document
organization under non-stationary environment conditions of growing document
quantity. They pointed to weaknesses of the original SOM approach (though SOM
is adaptive to some extent) and proposed a novel dynamic self-organizing neural model, so-called Dynamic Adaptive Self-Organizing Hybrid (DASH) model.
This model is based on an adaptive hierarchical document organization, supported

N. Lavrač, L. Todorovski, and K.P. Jantke (Eds.): DS 2006, LNAI 4265, pp. 65–76, 2006.
© Springer-Verlag Berlin Heidelberg 2006

by human-created concept-organization hints available in terms of WordNet. Other strategies like that of [4,14], attempt to capture the move of topics, enlarge dynamically the document map (by adding new cells, not necessarily on a rectangle map).

We propose a novel approach, addressing both the issue of topic drift and scalability. Section 3 explains in detail the concept of so-called contextual maps. While being based on a hierarchical (three-level) approach, it is characterized by three distinctive features. In our opinion, WebSOM-like clustering is inefficient, because its target 2D grid is too rigid. Hence, we propose first to use modified growing neural gas (GNG [6], section 2) clustering technique, which is also a structural one, but has a more flexible structure accommodating better to non-uniform similarity constraints. The GNG is then projected onto a 2D map, which is less time consuming than direct WebSOM like map creation. In fact, any other structural clustering like aiNet (AIS or artificial immune system approach) can be used instead of GNG[1]. The second innovation is the way we construct the hierarchy: first, we split the documents into (many) independent clusters (which we call "contexts"), then apply structural clustering within them, and in the end cluster structurally the "contexts" themselves. What we gain, is the possibility of drastic dimensionality reduction within the independent clusters (as they are more uniform topically) which accelerates the process and stabilizes it. The third innovation is the way we apply GNG technique. Instead of the classical global search, we invented a mixed global/local search especially suitable for GNG [11].

To evaluate the effectiveness of the novel map formation process, in experimental section 4 we compare it to the standard approaches. The crucial concept in contextual approach is the topic-sensitive representation of the vector space. In sections 4.3-4.5 we present experimental results which confirms its positive impact on clustering quality, processing complexity and meta-clustering graph structure. Final conclusions from our research work are given in section 5.

2 Growing Neural Gas for Document Clustering

Our baseline model for extracting topical groups was the Growing Neural Gas (GNG) network proposed in [6]. Like Kohonen (SOM) network, GNG can be viewed as topology learning algorithm, i.e. its aim can be summarized as follows: given some collection of high-dimensional data, find a topological structure which closely reflects the topology of the collection. In typical SOM the number of units and topology of the map is predefined. As observed in [6], the choice of SOM structure is difficult, and the need to define a decay schedule for various features is problematic.

GNG starts with very few units and new units are inserted successively every k iterations. To determine where to insert new units, local error measures are gathered during the adaptation process; new unit is inserted near the unit, which

---

[1] Actually, contextual aiNet model has also been implemented in BEATCA and proved to be an efficient alternative for a "flat" immune-based models [3].

has accumulated maximal error. Interestingly, GNG cells of the GNG network are joined automatically by links, hence as a result a possibly disconnected graph is obtained, and its connected components can be treated as different data clusters. The complete algorithm details can be found in [6].

2.1 GNG Extension with Utility Factor

Typical problem in web mining applications is that processed data is constantly changing - some documents disappear or become obsolete, while other enter analysis. All this requires models which are able to adapt their structure quickly in response to non-stationary distribution changes. Thus, we decided to adopt GNG with utility factor model [7].

A crucial concept here is to identify the least useful nodes and remove them from GNG network, enabling further node insertions in regions where they would be more necessary. The utility factor of each node reflects its contribution to the total classification error reduction. In other words, node utility is proportional to expected error growth if the particular node would have been removed. There are many possible choices for the utility factor. In our implementation, utility update rule of a winning node has been simply defined as $U_s = U_s + error_t - error_s$, where s is the index of the winning node, and t is the index of the second-best node (the one which would become the winner if the actual winning node would be non-existent). Newly inserted node utility is arbitrarily initialized to the mean of two nodes which have accumulated most of the error: $U_r = (U_u + U_v)/2$.

After utility update phase, a node k with the smallest utility is removed if the fraction $error_j /U_k$ is greater then some predefined threshold; where j is the node with the greatest accumulated error.

2.2 Ant Colony-Like Edge Utility Factor

We have developed additional edge utility function, which proved to make GNG model learning more robust in terms of convergence time of a single iteration. It is based on the one of the most prominent swarm intelligence approaches, so-called ant colony optimization (see e.g. [5]). Documents moving through graph edges are treated as artificial ants, which leave traces of artificial pheromone on the traversed path, in such a way that ants following them could behave more effectively. In our case, it means to shorten the document movement path by adding additional edges, linking starting node and the final node of the path traversed by a document. If such a node already exists, its utility factor is increased. When it eventually comes to edge removal, not only its age but also utility is taken into account; edges with utility greater than a predefined threshold are left unchanged. Contrary to the basic GNG algorithm, this approach exploits local information gathered during winner node search and reinforces search method with the memory of the model' dynamics in the recent phases of the learning process. Thus, it allows for temporal adaptation of the graph structure to facilitate accommodation of the streams of new documents within a certain subject or contextual group.

3   Contextual Maps

In our approach – like in many traditional IR systems – documents are represented as the points in term vector space by a standard term frequency/inverse document frequency (*tfidf*) weights:

$$tfidf\,(t_i, d_j) = w_{td} = f_{td} \cdot log\left(\frac{N}{f_t}\right) \qquad (1)$$

where w_{td} is the weight value for term t_i and document d_j, f_{td} is the number of occurrences of term t_i in document d_j, f_t is the number of documents containing term t_i and N is the total number of documents. As we have to deal with text documents represented in very high-dimensional space, first we apply entropy-based quality measures [2] to reduce the dimensionality of document-space.

It is a known phenomenon that text documents are not uniformly distributed over the space. Characteristics of frequency distributions of a particular term depend strongly on document location, while are similar for the neighboring documents. Our approach is based on the automatic division of the set of documents into a number of homogenous and disjoint subgroups each of which is described by unique (but usually overlapping) subset of terms.

We argue that after splitting documents in such groups, term frequency distributions within each group become much easier to analyze. In particular, it selection of significant and insignificant terms for efficient calculation of similarity measures during map formation step appears to be more robust. Such document clusters we call *contextual* groups. For each contextual group, separate maps are generated.

In the sequel we will distinguish between *hierarchical* and *contextual* model. In the former the set of terms, with *tfidf* weights (eq. (1)), is identical for each subgroup of documents, while in the later each subgroup is represented by different subset of terms weighted in accordance with the equation (3). Finally, when we do not split the entire set of documents and we construct a single, "flat", representation for whole collection – we will refer to a *global* model.

The contextual approach consists of two main stages. At first stage a hierarchical model is built, i.e. a collection D of documents is recurrently divided – by using Fuzzy ISODATA algorithm [1] – into homogenous groups consisting of approximately identical number of elements (by an additional modification in optimized quality criterion that penalizes for inbalanced splits, in terms of cluster size).

At the beginning, whole document set is split into a few (2-5) groups. Next, each of these groups is recursively divided until the number of documents inside a group meets required homogeneity or quantity criteria. After such process we obtain hierarchy, represented by a tree of clusters. In the last phase, groups which are smaller than predefined constraint, are merged to the closest group[2]. Similarity measure is defined as a single-linkage cosine angle between both clusters centroids.

---

[2] To avoid formation of additional maps which would represent only a few outliers in document collection.

The second, crucial phase of contextual document processing is the division of terms space (dictionary) into – possibly overlapping – subspaces. In this case it is important to calculate fuzzy membership level, which will represent importance of a particular word or phrase in different contexts (and implicitly, ambiguity of its meaning). Estimation of fuzzy within-group membership of the term m_{tG} is estimated as:

$$m_{tG} = \frac{\sum_{d \in G} (f_{td} \cdot m_{dG})}{f_G \cdot \sum_{d \in G} m_{dG}} \tag{2}$$

where f_G is the number of documents in the cluster G, m_{dG} is the degree of document d membership level in group G, f_{td} is the number of occurrences of term t in document d.

Next, vector-space representation of a document is modified to take into account document context. This representation increases weights of terms which are significant for a given contextual group and decrease weights of insignificant terms. In the extreme case, insignificant terms are ignored, what leads to the (topic-sensitive) reduction of representation space dimensionality. To estimate the significance of term in a given context, the following measure is applied:

$$w_{tdG} = f_{td} \cdot m_{tG} \cdot log \left(\frac{f_G}{f_t \cdot m_{tG}} \right) \tag{3}$$

where f_{td} is the number of occurrences of term t in document d, m_{tG} is the degree of membership of term t in group G, f_G is the number of documents in group G, f_t is the number of documents containing term t.

Main idea behind the proposed approach is to replace a single GNG model by a set of independently created contextual models and to merge them together into a hierarchical model. Training data for each model is a single contextual group. Each document is represented as a standard referential vector in term-document space. However, $tfidf$ measure (1) of vector components is replaced by w_{tdG}.

To represent visually similarity relation between contexts (represented by a set of contextual models), additional "global" map is required. Such model becomes a root of contextual maps hierarchy. Main map is created in a manner similar to previously created maps, with one distinction: an example in training data is a weighted centroid of referential vectors of the corresponding contextual model: $x_i = \sum_{c \in M_i} (d_c \cdot v_c)$, where M_i is the set of cells in i-th contextual model, d_c is the density of the cell and v_c is its referential vector.

Learning process of the contextual model is to some extent similar to the classic, non-contextual learning. However, it should be noted that each constituent model (and the corresponding contextual map) can be processed independently, in particular it can be distributed and calculated in parallel. Also a partial incremental update of such models appears to be much easier to perform, both in terms of model quality, stability and time complexity. The possibility of incremental learning stems from the fact that the very nature of the learning process is iterative. So if new documents come, we can consider the learning process as having been stopped at some stage and it is resumed now with all the documents. We claim that it is not necessary to start the learning process from scratch neither in the case that the new documents "fit" the distribution of the previous ones nor when their term distribution is significantly different [11]. In the section 4.3 we present some thoughts on scalability issues of contextual approach. Finally, we evaluate meta-clustering structure and compare contextual representation with standard $tfidf$ weights.

4   Experimental Results

To evaluate the effectiveness of the presented contextual map formation approach, we compared it to the global and hierarchical approaches. The architecture of our system supports comparative studies of clustering methods at the various stages of the process (i.e. initial document grouping, initial topic identification, incremental clustering, model projection and visualization, identification of topical areas on the map and its labeling). In particular, we conducted series of experiments to compare the quality and stability of GNG, AIS and SOM models for various model initialization methods, winner search methods and learning parameters [11,3]. In this paper we focus on evaluation of the proposed GNG-based model: its quality, learning complexity and stability as well as the graph structure of the resulting clustering with respect to the topic-sensitive learning approach.

In this section we describe the overall experimental design, quality measures used and the results obtained. The scalability study in section 4.3 was based on a collection of more than one million Internet documents, crawled by our topic-sensitive crawler, while the graph structure evaluation and contextual representation studies in sections 4.4-4.5 were based on a widely-used "20 Newgroups" benchmarking document collection.

4.1   Quality Measures for the Document Maps

A document map may be viewed as a special case of the concept of clustering. One can say that clustering is a learning process with hidden learning criterion. The criterion is intended to reflect some esthetic preferences, like: uniform split into groups (topological continuity) or appropriate split of documents with known a priori categorization. As the criterion is hidden, in the literature [15,8] a number of clustering quality measures have been developed, checking how the clustering fits the expectations. We selected the following ones for our study:

– **Average Map Quantization:** The average cosine distance between each pair of adjacent nodes. The goal is to measure topological continuity of the model (the lower this value is, the more "smooth" model is): $AvgMapQ = \frac{1}{|N|} \sum_{n \in N} \left(\frac{1}{|E(n)|} \sum_{m \in E(n)} c(n,m) \right)$, where N is the set of graph nodes, $E(n)$ is the set of nodes adjacent to the node n and $c(n,m)$ is the cosine distance between nodes n and m.

– **Average Document Quantization:** Average distance (according to cosine measure) for the learning set between the document and the node it was classified into. The goal is to measure the quality of clustering at the level of a single node: $AvgDocQ = \frac{1}{|N|} \sum_{n \in N} \left(\frac{1}{|D(n)|} \sum_{d \in D(n)} c(d, n) \right)$, where $D(n)$ is the set of documents assigned to the node n.

Both measures have values in the [0,1] interval, the lower values corresponds respectively to more "smooth" inter-cluster transitions and more "compact" clusters. To some extent, optimization of one of the measures entails increase of the other one.

4.2   Quality Measures for the Graphs of Clusters

Beside the clustering structure represented by nodes, GNG graph is also a meta-clustering model. Similarity between individual clusters is given by graph edges, linking referential vectors in nodes. Thus, there is a need to evaluate quality of the structure of the edges.

There are many possible ways to evaluate GNG model structure, here we present one which we have found to be the most be the most clear for the interpretation. This approach is based on the comparison of GNG model structure with two referential graphs: clique (complete graph) and minimal spanning tree (MST).

A 2x2 contingency matrix of the graph edges is built. Each edge of the clique falls into one of the four categories: common for both GNG and MST, present only in GNG, present only in MST, absent in both graphs. Next, we can calculate various statistics for edges within each contingency group, starting from the simplest ones (such as the total number of edges, average similarity) to the more complex one (e.g. fractal dimension for the induced subgraph).

4.3   Scalability Issues

To evaluate scalability of the proposed contextual approach (both in terms of space and time complexity), we built a model for a collection of more than one million documents crawled by our topic-sensitive crawler, starting from several Internet news sites (cnn, reuters, bbc).

Resulting model consisted of 412 contextual maps, which means that the average density of a single map was about 2500 documents. Experimental results in this section are presented in series of box-and-whisker plots, which allows to present a distribution of a given evaluation measure (e.g. time, model smoothness or quantization error) over all 412 models, measured after each iteration of the learning process (horizontal axis). Horizontal lines represent median values, area inside the box represents 25% - 75% quantiles, whiskers represent extreme values and each dot represents outlier values.

Starting with the initial document clustering/context initialization via hierarchical Fuzzy ISODATA (see section 3), followed by GNG model learning

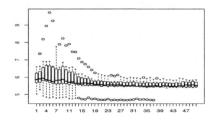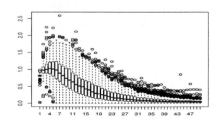

Fig. 1. Contextual model computation complexity (a) execution time of a single iteration (b) average path length of a document

(see section 2) and GNG-to-SOM projection (model visualization), the whole cycle of map creation process took 2 days. It is impressing result, taking into account that Kohonen and his co-workers reported processing times in order of weeks [13]. It should also be noted that the model was built on a single personal computer[3]. As it has been stated before, contextual model construction can be easily distributed and parallelized, what would lead to even shorter execution times.

The first observation is the complexity of a single iteration of the GNG model learning (Figure 1(a)), which is almost constant, regardless of the increasing size of the model graph. It confirms the observations from section 4, concerning efficiency of the tree-based winner search methods. One can also observe the positive impact of homogeneity of the distribution of term frequencies in documents grouped to a single map cell. Such homogeneity is - to some extent - acquired by initial split of a document collection into contexts. Another cause of the processing time reduction is the contextual reduction of vector representation dimensionality, described in the section 3.

In the Figure 1(b), the dynamic of the learning process is presented. The average path length of a document is the number of shifts over graph edges when documents is moved to a new, optimal location. It can be seen that model stabilizes quite fast; actually, most models converged to the final state in less than 30 iterations. The fast convergence is mainly due to topical initialization. It should be stressed here that the proper topical initialization can be obtained for well-defined topics, which is the case in contextual maps.

The Figure 2 presents the quality of the contextual models. The final values of average document quantization (Figure 2(a)) and the map quantization (Figure 2(b)) are low, which means that the resulting maps are both "smooth" in terms of local similarity of adjacent cells and precisely represent documents grouped in a single node. Moreover, such low values of document quantization measure have been obtained for moderate size of GNG models (majority of the models consisted of only 20-25 nodes - due to their fast convergence - and represented about 2500 documents each).

---

[3] Pentium IV HT 3.2 GHz, 1 GB RAM.

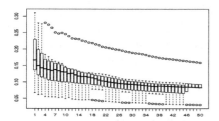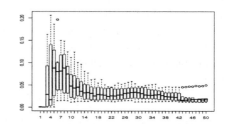

Fig. 2. Contextual model quality (a) Average Document Quantization (b) Average Map Quantization

4.4   Meta-clustering Structure Evaluation

To compare robustness of various variants of GNG-based models, graph structure quality measures have been applied. For each model, contingency matrices described in section 4.2 were computed. Due to the lack of space, we only briefly describe major results, comparing global GNG model, hierarchical model exploiting $tfidf$ representation 1 and contextual model using topic-sensitive representation 3. Experiments were executed on 20000 documents from 20 Newsgroups collection.

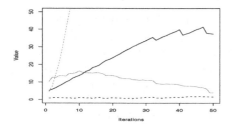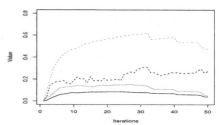

Fig. 3. Graph structure quality (a) Total Number of Edges (b) Average Edge Length (similarity)

As we expected, in all three cases, GNG graph structure has a lot in common with minimal spanning tree. Figure 3(a) shows the the total number of edges in each cell of the contingency matrix[4]. Graph contains relatively few edges comparing to the clique (complete graph, represented by steep, dashed grey line) and most of the edges are shared by MST and GNG graphs (black solid line). For this edges, the criterion of maximizing the similarity between linked clusters is met.

On the Figure 3(b), we can see that this criterion is met also for the edges present only in GNG graph (solid grey line). The figure present average similarity (i.e. edge length) between linked clusters within each contingency category.

---

[4] This and the following plots present averaged values for 20 contextual GNG graphs.

As one can see, the average length for GNG-only edges (solid grey line) is not significantly higher than for the edges common for MST and GNG . The purpose of this additional edges is to support the quick convergence of the model, so it can be observed on Figure 3(a) that there are much more of them at the beginning and in the middle phases of the learning process than in the final model.

The last category consists of the edges which are present only in MST and absent in GNG (dashed black line). They are quite scarce and noticeably longer than edges in two former categories. If we recall that GNG graph is usually unconnected, while MST is connected by definition, it is straightforward to conclude that these edges connect separated (and distant) meta-clusters (i.e. clusters of nodes).

To wrap up, we briefly mention that the number of edges present only in GNG graph and absent in MST is higher in case of the model exploiting contextual representation than in case of $tfidf$ representation. The same is observed when examining the model using ant colony edge utility factor (pheromone paths described in section 2.2). This observation is consistent with significant reduction of the processing time of the single learning iteration in those two models - enriched graph structure allows for faster and well-guided relocation of documents.

4.5   Convergence wrt Contextual Representation

The next series of experiments compared contextual GNG model with hierarchical GNG model (exploiting standard $tfidf$ representation). Figure 4(a) presents convergence (wrt Average Document Quantization measure) of the contextual model, represented by black line. In each iteration, we also calculated $AvgDocQ$ measure for the learning dataset (20 Newsgroups), but represented via $tfidf$ weights in vector space (grey line). In the opposite case, Figure 4(b), we have used $tfidf$ representation for learning (black line) and contextual weights for testing (grey line).

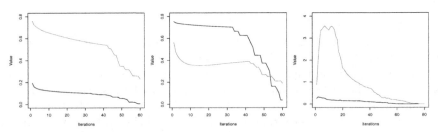

Fig. 4. Comparison of vector-space representations (a) convergence wrt contextual learning with $tfidf$ test dataset (b) convergence wrt $tfidf$ learning with contextual test dataset (c) processing time of a single iteration

In both cases it can be seen that $tfidf$ representation is characterized by a long, almost stable learning phase, followed by a rapid convergence. Moreover, when $tfidf$ is used for learning then contextual representation is divergent in the middle phase[5]. It can also be noted that the further convergence of test (contextual) representation is associated with the start of the rapid convergence of $tfidf$. There's no room for all the details here, but it can be shown that this undesirable behavior is the result of the noised information brought by additional terms, which finally appears to be not meaningful in the particular context (and thus are disregarded in contextual weights w_{dtG}).

What can not be seen on the plot is that convergence to the stable state requires less iterations in case of contextual representation than in case of $tfidf$ representation. The last figure, 4(c), presents the average time complexity (in seconds) of a single learning iteration for both models. For 20000 documents in 20 Newsgroups, the total processing time accounted for 1861.81 seconds for hierarchical GNG model using $tfidf$ representation against just 172.24 seconds for contextual GNG model. For juxtaposition, learning of a flat, global GNG model required more than two hours (and even more in case of the SOM model).

5   Concluding Remarks

We presented the contextual approach, which proved to be an effective solution to the problem of massive data clustering. It is mainly due to: (1) replacement of a flat, global, graph-based meta-clustering structure with a hierarchy of topic-sensitive models and (2) introduction of contextual term weighting instead of standard $tfidf$ weights so that document clusters can be represented in different subspaces of a global vector space. With these improvements we proposed a new approach to mining high dimensional datasets. The contextual approach appears to be fast, of good quality (in term of indices introduced in sections 4.1 and 4.2) and scalable (with the data size and dimension).

Clustering high dimensional data is both of practical importance and at the same time a big challenge, in particular for large collections of text documents. Still, it has to be stressed that not only textual, but also other high dimensional datasets (especially characterized by attributes of heterogeneous or correlated distributions) may be clustered using the presented method.

Contextual approach leads to many interesting research issues, such as context-dependent dictionary reduction and keywords identification, topic-sensitive document summarization, subjective model visualization based on particular user's information requirements, dynamic adaptation of the document representation and local similarity measure computation. Especially, the user-oriented, contextual data visualization can be a major step on the way to information retrieval personalization in search engines. We plan to tackle these issues in our future work.

---

[5] Since it is an averaged plot, so we note that in some of the contextual groups the divergence was highly significant.

References

1. J.C. Bezdek, S.K. Pal, Fuzzy Models for Pattern Recognition: Methods that Search for Structures in Data, IEEE, New York, 1992
2. K. Ciesielski, M. Draminski, M. Klopotek, M. Kujawiak, S. Wierzchon, Mapping document collections in non-standard geometries. B. De Beats, R. De Caluwe, G. de Tre, J. Fodor, J. Kacprzyk, S. Zadrony (eds): Current Issues in Data and Knowledge Engineering. Akademicka Oficyna Wydawnicza EXIT Publishing, Warszawa 2004, pp.122-132
3. K. Ciesielski, S. Wierzchon, M. Klopotek, An Immune Network for Contextual Text Data Clustering, in: Proceedings of the International Conference on Artificial Immune Systems (ICARIS-2006), Oeiras, Lecture Notes on Artificial Intelligence, LNAI 4163, Springer-Verlag, 2006
4. M. Dittenbach, A. Rauber, D. Merkl, Uncovering hierarchical structure in data using the Growing Hierarchical Self-Organizing Map. Neurocomputing 48 (1-4)2002, pp. 199-216.
5. M. Dorigo, G. Di Caro, The Ant Colony Optimization Meta-Heuristic, in: D. Corne, M. Dorigo, F. Glover (eds.), New Ideas in Optimization, McGraw-Hill, 1999, pp. 11-32.
6. B. Fritzke, A growing neural gas network learns topologies, in: G. Tesauro, D.S. Touretzky, and T.K. Leen (Eds.) Advances in Neural Information Processing Systems 7, MIT Press Cambridge, MA, 1995, pp. 625-632.
7. B. Fritzke, A self-organizing network that can follow non-stationary distributions, in: Proceeding of the International Conference on Artificial Neural Networks '97, Springer, 1997, pp.613-618
8. M. Halkidi, Y. Batistakis, M. Vazirgiannis, On clustering validation techniques. Journal of Intelligent Information Systems, 17(2-3), pp.107-145, 2001
9. C. Hung, S. Wermter, A constructive and hierarchical self-organising model in a non-stationary environment, International Joint Conference in Neural Networks, 2005
10. M. Klopotek, M. Draminski, K. Ciesielski, M. Kujawiak, S.T. Wierzchon, Mining document maps, in Proceedings of Statistical Approaches to Web Mining Workshop (SAWM) at PKDD'04, M. Gori, M. Celi, M. Nanni eds., Pisa, 2004, pp.87-98
11. M. Klopotek, S. Wierzchon, K. Ciesielski, M. Draminski, D. Czerski, Conceptual maps and intelligent navigation in document space, monography to appear in: Akademicka Oficyna Wydawnicza EXIT Publishing, Warszawa, 2006
12. T. Kohonen, Self-Organizing Maps, Springer Series in Information Sciences, vol. 30, Springer, Berlin, Heidelberg, New York, 2001
13. T. Kohonen, S. Kaski, P. Somervuo, K. Lagus, M. Oja, V. Paatero, Self-organization of very large document collections, Helsinki University of Technology technical report, 2003, http://www.cis.hut.fi/research/reports/biennial02-03
14. A. Rauber, Cluster Visualization in Unsupervised Neural Networks. Diplomarbeit, Technische Universitt Wien, Austria, 1996
15. Y. Zhao, G. Karypis, Criterion functions for document clustering: Experiments and analysis, available at http://www-users.cs.umn.edu/~karypis/publications/ir.html

Automatic Water Eddy Detection in SST Maps Using Random Ellipse Fitting and Vectorial Fields for Image Segmentation

Armando Fernandes[1] and Susana Nascimento[1,2]

[1] Centro de Inteligência Artificial - CENTRIA
[2] Departamento de Informática, Faculdade Ciências e Tecnologia,
Universidade Nova Lisboa, Quinta da Torre, 2829-516 Caparica, Portugal
arm.fernandes@gmail.com, snt@di.fct.unl.pt

Abstract. The impact of water eddies off the Iberian coast in the chemistry and biology of the ocean ecosystems, on the circulation of ocean waters and on climate still needs to be studied. The task of identifying water eddies in sea surface temperature maps (SST) is time-consuming for oceanographers due to the large number of SST available. This motivates the present investigation aiming to develop an automatic system capable of performing that task. The system developed consists of a pre-processing stage where a vectorial field is calculated using an optical flow algorithm with one SST map and a matrix of zeros for input. Next, a binary image of the modulus of the vectorial field is created using an iterative thresholding algorithm. Finally, five edge points of the binary image, classified according to their gradient vector direction, are randomly selected and an ellipse corresponding to a water eddy fitted to them.

Keywords: Image Segmentation, Optical Flow, Random Ellipse Detection, Water eddy.

1 Introduction

Water eddies, commonly called eddies, are a fundamental feature of ocean circulation patterns providing a mean of mixing ocean waters as a consequence of the turbulent motion associated with them. Thus, several physical, chemical and biological parameters are likely to be affected by the dynamics of eddy formation. One issue still to be understood is the importance of eddies of Mediterranean waters off the Iberian coast in the global thermohaline circulation of the ocean and their impact on global climate models. A thorough study of eddies requires their identification in sea surface temperature maps (SST) collected by satellites. Due to the large number of SST maps available for study, automatic eddy detection systems are of great support to oceanographers. Consequently, the present investigation aim is developing a system for automatic detection of eddies in SST maps using new methodologies, namely, vectorial fields obtained with an optical flow algorithm for eddy segmentation, and random ellipse fitting for eddy detection. Firstly, the optical flow algorithm [1] is applied in image pre-processing, but a new approach is used to prevent the problem of image sequence unavailability. Afterwards, the points of the pre-processed image are

N. Lavrač, L. Todorovski, and K.P. Jantke (Eds.): DS 2006, LNAI 4265, pp. 77–88, 2006.
© Springer-Verlag Berlin Heidelberg 2006

selected according to their gradient vectors and five points randomly chosen among the selected points. An ellipse is then fitted to those five points. The system is able to recognize images with or without eddies because, when executing the eddy detection algorithm several times, the centers of the ellipses obtained are concentrated in smaller areas in images with an eddy.

Several studies describing methods for eddy detection in SST were found in the scientific literature. Research on eddy detection with neural networks resulted in misdetection percentages of 28% for warm eddies and 13% for cold eddies [2]. Some neural networks developed for detection of eddies off the Iberian coast were able to correctly classify 91% of the eddies and 96% of the non-eddy structures [3]. The main drawback of neural network based systems is that the decision-making policy is not clear to the user. This inhibits the use of neural networks when oceanographers need to study and control that policy, which is the present case. Some eddy detection methods were based on the determination of velocity fields using temperature gradients and image averaging for periods of two days [4]. This averaging is not possible in the present work due to the frequent cloud coverage of the area being studied. In addition, SST isotherms are poorly defined making unfeasible velocity field determination based on temperature gradients. A method using isotherm curvature combined with texture analysis was proven to be inappropriate for eddy detection because it leads to a large number of false alarms [5]. Eddies with diameters of hundreds of kilometers were detected using optical flow approach which consists of utilizing temporally consecutive images to determine velocity fields [1]. However, the eddies one aims to detect have approximate diameters of only 100 kms. On the other hand, the optical flow approach has the disadvantage of using sequences of images, which are difficult to obtain due to the already mentioned cloud coverage phenomena of the zone under analysis. Hough transform and several circle fitting algorithms were applied to images pre-processed with the cluster-shade edge detector (CSED) which was based on pixel intensity co-occurrence matrix [6]. These algorithms have been tested in determining the radius and center position of eddies in the Gulf Stream area of the North Atlantic [7]. The present study uses different pre-processing and ellipse fitting methods and the area under investigation is the Iberian Coast whose SST maps present different features from the SST maps of the Gulf Stream area.

The literature search revealed several works on circle and ellipse detection whose similarities and differences to the present investigation will next be described. In the present study, eddy detection is performed by fitting an ellipse to edge points of the eddy. However, these eddies present shapes that can be approximated by ellipses only with some error. The main cause is that eddies are originated by turbulent fluid flow which makes eddy shapes and internal structures to vary. The majority of works for automatic ellipse detection are usually performed on images containing exact ellipse shapes like the ones of a dish, or a traffic sign viewed with a certain inclination [8, 9]. In [10], gradient information similar to the one used in the current work, is applied in choosing pairs of points for circle fitting in images containing complete circles. The present investigation consists of fitting ellipses in images containing ellipse arcs instead of complete ellipses. In [9] three points are chosen and an ellipse fitted to them. The points are chosen taking into consideration their intensity values and not their gradient vector direction. In [9] and [10] good results are shown in the analysis of binary images where almost all points belong to circles or ellipses. In the present study the percentage of points belonging to an ellipse is extremely reduced comparatively to the total number of points contained within the binary image obtained using a new segmentation procedure. In [8] traffic signs with elliptical shapes due to the angle of observation, are detected among a large number of points that do not belong to the signs. However, the detection algorithm starts by grouping points into arcs, a task that is not included in the proposed algorithm, in order to maintain the detection algorithm as simple as possible.

2 The Analyzed Images

Sixteen images containing eddies were selected for the experimental study. These images are relative to a zone located about 300 km to the south of the Iberian Peninsula and each pixel corresponds to an area of 1x1 km^2. The images used were those among the images available that showed large areas free of clouds and contained some of the largest eddies appearing in the Iberian Peninsula region. The eddies analyzed present blurred edges, and the temperature variation between the zones inside and outside the eddy is approximately 3 degrees Celsius. Two of the images containing eddies are shown in Fig. 1. Five images without eddies were also selected with the aim of testing the possibility of distinguishing images containing eddies from those that do not contain eddies. Only few images do not contain eddies.

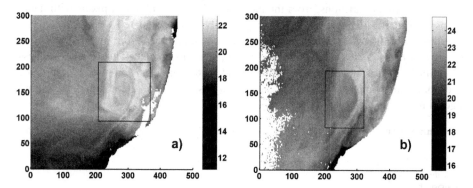

Fig. 1. Images containing two eddies. The rectangles show the eddy position. The left eddy (a) presents a spiral with a better outline than the right eddy (b).

3 The Proposed Algorithm

The algorithm for eddy detection consists of three stages. In the first stage the original SST is pre-processed using an optical flow algorithm and the modulus of the vectorial field obtained is given as input to an iterative thresholding algorithm. The resulting threshold is used to create a binary image from the vectorial field modulus. In the second stage, the groups of points resembling ellipses are detected in the binary image, and in the third stage those ellipses corresponding to eddy are selected according to an ellipse coverage criterion.

3.1 Segmentation Stage

The application of optical flow algorithms to temporal sequences of images with the aim of detecting motion is common in the scientific literature [11]. In the present work an optical flow algorithm [1] is used for edge detection instead. This algorithm determines the vectorial field minimizing the pixel intensity variation between two consecutive images. However, in the present study, it is difficult to gather temporal sequences of SSTs due to the existence of clouds between satellites and sea surface. To overcome this problem, the authors apply the optical flow algorithm in an innovative way. Traditionally, the two SSTs used as input to the optical flow algorithm are visualizations of the same area in two consecutive time instants. In the present work, only one SST map is required because the second image is composed of pixels with intensity values equal to zero. This type of image pre-processing has revealed to be powerful for eddy segmentation because it allows enhancing eddy edges even when they present low gradient values. The result of the optical flow algorithm is a vectorial field whose modulus is then used to create a binary image. The threshold value for the modulus of the vectorial field, below or above which a value of one or zero, respectively, is assigned to each pixel composing the binary image. This threshold is determined using an iterative thresholding algorithm [12], described next.

The iterative thresholding algorithm consists of the following steps. First, an initial threshold value, τ, that corresponds to the average pixel intensity value is defined. Second, two pixel partitions, from the background pixels and object pixels with mean intensity values $\mu Back$ and μObj, respectively, are calculated using the initial threshold. Third, a new threshold value, τ, corresponding to the average value of $\mu Back$ and μObj is determined and new pixel partitions created. The iterative process of estimation of new threshold values and pixel partitions continues until the threshold value changes become sufficiently small. The iterative thresholding algorithm is stated as follows:

```
τ = function ITERATIVE_THRESHOLD(Image)          /* Image is the
                                                    input SST
    τ_0= -∞;
    τ= mean(Image);
Repeat
    τ_0= τ;
    µObj= mean(Image < τ);
    µBack= mean(Image > τ);
    τ= (µObj + µBack)/2;
until (|τ-τ_0|≤ ε)
end
```

3.2 Automatic Ellipse Fitting Stage

The detection of structures resembling ellipses is performed over the binary images obtained in the segmentation stage. The ellipse detection algorithm randomly selects five edge points and then fits an ellipse to those five points using a least mean squares procedure [13]. The five points used for the ellipse fitting are chosen in such a way that the distance between them is neither smaller than a minimum distance, *MinDis*, neither larger than a maximum distance, *MaxDis*, with both distances *MinDis* and *MaxDis* being parameters defined by the user. The maximum distance should be larger than the maximum expected eddy diameter, but as small as possible in order to minimize the number of edge points that can be chosen for the ellipse fitting. On the other hand, the minimum point distance should be chosen so that the small islands of ones usually existing in the binary images are not considered for ellipse detection. The number of five point groups possible to build for each binary image, which contains approximately 10000 edge points, is about 8×10^{17}. However, not every five point group is relevant for an ellipse detection. In fact, only the points with gradient vectors (calculated in the binary image) pointing inside the ellipse are considered. The scheme presented in Fig. 2 illustrates that. The gradient vectors are required to point inside the ellipse because the region to be detected contains ones and is surrounded by zeros.

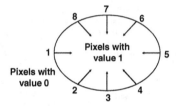

Fig. 2. Indexes for points according to their gradient vector directions and their optimal relative position in an ellipse

There are 8 possible gradient vector directions with $\pi/4$ difference in angle as represented in Fig. 2. Usually, the eddy edges presented in the binary image do not allow to form a complete ellipse. Therefore, the number of different gradient vector directions used in the ellipse detection algorithm is inferior to 8. Choosing five points with different vector directions allows coping with this requirement and simultaneously guaranteeing that the group of points represents more than half an ellipse. This is enough to help detecting the more relevant elliptical structures.

3.2.1 Choosing the Points for the Ellipse Fitting

The ellipse fitting algorithm starts by collecting all edge points of the binary image. Let *PointsToChoose* be that set of points. Each edge point has an index value associated that corresponds to an integer between 1 and 8, and which is assigned according to the gradient directions, as shown in Fig. 2. Each index assigned to an edge point does not depend on the point position. Next, choose randomly a point *P* with index *N* from the set of points in *PointsToChoose*, and include *P* in the set of points, *PointsToFit*. Remove from set *PointsToChoose* all the points with index *N*. Then, create two axes with origin in point *P*, with the *xx* axis following the same direction as the gradient vector at *P* and the *yy* axis being rotated 90° counter clockwise relatively to the *xx* axis. From there, discard all points in *PointsToChoose* whose coordinates correspond to negative values of *xx*. Keep the points with positive *xx* coordinates, positive *yy* coordinates and indexes equal to *N-1*, *N-2* and *N-3*, as well as the points with positive *xx* coordinates, negative *yy* coordinates and indexes equal to *N+1*, *N+2* and *N+3*. Keep also the points with positive *xx* coordinates, index equal to *N+4* and distance to *P* larger than *MinDis*. The index is labelled in the sequence 123456781234, so that index *N-3* corresponds to 7 when *N* is equal to 2. Continue by randomly selecting points among those remaining in set *PointsToChoose* until *PointsToFit* contains 5 points. After selecting randomly an edge point, it may only be added to *PointsToFit* if its distance to the points already included in *PointsToFit* is not smaller than *MinDis* neither larger than *MaxDis*. Finally, an ellipse is fitted to the five points contained in *PointsToFit* using the least mean square algorithm described in [13]. Fig. 3 shows two examples of applying the above described algorithm for two randomly selected points with indexes 1 and 6.

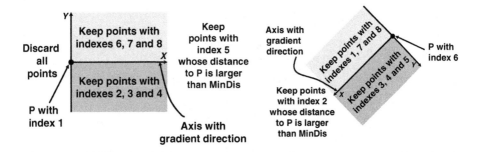

Fig. 3. Selection of edge points of a binary image for points *P* with indexes 1 and 6

3.3 Ellipse Selection Criterion

The ellipse fitting process is performed on five points collected randomly. Since it is not known *a priori* which points allow to determine ellipses corresponding to eddies, each 'run' for an eddy detection consists of iterating the process of selecting groups of five points and fitting one ellipse to each one of these groups of points. Consequently, it is necessary to define a criterion to select in each run the ellipse that best recognizes an eddy. Such criterion consists of choosing the ellipse with the largest perimeter percentage covered by the edge points of the binary image. To define which points cover an ellipse it is necessary to recall that every ellipse has a "characteristic distance", which according to Fig. 4a, corresponds to the sum of the distances *d1* and *d2* of any ellipse point to the ellipse foci *F1* and *F2*. An edge point is said to cover an ellipse when it is inside a band delimited by two ellipses whose "characteristic distances" are equal to the fitted ellipse "characteristic distance" minus and plus a percentage of the length of the smallest axis relative to the fitted ellipse. In addition, the edge points position and index must be compatible with Fig. 4b. The percentage of ellipse axis size is as small as possible in order to have each ellipse point covered by one edge point at most.

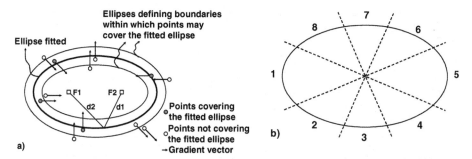

Fig. 4. a) Representation of point coverage of the fitted ellipse. The small circles represent edge points of the binary image. b) Ellipse zones where points with a certain index contribute to the ellipse perimeter coverage.

By analyzing the percentage of ellipse perimeter covered it is possible to compare ellipses of different dimensions. The ellipses obtained from the fitting procedure whose largest axis size is larger than *MaxDis* or semiminor axis size is smaller than *MinDis* are rejected and thus are not considered for eddy detection. The ellipse perimeter, *P*, is calculated, in pixels, according to the following equation

$$P \approx \pi\sqrt{2(a^2 + b^2)} \ , \tag{1}$$

where *a* and *b* represent the semimajor and semiminor ellipse axis length, respectively.

3.4   The Complete Eddy Detection Algorithm

The calculations are performed with the optical flow algorithm having as input one SST map and one image composed of pixels with intensity values equal to zero. The output of the optical flow algorithm is a vectorial field whose modulus serves as input to an iterative thresholding algorithm [12], which allows producing a binary image. Next, five edge points of the binary image are randomly selected among all edge points. A constraint is made on the random selection, the five points must present different gradient vector directions and all gradient vectors must aim inside the area delimited by those points. An ellipse is then fitted to these five points using an algorithm based on least squares minimization [13]. Afterwards, several ellipses are fitted to different groups of five points, which is called a 'run', and the ellipse presenting the largest percentage of its points close to the edge points of the binary image is finally chosen as being the one outlining an eddy. Only the edge points with gradient vectors pointing at the inner region of the ellipse are accounted for that percentage. The algorithm is the following:

```
BestEllipse= function EddyDetection(Image)     /* Image  is  the
                                               input image
  Niterations_Per_Run= 500;
  BestEllipse= 0;
  [U,V]= OpticalFlow(Image);
  τ= IterativeThreshold(Sqrt(U.U+V.V));
  BImage= BinaryImage(Image,τ);
```

```
L= EdgePoints(BImage);            /* L contains all edge points
Indexes= GradientDirections(L);         /* According to Fig. 2
For i=1 to Niterations_Per_Run
  PointsToChoose= {(P, N): P ∈ L ∧ N ∈ Indexes};
  PointsToFit= {};
  P= ChooseRandomly(PointsToChoose);             /*Choose one point
  While (|PointsToFit|≤5 ∧ MinDis≤ Dist(P,PointsToFit) ≤MaxDis)
      PointsToFit= PointsToFit ∪ {P};
      N= Index(P);
      PointsToChoose= PointsToChoose - {P': Index(P')==N};
      CreateAxesWithOrigin(P);/* xx axis with gradient direction
      PointsToChoose= PointsToChoose - {P':xxCoordinate(P')< 0};
      PointsToChoose= PointsToChoose -
            {P': xxCoordinate(P')> 0 ∧ yyCoordinate(P')> 0 ∧
            (Index(P')==N+1 ∨ Index(P')==N+2 ∨ Index(P')==N+3)
            };
      PointsToChoose= PointsToChoose -
            {P': xxCoordinate(P')> 0 ∧ yyCoordinate(P')< 0 ∧
            (Index(P')==N-1 ∨ Index(P')==N-2 ∨ Index(P')==N-3)
            };
      PointsToChoose= PointsToChoose -
            {P': xxCoordinate(P')> 0 ∧ Index(P')==N+4 ∧
            Dist(P',P)< MinDis
            };
      P= ChooseRandomly(PointsToChoose);         /*Choose one point
  endWhile
  If (|PointsToFit|== 5)
      Ellipse= EllipseFit(PointsToFit); /* Least-squares fit
      BestEllipse= ChooseBestEllipse(Ellipse,BestEllipse);
  endIf
 endFor
end
```

4   Experimental Study

The goal of the experimental study is to analyze the effectiveness of the algorithm described on Section 3 in recognizing eddies similar to those presented in Section 2.

4.1   Image Segmentation Stage

The image pre-processing is carried out executing the optical flow algorithm for 100 iterations, with the α parameter equal to 1 and β parameter equal to 0 [1]. The origin of parameters α and β is described in [1]. When β is equal to zero the equations in [1] are similar to the original equations reported in [14]. Two binary images resulting from applying the iterative thresholding algorithm to the modulus of the vectorial field obtained with the optical flow, as explained in Section 3.1, are shown in Fig. 5. These are the binary images resulting from the original eddy images shown in Fig. 1. In Fig. 5 a and b one can recognize several eddy edges similar to elliptical arcs around position with xx coordinates equal to 270 and yy coordinates equal to 150. In the 16 images studied, there are several obstacles for eddy detection. The arcs corresponding to the eddy edges usually do not form a complete ellipse when assembled together, and their size, number and position relatively to the others is quite variable. In addition, the binary images contain a large number of edge points that are not related to eddy recognition. However, the ellipse detection algorithm was developed to be able to work properly with these properties.

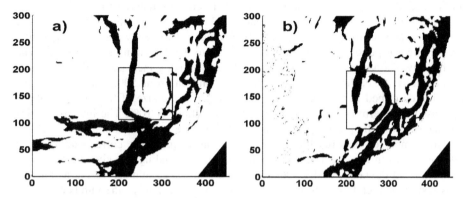

Fig. 5. Binary images of the images presented in Fig. 1. The rectangles show the eddy position.

4.2 Ellipse Fitting Stage

The ellipse detection algorithm is used with parameters *MinDis* and *MaxDis* set to 10 and 90 pixels, respectively. In each image, the number of groups of 5 points with the required gradient vectors decreases with *MinDis*. The detection of two ellipses of different eccentricities that correspond to eddies is presented in Fig. 6.

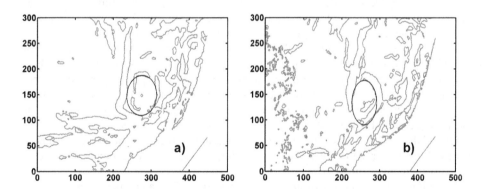

Fig. 6. Detection of the ellipses corresponding to eddies in the binary images of Fig. 5

The gray lines represent the edge points of the binary image. In Fig. 6a the arcs associated with the eddy, are almost flat and in some cases have small dimensions when compared to the eddy perimeter. However, the detection algorithm is effective in detecting a group of points belonging to those arcs, which allow to obtain one ellipse that corresponds to the eddy. In Fig. 6b the algorithm is able to detect two large, independent elliptical arcs corresponding to an eddy. The examples shown in

Fig. 6 illustrate the robustness of the ellipse detection algorithm in recognizing eddies, since it is able to detect eddies presenting different features in the binary images.

4.3 Performance Evaluation of the Ellipse Fitting Algorithm

In this section it is discussed the influence of the random nature of the ellipse fitting algorithm on eddy detection. For each one of the 16 images containing eddies, 30 runs were executed in order to analyze the effectiveness of the eddy detection algorithm. Each run corresponds to 500 iterations, and thus to the detection of 500 ellipses. The probability of finding an ellipse corresponding to an eddy increases with the number of iterations. In each run the best ellipse is chosen from 500 ellipses using the criterion described in Section 3.3. The percentage of the size of the smallest ellipse axis used in determining the points covering an ellipse is 10%. The increase of this parameter allows finding elliptical structures with more irregular boundaries.

The number of runs resulting in the detection of an ellipse corresponding to an eddy is shown in Fig. 7a. The average number of runs ending in eddy detection is 25 with standard deviation of 4.7. Only in the case of image number 10 the number of runs resulting in eddy detection is inferior to 15, i.e. to half of the number of runs. This result derives from the fact that image 10 contains a structure that is not an eddy but resembles an ellipse. Fig. 1 a and b correspond to images with index 1 and 12 in Fig. 7a, respectively. Images with index number ranging from 1 to 9 contain eddies similar to the eddy shown in Fig. 1a while images with index number ranging from 10 to 16 contain eddies similar to the eddy presented in Fig. 1b. The analysis of the graphic shown in Fig. 7a allows to conclude that the eddies similar to the eddy shown in Fig. 1a are easier to detect than the eddies similar to the one shown in Fig. 1b. The cause is that the former eddies present spiral patterns with a better outline than the latter eddies. Since the edges of these spiral patterns appear as ellipse arcs in the binary image, the former eddies have a wider variety of elliptical arcs associated with them than the latter eddies. This wider variety of arcs increases the number of different possibilities of fitting ellipses on the eddy zone, which eases the eddy detection process. Since not every run results in an ellipse corresponding to an eddy, only the best ellipse out of several runs is considered to correspond to an eddy. In the present study thirty runs are used. The best ellipse out of the 30 runs is chosen with the same criterion used in choosing the best ellipse out of 500 ellipses obtained in each run, i.e. it is the one with the largest percentage of its perimeter covered by edge points of the binary image. In 16 images containing eddies only the eddy in image with index number 10 is not detected after thirty runs. This corresponds to an eddy detection percentage of 94%.

The ellipse detection algorithm was also tested on 5 images that do not contain eddies. However, the algorithm described in Section 3 detects ellipses even if no eddy is present in the image. The images containing eddies are automatically distinguished from those that do not contain eddies by analyzing the sum of the standard deviation of xx and yy coordinates of the ellipse centers obtained in the 30 runs performed for each image. In Fig. 7b it is possible to observe that the sum of standard deviations is larger in images that do not contain eddies than in images containing eddies. The value of the horizontal line separating the two types of image is automatically determined using the iterative thresholding algorithm. In this case, the algorithm input is the sum of the standard deviation of xx and yy coordinates of the ellipse centers instead of the modulus of the vectorial field obtained with the optical flow. The labels, with or without eddies, shown in Fig. 7b are for display purposes only. They were not used as an input parameter of the iterative thresholding algorithm.

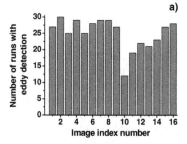 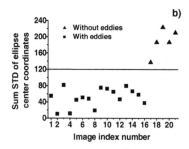

Fig. 7. a) Number of runs out of 30 that resulted in eddy detection for each one of the 16 images with eddies. b) Comparison of the sum of standard deviation of ellipse center coordinates in 30 run for images with and without eddies.

5   Conclusion

The present investigation focused on the automatic detection of water eddies in sea surface temperature (SST) maps. Due to their nature being the one of a turbulent flow, these eddies present highly variable internal structures and shapes. However, these shapes may be approximated by ellipses.

A new segmentation method, consisting in an optical flow algorithm using solely one SST map and an iterative thresholding algorithm, was proven to be robust to changes in SST and eddy characteristics, and efficient in creating binary images whose edge points outline eddies. The ellipse fitting algorithm developed, presents a good performance in discovering points belonging to elliptical arcs corresponding to eddies even though the total number of points in these arcs is small comparatively to the total number of points contained within the binary images. The use of the gradient direction of the edge points from the binary image for choosing points for fitting ellipses is essential to achieve this good performance. Moreover, this algorithm for automatic eddy detection is easily implemented computationally.

Thirty runs composed by 500 iterations of the automatic ellipse detection algorithm were carried out for each image containing one eddy. An eddy was recognized in 25 runs in average. Choosing among 30 ellipses (the best of each run), the ellipse with the largest perimeter percentage covered by edge points of the binary image allows detecting 15 eddies in 16 possible, which corresponds to a recognition efficiency of 94%. Only the edge points with gradient vectors pointing to the ellipse inner region are considered to cover the ellipse. It is possible to automatically distinguish 16 images containing eddies from 5 images that do not contain eddies just by analyzing the standard deviation of the ellipse center position, that is larger in the latter case.

In the future, a larger number of images with and without eddies are going to be included in the study. In addition, fuzzy clustering techniques will be used for eddy detection and their effectiveness compared to the one of the algorithm presented here.

Acknowledgments. The authors would like to acknowledge Instituto de Oceanografia, Faculdade de Ciências, Universidade de Lisboa, for providing the SST images and their support at the crucial stages of this work.

References

1. Yang, Q., Parvin, B., and Mariano, A.: Singular features in sea surface temperature data. 15th International Conference on Pattern Recognition, Vol 1, Proceedings - Computer Vision and Image Analysis International Conference on Pattern Recognition (2000) 516-520.
2. Arriaza, J.A.T., Rojas, F.G., Lopez, M.P., and Canton, M.: Competitive neural-net-based system for the automatic detection of oceanic mesoscalar structures on AVHRR scenes. IEEE Transactions on Geoscience and Remote Sensing 41(4) (2003) 845-852.
3. Castellani, M. and Marques, N.C.: Automatic Detection of Meddies Through Texture Analysis of Sea Surface Temperature Maps. EPIA - Portuguese Conference on Artificial Intelligence Covilhã, Portugal (2005) 359-370.
4. Alexanin, A.I. and Alexanina, M.G.: Quantitative analysis of thermal sea surface structures on NOAA IR-images. Proceedings of CREAMS Vladivostok, Russia (2000) 158–165.
5. Thonet, H., Lemonnier, B., and Delmas, R.: Automatic segmentation of oceanic eddies on AVHRR thermal infrared sea surface images. Oceans '95 Mts/IEEE - Challenges of Our Changing Global Environment, Conference Proceedings, Vols 1-3 (1995) 1122-1127.
6. Holyer, R.J. and Peckinpaugh, S.H.: Edge-Detection Applied to Satellite Imagery of the Oceans. IEEE Transactions on Geoscience and Remote Sensing 27(1) (1989) 46-56.
7. Peckinpaugh, S.H. and Holyer, R.J.: Circle Detection for Extracting Eddy Size and Position from Satellite Imagery of the Ocean. IEEE Transactions on Geoscience and Remote Sensing 32(2) (1994) 267-273.
8. Kim, E., Haseyama, M., and Kitajima, H. (2002). Fast and Robust Ellipse Extraction from Complicated Images, http://citeseer.ist.psu.edu/554140.html,
9. Ho, C.T. and Chen, L.H.: A high-speed algorithm for elliptical object detection. IEEE Transactions on Image Processing 5(3) (1996) 547-550.
10. Rad, A., Faez, K., and Qaragozlou, N.: Fast Circle Detection Using Gradient Pair Vectors. Proc. VIIth Digital Image Computing: Techniques and Applications, C. Sun, et al.(Ed) Sydney (2003).
11. Barron, J.L., Fleet, D.J., and Beauchemin, S.S.: Performance of Optical-Flow Techniques. International Journal of Computer Vision 12(1) (1994) 43-77.
12. Sezgin, M. and Sankur, B.: Survey over image thresholding techniques and quantitative performance evaluation. Journal of Electronic Imaging 13(1) (2004) 146-165.
13. Halir, R. and Flusser, J.: Numerically Stable Direct Least Squares Fitting of Ellipses. Proc. Int. Conf. in Central Europe on Computer Graphics, Visualization and Interactive Digital Media, V. Skala(Ed) (1998) 125-132.
14. Horn, B.K.P. and Schunck, B.G.: Determining Optical-Flow. Artificial Intelligence 17(1-3) (1981) 185-203.

Mining Approximate Motifs in Time Series

Pedro G. Ferreira,[1] Paulo J. Azevedo,[1] Cândida G. Silva,[2]
and Rui M.M. Brito[2],[*]

[1] Department of Informatics, University of Minho 4710-057 Braga, Portugal
{pedrogabriel, pja}@di.uminho.pt
[2] Chemistry Department, Faculty of Sciences and Technology, and Centre of
Neurosciences of Coimbra, University of Coimbra, 3004-517 Coimbra, Portugal
csilva@student.uc.pt, brito@ci.uc.pt

Abstract. The problem of discovering previously unknown frequent patterns in time series, also called motifs, has been recently introduced. A motif is a subseries pattern that appears a significant number of times. Results demonstrate that motifs may provide valuable insights about the data and have a wide range of applications in data mining tasks. The main motivation for this study was the need to mine time series data from protein folding/unfolding simulations. We propose an algorithm that extracts approximate motifs, i.e. motifs that capture portions of time series with a similar and eventually symmetric behavior. Preliminary results on the analysis of protein unfolding data support this proposal as a valuable tool. Additional experiments demonstrate that the application of utility of our algorithm is not limited to this particular problem. Rather it can be an interesting tool to be applied in many real world problems.

1 Introduction

The mining of time series data has gathered a great deal of attention from the research community in the last 15 years. These studies have impact in many fields, ranging from biology, physics, astronomy, medical, financial and stock market analysis, among others. The research in mining time series has been mainly focused in four problems [9]: *indexing* or *query by content, clustering, classification* and *segmentation*. Lately, the problem of mining unusual and surprising patterns [10] has also been enthusiastically studied. Other challenging and recently proposed problem in the context of time series is the mining of previously unknown patterns. These patterns, here referred as *episodes*, consist of subsequences that appear, in a unique and longer sequence [15,7,5] or are subsequences that occur simultaneously in more than one sequence from a set of related sequences, in this case called *motifs* or *sequence patterns*. These motifs have a wide range of applications. They can be used in the clustering and classification of time series. They can also be applied in the generation of sequence rules and in the detection

---

[*] The authors acknowledge the support of the "Fundação para a Ciência e Tecnologia" and the program FEDER, Portugal, POCTI/BME/49583/2002 (to RMMB) and the Fellowships SFRH/BD/13462/2003 (to PGF) and SFRH/BD/16888/2004 (to CGS).

N. Lavrač, L. Todorovski, and K.P. Jantke (Eds.): DS 2006, LNAI 4265, pp. 89–101, 2006.

of interesting behaviors, which can give the user/domain expert valuable insights about the problem that is being studied.

In this work we are interested in the extraction of time series motifs. We present an algorithm that given as input the symbolic representation of a set of comparable time sequences, it finds all the patterns that occur a number of times equal or greater than a threshold value. A sequence pattern or motif consists in a set of subsequences that share among them a similarity greater than an user defined value. The definitions adopted in this work and the development of the algorithm were mainly motivated by the specificities of the mining of time series data from protein folding/unfolding simulations, as we will discuss in section 4. However, as we will see in the same section, these ideas can also be applied to many different domains and application contexts.

2   Definitions and Notations

In this section we will present some definitions that will be used throughout this paper.

Definition 1. (Time Sequence) *A time sequence T is an ordered set of values $(t_1, t_2, \ldots, t_n)$, where $t_i \in \mathbb{R}$; $(t_p, t_{p+1}, \ldots, t_q)$ is a subsequence of T starting at position p and ending at position q, where $1 \leqslant p$ and $q \leqslant n$. The length of sequence, $|T|$ is equal to n.*

In this work we are interested in finding patterns over a set of temporal sequences that for a certain period of time reflect a similar and/or a symmetric tendency. This trend or tendency reflects a measure of interest of the motif and is called here *approximate similarity.*

According to our notion of approximate similarity, several measures and coefficients appear as candidates for similarity functions. The most popular measures of distance appearing in literature are the Euclidean distance and the Dynamic Time Warping (DTW) measure [6,11]. For this particular problem, the drawback of these two measures is that they are not sensitive to the association linearity between the elements of the subsequences. We are interested in finding patterns based on an approximate similarity, thus a more suitable measure is required.

The Pearson Correlation Coefficient [20], r, measures the magnitude and the direction of the association between the values of the subsequences. The correlation coefficient ranges from -1 to +1 and reflects the linear relation between the values of the subsequences. The Pearson correlation is a metric measure that satisfies the three following properties: *positivity*($r(x, y) \geq 0$ and $r(x, y) = 0$, if $x = y$), *symmetry* ($r(x, y) = r(y, x)$) and *triangle inequality* ($r(x, z) \leq r(x, y) + r(y, z)$)). This last property as we will see, will be particularly useful to cluster pairs of similar subsequences to form the motifs. In fact, Pearson correlation tends to be robust to small variations.

Definition 2. (Match) *Given the similarity function between two subsequences X and Y, $sim(X, Y)$, we say that subsequence X matches Y if $|sim(X, Y)| \geqslant R$, where R is a user supplied positive real number.*

The absolute similarity value used in definition 2 handles the notion of approximate similarity, i.e. two subsequences may have an inverted behavior and still be considered a match.

Definition 3. (Instance) *A subsequence X is an instance of a subsequence Y, where $|X| = |Y|$, if $|sim(X,Y)| \geqslant R$.*

Definition 4. (Overall Similarity) *The overall similarity for a set of subsequences corresponds to the average value of similarity between all the pairs of subsequences in the set.*

Therefore, a motif can informally be defined as a set of interesting subsequences. The interestingness is defined by its overall similarity and by the frequency of appearance, i.e. how much recurrent are the subsequences in the set of the input time sequences. The frequency is provided by the cardinality of the set and is usually called as *support*. Hence, formally a motif can be defined as follows:

Definition 5. (Approximate Motif) *Given a database D of time sequences, a minimum support σ and a minimum value of similarity/correlation R_{min}. We consider that k subsequences consist of an approximate motif, if $k \geq \sigma$ and all subsequences pairwisely match for a value of R_{min}.*

Definition 6. (K-Cluster) *We denote a group of k (related) instances as k-cluster.*

Definition 7. (Cluster Containment) *A cluster C_1 is contained in a cluster C_2, if all instances of C_1 are in C_2.*

Definition 8. (Overlap Degree) *The degree of overlap between two sequences X and Y is defined as $Od = \frac{|X \cap Y|}{w} \times 100\%$, where $\cap$ is the intersection operation and w the length of the sequences. $|X \cap Y|$ gives the overlap region between X and Y.*

3   Algorithm

In this section we start by formulating the described problem. Next, we give an overview of the algorithm that we propose to tackle this problem.

   Problem Formulation: Given a database D of time sequences, a minimum support σ, a minimum similarity R_{min}, a window length w and a window frame length deltaW [1] *find all the approximate motifs.*

   Typically, before the algorithm is applied, a pre-processing step is performed. We adopted a two step approach called SAX [14]. In the first step, a complexity/dimensionality reduction on the data may be necessary and desirable. This reduction can be achieved through means of a *scaling* operation, also called PAA

---

[1] Concept introduced in subsection 3.1.

(Piecewise Aggregate Approximation) [16]. This operation basically consists in a reduction on the number of intervals in the time axis. A group of successive points is replaced by their average value. The second part of pre-processing step is the *symbolization*. The amplitude axis is scaled and divided into a finite number of intervals of equal size. Each interval is represented by a symbol. Sequence values are them mapped into the respective symbol of belonging interval. In our particular case all the sequence values are mapped into integer values (called *alphabet*, and denote it as Σ), therefore a symbolic representation of the time sequences is used. This approach brings two benefits to the data analysis, which are the robustness to small variations and to noise [14]. Both scaling and symbolization are performed in linear time. We continue by outlining the proposed algorithm.

3.1   Phase One: Motif Detection

This phase of the algorithm is divided into two steps. In step 1, all the subsequences of length w in D are scanned and compared against each other, in order to find similarities among them. If two subsequences match, they form a 2-cluster. In the second step of this phase, the pairs of subsequences are successively clustered with the goal of finding longer clusters of subsequences. If the overall similarity is above the minimum similarity threshold R_{min}, i.e. all instances pairwisely match, and its cardinality is greater than the minimum support σ, the cluster is then considered a motif. Our technique is a bottom-up or agglomerative method and it resembles the hierarchical agglomerative clustering technique. We start with all the clusters of size two and we keep merging the most similar ones until we obtain one cluster that can no longer be extended. If the cluster satisfies the motif definition 5, then it is considered a motif.

```
input  : w, R_min
output: ClusterInfo : List with the 2clusters
1  cnt = 0;
2  foreach S in D do
3      for i = 0 to |S| − w + 1 do
4          ss = S(i, i + w);
5          lstFSS = findFollowSS(D, ss);
6          foreach fss in lstFSS do
7              if |sim(fss, ss)| ≥ R_min then
8                  clusterInfo[cnt] = {ss, fss};
9                  cnt = cnt + 1;
10             end
11         end
12     end
13 end
```

Algorithm 1: *2clustersEnumeration* function

Step 1: 2-Cluster Enumeration. The pseudo-code in algorithm 1 shows the application of the sliding window methodology. Each subsequence is defined by the tuple $< seq, start >$, that represents the sequence identifier and the start where the subsequence occurs in D. It scans all the sequences in the database (line 2) and for each sequence it also scans all its subsequences of size w (line 3 and 4). For each subsequence (line 5) the respective following subsequences are obtained. If any of these subsequences match ss, the information for this pair of subsequences is saved (line 8) in the clusterInfo list. Function findFollowSS retrieves all the subsequent subsequences of ss. It basically consists in two for loops that scan all the windows that occur after ss in D. At this point two scenarios are possible. The algorithm may look for all the occurrences in D of the motif (one sequence may contribute with more than one instance for the counting) or only for the occurrences in different sequences. In the first case, the function starts to scan D in the sequence $ss.seq$ at position $ss.start + 1$. In the second case, it starts in the sequence $ss.seq + 1$ at position 0. This last case eliminates the existence of trivial matches, i.e. matches between subsequences that are apart from each other only few points. In this work we are only interested in the second case.

In the motif discovery process, we have decided for an agglomerative approach to cluster all the instances of a motif. Another possibility would be to use a subsequence oriented approach, where for each scanned subsequence (reference) and respective matches, if the motif definition is verified it would be immediately reported as a motif. Although simpler, this approach has two problems. First too many repeated motifs would be reported. For example, for a motif with instances {A,B,C} the motifs {A,C,B}, {C,B,A} and so on would be reported. The second problem is that when some instances match with the reference subsequences but not with the other matches of this subsequence all the possible combinations have to be tried. Consider the example in figure 1(a). Instance A (reference) matches B,C,D but instance B does not match C and D. Thus, we have at least two clusters {A,B} and {A,C,D}. When the number of non-matching instances is greater, the number of possible clusters combinations increases, which become more difficult to manage. The agglomerative approach prevents these two problems by avoiding most of the motif redundancy and making a clear separation of the clusters.

Concerning the enumeration of 2-clusters, its dispersion along the time line arises as an interesting question. Suppose a motif that has an instance that occurs in the initial part of a time sequence and other in the end of other sequence. A significant distance between the instances may invalidate the meaning of the motif. Therefore, we incorporate the option to enumerate motif instances that occur only within a certain window frame.

Figure 1 (b) shows an example of the application of the window frame of size $deltaW$. Although, instances A, B, C may have an overall similarity greater than

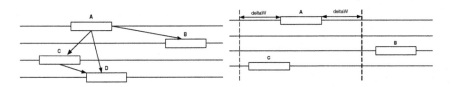

Fig. 1. (a) Example where a subsequence has several matches in more than one cluster; (b) Example of the application of a window frame to the enumeration of motif instances

R_{min}, instance B is not within the frame $[A.start - deltaW; A.end + deltaW]$. Thus, B should not be considered as a motif instance. Note, that we could ignore this window frame option and still consider the sequences in all their extension. This feature can be easily handled by introducing in line 7 of algorithm 1, the condition $|ss.start - fss.start| < deltaW$ and in line 8 of algorithm 3 the condition $\forall i, j \ |ss_i.start - ss_j.start| < deltaW$.

One problem that typically arises in heuristic methods that find probabilistic motifs in biological data like in [18,2], is the "phase" problem. This means that when looking for a pattern, the algorithm may get locked in a local optimum and not in the global optimum of the pattern. For example, consider that the best solution of a pattern starts at position 7, 19, 13, ... within the given sequences. If after some iterations the algorithm chooses positions 9, 21 in the first two sequences it probably will consider the position 15 in the third sequence and so on. It means that the algorithm got trapped in a local optimum and the obtained solution is shifted by a few positions from the correct solution. The sliding-window based methods may also suffer from this problem, as is the case of our algorithm. In order to solve it we analyze the 2-cluster list, where each cluster is compared to all the other 2-clusters that are within a certain neighborhood range. Only the 2-cluster which maximize the similarity value is kept. We consider that two sequences can be considered different when they differ in more than $1/3$ of its length. Thus, two 2-cluster are in the same neighborhood if the overlap degree between the respective subsequences is greater than $2/3$ of their length.

Step 2: Cluster Extension. At the end of step 1 we obtain a list of 2-clusters that will form the basis of the potential motifs. In an agglomerative way the similar pairs of subsequences (2-clusters) can be merged, until one big cluster is found. The criteria used to merge one 2-cluster into a N-cluster ($N \geq 2$) is based on the triangle inequality property of our similarity measure. The idea is that if instance x is similar to y and, y is similar to z, then it is expected that x is also similar to z. Thus, the cluster $\{y, z\}$ is merged with the cluster $\{x, y\}$ when all the pairs of instances from $\{x, y, z\}$ match for a value of R_{min}. In fact, the triangle inequality property allows us to define a linkage method to merge the clusters. The extension of a cluster C stops when no more 2-clusters exist to merge or when a in the result of a cluster extension a pair of instances does not match. Function *seedExtension* summarizes the described ideas.

In algorithm 2 the list of 2-clusters (clusterInfo) is traversed in order to try the extension of one 2-cluster with another 2-cluster. If the cluster can not be

```
input  : ClusterInfo, R_min, σ
output: motifList : List with motifs
1  for i = 0 to clusterInfo.size − 1 do
2      for j = i + 1 to clusterInfo.size do
3          status = extendCluster(clusterInfo[i], clusterInfo[j], j, R_min, σ);
4          if (σ == 2)AND(status == NotExtended) then
5              motifList.add(clusterInfo[i]);
6          end
7      end
8  end
```

Algorithm 2: *seedExtension* function

```
input  : ClusterInfo, clusterExt, clusterPair, index, R_min, σ
output: motifList :  List with motifs
1  if (index == clusterInfo.size) then
2      return NotExtended;
3  end
   /* Intersect the subsequences of the clusters                              */
4  IS = intersect(clusterExt, clusterPair);
5  if (index == 0) then  return NotExtended;
6  else  clusterExt = join(clusterExt, clusterPair);
   /* Find the new similarity value                                           */
7  newSimil = avgSimil(clusterExt);
8  if |newSimil| ≥ R_min then
9      for j = index + 1 to clusterInfo.size do
10         status = extendCluster(clusterExt, clusterInfo[j], i, R_min, σ);
11         if (clusterExt.size ≥ σ)AND(status == NotExtended) then  motifList.add(clusterInfo[i]);
12         return NotExtended;
13         else  return Extended;
14     end
15 end
```

Algorithm 3: *clusterExtension* function

extended and the minimum support is 2, then it can be considered as a motif (line 5). In order to avoid redundant motifs (motifs contained in other motifs) we only consider a cluster to be a motif when it can no longer be extended.

In procedure *clusterExtension*, the input consists of a cluster to be extended, clusterExt, and a cluster which is used to try an extension, clusterPair. If one subsequence in clusterPair is present in clusterExt (verified with function intersect in line 4), clusters are joined (line 6) and an extension is tried (lines 7 to 15). If all pairs of instances match, the new similarity value of the cluster is above R_{min}, the cluster can be extended. If its cardinality is equal or greater than σ and it is the last possible extension then it is considered a motif (lines 11 to 12). Otherwise, no extension is performed and the status *NotExtended* is returned. Note, that since this procedure is applied recursively, line 1 is used to test if no more 2-clusters exists when trying a new extension. At the end of this step the algorithm already retrieved the list of all possible motifs in D. Eventually, due to the extension process, some of these motifs are non-maximal, i.e. they are contained in other motifs. In order to eliminate this redundancy we apply a procedure that removes non-maximal motifs from this list. This is a time-consuming operation since each motif in the list has to verified whether it contains or is contained in another motif.

3.2   Phase2: Length Extension

In phase 2, a length extension of all the motifs extracted in phase 1 is tried. Since the subsequences of a motif are still a motif, we need to find its longest length. Thus, for a given motif all its instances are extended until the overall similarity drops below R_{min}. When the length extension is performed, a previous test has to be done since it is not know in advance the extension direction. For each pair of instances in the motif, all the four possible direction combinations are tried. The information, namely the extension direction and gainOfExtension = newValue - oldValue of each instance, is saved when the tested extension maximizes the gain. When all the pairs of subsequences are verified, a list with the directions of the extension of each subsequence is obtained. Next, based on these directions, an extension is performed one event at a time, until the overall similarity of the motif drops below R_{min}.

3.3   Motif Features and Statistical Significance

Two types of patterns can be distinguished. Motifs that contain only positive correlations and mixed motifs that contain both positive and negative correlations. If a sequence Y regresses on X, the equation $Y \simeq \beta_0 + \beta_1 X + u$, where u is the model error, may model such regression. In order to provide the scale in which all the instances of the motifs are related we calculate the average value of β_1 (refer to [20] for this calculation).

To assess the statistical significance of a motif, two measures are provided for each motif: Information Gain [19] and Log-Odds significance [12]. The information content I of a pattern, measures how likely a pattern is to occur, or equivalently, what is the amount of "surprise" when the pattern occurs. The odds score of a sequence measures the degree of surprise of the motif by comparing its probability of occurrence with the expected probability of occurrence according to the background distribution.

4   Experimental Evaluation

In our experiments we used a prototype developed in the C++ language. All the tests were done in 3.0GHz Pentium4 machine with 1GB of main memory, running windows XP Professional. Our algorithm works in "in-memory" way since it first loads to main memory the entire dataset and then starts the mining. Nevertheless, in all the experiments the maximum memory usage was 20MB. Each motif graphic contain the correspondent Overall Similarity (S), average β_1(B) and the type correlation of the pattern (T): only positive (1) and both (0).

4.1   Protein Unfolding

Protein folding is the process of acquisition of the native three-dimensional structure of a protein. The 3D structure, ultimately determined by its linear sequence of amino-acids, is essential for protein function. In recent years, several human and animal pathologies, such as cystic fibrosis, Alzheimer's and mad cow disease, among others, have been identified as protein folding or protein unfolding disorders. Over the years, many experimental and computational approaches have been used to study protein folding and protein unfolding. Here we use the proposed algorithm to assist in the study of the unfolding mechanisms of Transthyretin (TTR), a human plasma protein involved in amyloid diseases such as Familial Amyloid Polyneuropathy. Our goal is to find approximate motifs, i.e. simultaneous events on variations of molecular properties characterizing the unfolding process of TTR. The data analyzed is constituted by changes observed in molecular properties calculated from Molecular Dynamics (MD) protein unfolding simulations of TTR [1,3]. In the present case, the dataset consists of 127 time series, each representing the variation over time of the Solvent Accessible

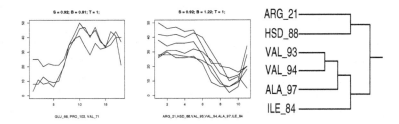

Fig. 2. (a) Example of a motif with an increasing SASA; (b) Example of a motif with a decreasing SASA and (c) respective cluster of Amino-Acids

Surface Area (SASA) of each amino acid in the protein. Each time series is a collection of 8000 data points, one data point per picosecond (ps) for a total of 8 nanoseconds (ns) of simulation. Before the algorithm is applied, pre-processing of the data is performed. The 8000 data points were scaled to 160 intervals. Each interval represents the average variation of SASA over 50ps. Symbolization was performed by rounding each value to its closest integer.

In the unfolding process of a protein, it is expected that the SASA increases for most amino-acids, i.e. they tend to become more exposed to the solvent upon protein unfolding. However, identifying how and when the SASA increases, and which amino-acids have similar (positive correlation) or opposite behavior (negative correlation), may reveal important details of the unfolding mechanism and which amino-acids constitute structural intermediates essential in the unfolding process. Figure 2 (a) depicts a motif that represents a synchronized increase of the SASA values for three amino-acids. Around data point 5, an unidentified event triggered the increase of solvent exposure. This is an example of a motif that is worthwhile to be investigated. Figure 2 (b) shows a motif with an overall tendency of SASA decreasing. Since this motif opposes to the expected behavior, it is also interesting to be further investigated. The motifs in figure 2 were obtained respectively with the parameters $\sigma = 3$, $w = 15(750ps)$ and $\sigma = 4$, $w = 10(500ps)$, with $deltaW = 5(250ps)$.

4.2   Stock Market Analysis and Synthetic Control Charts

The use of stock market datasets for the evaluation of time series algorithms is almost classic. We analyzed the Standard and Poor (S&P) 500 index historical stock data to demonstrate another possible application domain of our algorithm. This dataset contains 515 stocks with daily quotes fluctuations recorded over the course of the year 2005. It was obtained from http://kumo.swcp.com/stocks. We have analyzed the volume data for each day. The data contained a variable length (between 50 and 252) of points, since no scaling was done in this case. The size of the symbols alphabet was 100. We have made several runs on data, and we choose the results obtained with the following parameters: $\sigma = 0.01\%$; $R_{min} = 0.95$; w = 15 and deltaW = 5. It resulted in 9 motifs. From these we choose the 4 motifs with the highest statistical significance, according to the LogOdds and

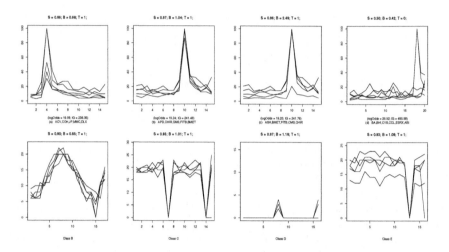

Fig. 3. (Upper row) Example motifs from the SP500 dataset; (Lower row) Selected motifs from the synthetic control dataset from class B, C, D, E

Information Gain measures. For each motif in figure 3 (upper row) it is presented the identifier of the company where the motif occurs and the respective statistical scores. Note that motif (d) has a overall similarity below R_{min}. This happen since this is a motif with two types of correlation (T=0).

In order to test our algorithm in a set of time series with well differentiated characteristics we applied it to the SCC dataset. This dataset was obtained from the UCI repository [8] and contain 600 examples of control charts synthetically generated, divided in six different classes of control charts. We mined each class individually and according to the parameters: $\sigma = 0.05\%$, $R_{min} = 0.9$, w=15, $deltaW = 15$. The average runtime was 3.67 secs and average number of motifs: 12.8. Figure 3 (lower row) shows examples of motifs with the highest similarity value for each of the four classes.

4.3   Performance Evaluation

Due to the lack of larger datasets we used synthetic random walk sequences to evaluate the performance of our algorithm in several dimensions. The formula used to generate the points of the sequences was: $S(j + 1) = S(j) + RandNum(10)$.

Due to its different phases and steps our algorithm has also different levels of complexity. Step 1 of Phase 1 is quadratic with respect to the length of the sequences. For N possible windows in database, this step requires approximately N^2 calls to the similarity function. Depending on the size of the window frame the number of tested pairs of windows can be greatly reduced with a consequently diminution in the runtime (see figure 4(b)). The complexity of Step 2 depends on the output of Step 1, namely it is proportional to number of 2-Clusters. This number directly depends on the R_{min} and on w values (see figure 4(a)). The

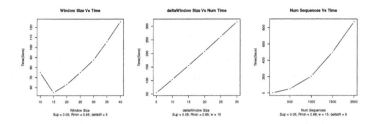

Fig. 4. Performance of the algorithm with different databases and parameter settings: Runtime w.r.t: (a) window size;(b) deltaW size;(c) number of sequences

initial value in the (see figure 4(a)) is explained by the significative difference in the number of 2-clusters obtained with $w = 10$ and with $w = 15$. The Phase 2, which corresponds to the length extension of the motifs is the less demanding. It typically represents an insignificant part of the overall time. Finally, we give some insights about the algorithm scalability. Figure 4(c) shows that our algorithm scales nearly in quadratic time in relation to the database size, which in practice is a is a reasonable performance for small or medium size datasets.

As it is expected in this type of mining applications, the interest of the results is a direct outcome of parameters values. Therefore, an iterative and interactive application of the algorithm is necessary. Heuristic or statistical methods, like the MDL principle used in [17], may provide initial values for the parameter setting.

5   Related Work

Our work can be viewed as a fusion of two research areas of data mining. The sequence and motif mining area where we emphasize the motivation provided by the algorithms in bioinformatics for mining overrepresented patters in a set of related and comparable biological sequences (proteins/dna) [18,2]. A second area corresponds to all the research made in time series, where particular attention is given to recent advances in the algorithms for pattern extraction. In the context of time series, we start by emphasizing some earlier work that have been done in the mining of recurrent subseries/patterns throughout a particular sequence [4,7]. Recently, Keogh et al. has introduced the issue of mining motifs in time series [16] and proposed algorithms [16,5] for this task. In [5] is described a probabilistic algorithm inspired on another algorithm from bioinformatics, called random projection. The difference from our work to the previous work starts with the definition of a motif. We are interested in discovering groups of subsequences that reveal the same trend, possibly in a symmetric way (approximate motifs). Additionally, we search for motifs in a set of related time sequences, eventually confined to a certain window frame, and not only in one time sequence. One work that has also inspired us is [13]. Here, the Pearson's correlation is used in the context of linear regression analysis in order to cluster time series.

6 Conclusions

In this work we have formulated the problem of mining approximate motifs in a set of related sequences. We also propose an algorithm that allows discovering all the motifs in the database, thus ensuring no false dismissals. The application of the algorithm to the problem that first motivated its development has already proved to be a valuable tool to assist biologists. However, the application of this method is not limited to this case study. As we have demonstrated it is an interesting method to be applied in other application domains.

References

1. P. Azevedo, C. Silva, J. Rodrigues., N. Loureiro-Ferreira, and R. Brito. Detection of hydrophobic clusters in molecular dynamics protein unfolding simulations using association rules. *LNCS*, (3745), 2005.
2. T. Bailey and C. Elkan. Fitting a mixture model by expectation maximization to discover motifs in biopolymers. In *Proc. of the 2th ISMB*, 1994.
3. R. Brito, W. Dubitzky, and J. Rodrigues. Protein folding and unfolding simulations: A new challenge for data mining. *OMICS: A Journal of Integrative Biology*, (8):153–166, 2004.
4. J. Caraca-Valente and I.Lopez-Chavarrias. Discovering similar patterns in time series. In *Proc. of the 6th ACM SIGKDD*, 2000.
5. B. Chiu, E. Keogh, and S. Lonardi. Probablistic discovery of time series motifs. In *Proceedings of the 9th ACM SIGKDD*, Washington DC, USA, August 24-27 2003.
6. D. Gunopulos and G. Das. Time series similarity measures (tutorial pm-2). In *Tutorial notes of the 6th ACM SIGKDD*, 2000.
7. J. Han, G. Dong, and Y. Yin. Efficient mining of partial periodic patterns in time series database. In *Proc. of the 15th ICDE*, 1999.
8. S. Hettich and S. D. Bay. The uci kdd archive irvine,[http://kdd.ics.uci.edu], ca: University of california, department of information and computer science, 1999.
9. E. Keogh and S. Kasetty. On the need for time series data mining benchmarks: A survey and empricial demonstration. In *Proc. of the 8th ACM SIGKDD*, 2002.
10. E. Keogh, J. Lin, and A. Fu. Hot sax: Efficiently finding the most unusual time series subsequence. In *Proc. of the 5th IEEE ICDM*, 2005.
11. E. Keogh and M. Pazzani. Scaling up dynamic time warping for datamining applications. In *Proc. of the 6th ACM SIGKDD*, 2000.
12. Anders Krogh. *An Introduction to Hidden Markov Models for Biological Sequences*, chapter 4, pages 45–63. Elsevier, 1998.
13. H. Lei and V. Govindaraju. Grm: A new model for clustering linear sequences. In *Proc. of SIAM Int'l Conference on Data Mining*, 2004.
14. J. Lin, E. Keogh, S. Lonardi, and B. Chiu. A symbolic representation of time series, with implications for streaming algorithms. In *Proc. of the 8th ACM SIGMOD workshop DMKD'03*, 2003.
15. H. Mannila, H. Toivonen, and A. Verkamo. Discovery of frequent episodes in event sequences. *Data Mining and Knowledge Discovery*, 1(3):259–289, 1997.
16. P. Patel, E.Keogh, J.Lin, and S.Lonardi. Mining motifs in massive time series databases. In *Proc. of 2th IEEE ICDM*, December 2002.

17. Y. Tanaka and K. Uehara. Discover motifs in multi-dimensional time-series using the principal component analysis and the mdl principle. In *Proc. of 3th MLDM*, 2003.

18. W. Thompson, E. Rouchka, and C. Lawrence. Gibbs recursive sampler: finding transcription factor binding sites. *Nucleic Acids Research*, 31(13):3580–3585, 2003.

19. J. Yang and P.S. Yu W. Wang. Mining surprising periodic patterns. In *Proc. of the 7th ACM SIGKDD*, 2001.

20. J.H. Zar. *Biostatistical Analysis (4th Edition)*. Prentice Hall, 1998.

Identifying Historical Period and Ethnic Origin of Documents Using Stylistic Feature Sets

Yaakov HaCohen-Kerner[1], Hananya Beck[1], Elchai Yehudai[1], and Dror Mughaz[1,2]

[1] Department of Computer Science, Jerusalem College of Technology (Machon Lev)
21 Havaad Haleumi St., P.O.B. 16031, 91160 Jerusalem, Israel
{kerner, hananya, yehuday}@jct.ac.il, myghaz@cs.biu.ac.il
[2] Department of Computer Science, Bar-Ilan University, 52900 Ramat-Gan, Israel

Abstract. Text classification is an important and challenging research domain. In this paper, identifying historical period and ethnic origin of documents using stylistic feature sets is investigated. The application domain is Jewish Law articles written in Hebrew-Aramaic. Such documents present various interesting problems for stylistic classification. Firstly, these documents include words from both languages. Secondly, Hebrew and Aramaic are richer than English in their morphology forms. The classification is done using six different sets of stylistic features: quantitative features, orthographic features, topographic features, lexical features and vocabulary richness. Each set of features includes various baseline features, some of them formalized by us. SVM has been chosen as the applied machine learning method since it has been very successful in text classification. The quantitative set was found as very successful and superior to all other sets. Its features are domain-independent and language-independent. It will be interesting to apply these feature sets in general and the quantitative set in particular into other domains as well as into other.

1 Introduction

Text classification (TC) is the supervised learning task of assigning natural language text documents to one or more predefined classes (also called categories) [19]. The meaning of supervised in this definition is that all the documents in a training set are pre-assigned a class before the training process starts.

TC is applied in many tasks, such as: clustering, document indexing, document filtering, information retrieval (IR), information extraction (IE), word sense disambiguation (WSD), text filtering, and text mining [13, 22]. Current-day TC presents challenges due to the large number of features present in the text set, their dependencies and the large number of training documents. The main difficulty with having a large number of features is finding out if some of them are redundant so that they can be ignored.

The most frequent TC task is to classify documents to one or more predefined categories according to their content. Another type of classification is stylistic classification, i.e.: classifying documents according to their author's style.

Classification according to categories is usually based on content words and collocations. For instance, texts about economics are different from texts about army by

N. Lavrač, L. Todorovski, and K.P. Jantke (Eds.): DS 2006, LNAI 4265, pp. 102–113, 2006.

their content words and collocations. In contrast, stylistic classification is usually based on linguistic features, e.g.: Argamon et al [1] on news stories and Koppel et al [14] on gender.

Hebrew-Aramaic documents present various interesting problems for stylistic classification. Firstly, these documents include words from both languages. Secondly, Hebrew and Aramaic are richer than English in their morphology forms. Thirdly, these documents include a relatively high rate of abbreviations [9].

The corpus that was analyzed in this paper includes responsa (answers written in response to Jewish legal questions) authored by rabbinic scholars. These documents are taken from a widespread variety of Jewish domains, e.g.: laws, holidays, customs, kosher food, economics and army. Each answer is based on both ancient Jewish writings and answers given by previous rabbinical authorities over the years. More so, arguments contradicting the author's answer should also be referred to. The author should give an acceptable explanation to solve such arguments.

In this research, the following classification tasks are investigated: Jewish ethnic origin of the authors (Sephardim or Ashkenazim) and / or historical period when the responsa were written. To the best of our knowledge, identifying the ethnic origin of documents' authors using stylistic feature sets is the first proposed.

The proposed model chooses the best combination of sets of stylistic features. The results are rather successful for all investigated classification tasks. The quantitative feature set was found as very successful and superior to all other sets. Its features are language-independent and domain-independent. In addition, this research can yield results of great use to scholars in the humanities, e.g.: identifying the differences in writing-style, culture and customs between writers who belong to different ethnic origin and / or historical period.

This paper is organized as follows: Section 2 gives background concerning the Hebrew and Aramaic languages. Section 3 describes previous classification of Hebrew-Aramaic texts. Section 4 presents feature sets for classification. Section 5 describes the proposed model. Section 6 presents the results of the experiments and analyzes them. Section 7 concludes and proposes future directions.

2 The Hebrew and the Aramaic Languages

2.1 The Hebrew Language

Hebrew is a Semitic language. It uses the Hebrew alphabet and it is written from right to left. Hebrew words in general and Hebrew verbs in particular are based on three (sometimes four) basic letters, which create the word's stem. The stem of a Hebrew verb is called $pl^{1,2}$ (פעל, "verb"). The first letter of the stem p (פ) is called $pe\ hapoal$; the second letter of the stem (ע) is called $ayin\ hapoal$ and the third letter of the stem l

---

[1] The Hebrew Transliteration Table that has been used in this paper, is taken from the web site of the Princeton university library.

[2] In this Section, each Hebrew word is presented in three forms: (1) transliteration of the Hebrew letters written in italics, (2) the Hebrew letters, and (3) its translation into English in quotes.

(ל) is called *lamed hapoal*. The names of the letters are especially important for the verbs' declensions according to the suitable verb types.

Except for the word's stem, there are other components which create the word's declensions, e.g.: conjugations, verb types, subject, prepositions, belonging, object and terminal letters. In Hebrew, it is impossible to find the declensions of a certain stem without an exact morphological analysis based on these features, as follows:

Conjugations: The Hebrew language contains seven conjugations that include the verb's stem. The conjugations add different meanings to the stem such as: active, passive, cause, etc. For example the stem *hrs* (הרס, "destroy") in one conjugation hrs means destroy but in another conjugation *nhrs* (נהרס, "being destroyed").

Verb types: The Hebrew language contains several verb types. Each verb type is a group of stems that their verbs are acting the same form in different tenses and different conjugations. There is a difference in the declensions of the stem in different verb types. In English, in order to change the tense, there is a need to add only one or two letters as suffixes. However, In Hebrew, for each verb type there is a different way that the word changes following the tense.

To demonstrate, we choose two verbs in past tense of different verb types: (1) *ktv* (כתב, "wrote") of the *shlemim* verb type (strong verbs - all three letters of the stem are apparent), and (2) the word *nfl* (נפל, "fell") of the *hasrey_pay_*noon verb type (where the first letter of the stem is the letter n and in several declensions of the stem this letter is omitted). When we use the future tense, the word *ktv* (כתב, "wrote") will change to *ykhtv* (יכתב, "will write") while the second word nfl will change to *ypl* (יפל, "will fall") which does not include the letter n. Therefore, in order to find the right declensions for a certain stem, it is necessary to know from which verb type the stem comes from.

Subject: Usually, in English we add the subject as a separate word before the verb. For example: I ate, you ate; where the verb change is minimal if at all. However, in Hebrew the subject does not have to be a separated word and it can appear as a suffix.

Prepositions: Unlike English, which has unique words dedicated to expressing relations between objects (e.g.: at, in, from), Hebrew has 8 prepositions that can be written as a concatenated letter at the beginning of the word. Each letter expresses another relation. For example: (1) the meaning of the letter *v* (ו) at the beginning of word is identical to the meaning of the word "and" in English and (2) the meaning of the letter *l* (ל) at the beginning of word is similar to the English word "to".

Belonging: In English, there are some unique words that indicate belonging (e.g.: my, his, her). This phenomenon exists also in Hebrew. In addition, there are several suffixes that can be concatenated at the end of the word for that purpose. The meaning of the letter *y* (י) at the end of word is identical to the meaning of the word "my" in English. For example, the Hebrew word *ty* (עטי) has the same meaning as the English words "my pen".

Object: In English, there are some unique words that indicate the object in the sentence, such as: him, her, and them. This is also the case in Hebrew. In addition, there are several letters that can be concatenated at the end of the word for that purpose. The letter *v* (ו) at the end of a word has the same meaning as the word him in English. For example, the Hebrew word *r'ytyv* (ראיתיו) has the same meaning as the English words "I saw him".

Terminal letters: In Hebrew, there are five letters: *m* (מ), *n* (נ), *ts* (צ), *p* (פ), *kh* (כ) which are written differently when they appear at the end of word: *m* (ם), *n* (ן), *ts* (ץ), *p* (ף), *kh* (ך) respectively. For example, the verb *ysn* (ישן, "he slept") and the verb *ysnty* (ישנתי, "I slept"). The two verbs have the same stem *ysn*, but the last letter of the stem is written differently in each one of the verbs.

The English language is richer in its vocabulary than Hebrew. The English language has about 40,000 stems while Hebrew has only about 3,500 and the number of lexical entries in the English dictionary is 150,000 compared with only 35,000 in the Hebrew dictionary [3].

However, the Hebrew language is richer in its morphology forms. The Hebrew language has 70,000,000 valid (inflected) forms while English has only 1,000,000 [3]. For instance, the single Hebrew word וכשיכוהו is translated into the following sequence of six English words: "and when they will hit him".

In Hebrew, there are up to seven thousand declensions for only one stem, while in English there is only a few declensions. For example, the English word eat has only four declensions (eats, eating, eaten and ate). The relevant Hebrew stem אכל (eat) has thousands of declensions.

Hebrew in general is very rich in its vocabulary of abbreviations. The number of Hebrew abbreviations is about 17,000 not including unique professional abbreviations, relatively high comparing to 40,000 lexical entries in the Hebrew language. About 35% of them are ambiguous. That is, about 6000 abbreviations have more than one possible extension for each abbreviation in particular contain a high frequency of abbreviations. Moreover, Jewish Law articles written in Hebrew-Aramaic include a high rate of abbreviations. About 20% of all the words in the documents, while more then one third of them (about 8%) are ambiguous.

2.2 The Aramaic Language

Aramaic is another Semitic language. It is particularly closely related to Hebrew, and was written in a variety of alphabetic scripts. Aramaic was the language of Semitic peoples throughout the ancient Near East. It is spoken for at least three thousand years. Aramaic is still spoken today in its many dialects, especially among the Chaldeans and Assyrians. (more details can be found in [27]).

Although Aramaic and Hebrew have much in common, there are several major differences between them. The main difference in grammar is that while Hebrew uses aspects and word order to create tenses, Aramaic uses tense forms. Another important difference is that there are several types of changes in one particular letter in many words. In some cases a Hebrew prefix is replaced in Aramaic by a suffix. More details can be found in Melamed [18].

3 Previous Classification of Hebrew-Aramaic Texts

Automatic classification of Hebrew-Aramaic texts is almost an uninvestigated research domain. CHAT, a system for stylistic classification of Hebrew-Aramaic texts is presented in [15, 16, 20]. It presents applications of several TC tasks to Hebrew-Aramaic texts:

1. Which of a set of known authors is the most likely author of a given document of unknown provenance?
2. Were two given corpora written/edited by the same author or not?
3. Which of a set of documents preceded which and did some influence others?
4. From which version (manuscript) of a document is a given fragment taken?

CHAT uses as features only single words, prefixes and suffixes. It uses simple ML methods such as Winnow and Perceptron. Its datasets contain only a few hundreds of documents. CHAT does not investigate the various tasks proposed at the end of Section 1.

Classification of Biblical documents has been done by Radai [24, 25, 26]. However, he did not implement any ML method.

4   Classification Features

Various kinds of stylistic features have been proposed during the years. For example: Quantitative features such as word and sentence length and punctuation–signs proposed by Yule [34]. Function words (e.g.: are, at, is, of, on, then and will) proposed by Mosteller and Wallace [21].

Many kinds of stylistic features have been applied for automatic classification of documents. Karlgren and Cutting [12] developed 20 features that belong to the three following sets: POS counts (e.g.: noun and adverb), lexical counts (e.g.: "that" and "we"), and textual counts (e.g.: characters per word and words per sentence). A set of orthographic features (abbreviations, acronyms, various spellings of the same words) has been proposed by Friedman [7]. Stamatatos et al [30] applied syntactic features (e.g.: frequencies and distribution of parts of speech tags, such as: noun, verb, adjective, adverb; basic syntactic sequences, such as noun-adjective and subject-verb relations and active and passive sentences).

Lim et al [17] used five different sets of features for automatic genre classification of web documents. Two of them were web-oriented: URL tags (e.g.: depth of URL, document type and domain area), and HTML tags (frequencies of various types of links). The other three sets were token information (e.g.: average number of characters per word and average number of words per sentence), lexical information (e.g.: frequency of content words, frequency of function words and frequency of punctuation marks) and structural information (e.g.: number of declarative sentences and number of question sentences).

5   The Proposed Model

As mentioned at Section 1, Hebrew-Aramaic documents present interesting problems for stylistic classification. Therefore, methods already used in text classification require adaptation to handle these problems. Firstly, the definition of suitable feature sets is required. Secondly, the proper combination of feature sets is needed to be investigated for a variety of classification experiments.

5.1 Feature Sets for Classification of Hebrew-Aramaic Texts

Several sets of linguistic features mentioned previously found to be applicable for this research. Unfortunately, no parts-of-speech tagger for Hebrew was available to us. Therefore, neither morphological features (e.g.: adjectives and verbs) nor syntactic features (frequencies and distribution of parts of speech tags, such as: noun, verb, adjective, adverb) are used.

Forty two baseline stylistic features appropriate for Hebrew-Aramaic texts have been defined and programmed. All features are normalized by the number of words in document. Features regarding sentences have two versions (relating or not relating to a comma as an end of a sentence). These features are detailed in the six following sets:

1. Lexical features: normalized frequencies of 958 religious help words (e.g.: bible, responsa, rabbi), 533 general help words (e.g.: of, at, on, no), both religious and general help words and 307 summarization words (e.g.: conclusion, to conclude, summary, to sum up).
2. Orthographic features: normalized frequencies of acronyms and abbreviations.
3. Topographic features: first n (10, 20) words, last n words, first n/2 words and last n/2 words, words in first n (2, 4) sentences, words in last n sentences, words in first n/2 sentences and words in last n/2 sentences, words in first n (1, 2) paragraphs, words in last n paragraphs, words in first n/2 paragraphs and words in last n/2 paragraphs.
4. Quantitative features: average number of characters in a word / sentence / paragraph / file, average number of words in a sentence / paragraph / file, average number of sentences in a paragraph / file, average number of paragraphs in a file, and average number of punctuation-signs (e.g.: !. ?, :, ;) in a file.
5. Function features: normalized frequencies of sentences appeared in brackets and normalized frequency of 11 pronouns (e.g.: I, we, you, he, she).
6. Vocabulary's richness: size of author's vocabulary in words.

The lexical, orthographic and pronoun features are domain-dependent and language-dependent. They have been especially fitted to the Hebrew and Aramaic languages. Dictionaries containing religious help words, general help words, summarization words, abbreviations and acronyms have been built. Moreover, they have been combined with various kinds of prepositions, belonging, terminal letters, etc. The quantitative, function and vocabulary's richness features are domain-independent and language-independent.

In order to find the best combination of sets for any particular classification task, one should try all possible combinations of sets. One of the known properties of the binomial coefficients is that the total number of distinct k-subsets on a set of n elements (i.e., the number of possible subsets) is $\sum_{k=0}^{n} \binom{n}{k} = 2^n$ where a k-subset is a subset of a set on n elements containing exactly k elements.

In this research, there are six feature sets. That is, there are only $2^6 = 64$ combinations (including the empty set). The combination with the highest classification rate is selected. Investigating whether the accuracy can be improved if the feature selection process will be applied to every single feature not to entire feature set is left for future research.

5.2 Support Vector Machines

Various machine learning methods have been applied for TC [29], e.g.: Naïve Bayes [28], C4.5 [8] and Winnow [15, 16, 20].

However, the Support Vector Machines (SVM) method [4, 31] has been chosen to be applied in this model since it seems to be the most successful for TC [5, 6, 10, 11, 33]. There are various variants of SVMs. One simple and fast method is Sequential Minimal Optimization (SMO) developed by Platt [23]. In this setting multi-class problems are solved using pairwise classification.

6   Experimental Results

To test the accuracy of the model, the 5-fold cross-validation is used. The SVM applied version was the SMO implementation of Weka [32] using a linear kernel, default values and no feature normalization. Model tuning is left for future research.

Three experiments have been applied on each one of two different data sets. These data sets include responsa that were downloaded from The Global Jewish Database (The Responsa Project[3]) at Bar-Ilan University. As mentioned above, these documents are taken from a widespread variety of domains relevant to Jewish life, e.g.: laws, holidays, customs, kosher food, economics and army.

The history of Jewish responsa covers a period of 1,700 years. The responsa investigated in this research were authored by Sephardic rabbis (Mediterranean) and Ashkenazic rabbis (eastern or central Europe) in the last 500 years (16th - 20th centuries). This era is called the *Acharonim* era. *Acharonim* (אחרונים) literally "the later ones", is a term used in Jewish law and history, to signify the leading rabbis living from roughly the 16th century to the present. The experts in the responsa domain think that the 20th century's responsa are much different from the previous ones for various reasons. In the 20th century there were major changes in the world in general and in the Jewish world in particular. Jewish rabbis from different ethnic origin (Sephardim and Ashkenazim) met each other. Some of them even immigrated to live within the communities of the other ethnic origin. Since the means of communication have developed so much Jewish rabbis read much more responsa written by other rabbis including from the other ethnic origin, and were influenced to one degree or another.

The accuracy that was measured in all experiments is the fraction of the number of documents correctly classified to the total number of possible documents to be classified.

6.1 Experiment # 1

The data set is a collection of 10,504 responsa authored by 60 Jewish rabbinic scholars, with about 175 documents for each scholar. These scholars belong to the two major Jewish ethnic groups (Sephardim and Ashkenazim). The responsa were written in the 16th - 20th centuries either. The total number of words in all files is about 18.5M words, while a document contains on average about 1763 words.

---

[3] http://www.biu.ac.il/ICJI/Responsa

This dataset can be classified into three different classifications: (1) ethnic Jewish groups (Sephardim or Ashkenazim) of their authors, (2) historical period when they were written (in the 16th - 19th centuries or in the 20th century) and (3) combination of classification (according to ethnic and time). For each classification task the dataset was divided into equal-sized suitable categories. That is the number of documents in the positive and in the negative sets is the same in each sub-experiment. The main results of all these three sub-experiments are presented in Table 1.

Table 1. Classification results of Dataset # 1

Kind of Classif-ication	Classification by each set alone						Classification by the best combination of sets	Std-dev
	1 Lexical	2 Ortho-graphic	3 Topo-graphic	4 Quan-titative	5 Function	6 Vocab-ulary richness		
ethnic	81.79	57.69	58.18	**97.97**	74.28	49.76	{1,2,3,4,5,6} **98.67**	0.3
time	83.6	58.61	62.25	**96.94**	80.05	53.97	{1,2,3,4,5,6} **98.95**	0.15
ethnic / time	62.83	41.92	38.13	**86.33**	57.04	28.8	{1,2,3,4,5} **92.81**	0.43

The first / second / third results' rows in Table 1 present the results of the classification to ethnic / time / both ethnic and time, respectively. The first results' row describes the classification according to the Jewish ethnic groups (Sephardim or Ashkenazim). The second results' row describes the classification according to the two historical periods when the responsa were written (Old or New). The last results' row describes the classification to the four possible categories (Old Sephardim, Old Ashkenazim, New Sephardim and New Ashkenazim).

In all sub-experiments, the best result by a unique set was always achieved by the quantitative set. The best result achieved by a combination of feature sets was achieved by a combination of all / almost all sets.

Several general conclusions can be drawn from Table 1:

1. All three classification tasks were highly successful.
2. The quantitative set (set # 4) was superior to all other sets and in the first two experiments it was enough for an excellent classification.
3. Only in the last sub-experiment, the most complex one (both ethnic and time), there is a meaningful contribution of other sets to the quantitative set.
4. The more complex is the classification task, the lowest classification results we achieve.
5. The lexical (# 1) and function (# 5) sets achieve reasonable results and are always part of the best combination of sets.
6. The orthographic, topographic and the vocabulary richness sets were rather poor, especially the last one.
7. The more relevant features we use the higher results we achieve.
8. The low values of the standard deviations indicate that the classification results are stable.

6.2 Experiment # 2

The second experiment is similar to the previous one, but now applied on a larger dataset containing responsa written only in the 19th and 20th centuries. This dataset includes 12,020 responsa authored by 48 Jewish rabbinic scholars, with about 250 documents for each scholar. These scholars belong to the two major Jewish ethnic groups (Sephardim or Ashkenazim). The responsa were written in the 19th and 20th centuries). The total number of words in all files is about 19M words, while a document contains on average about 1579 words.

This dataset can be also classified into three classifications: (1) ethnic Jewish groups (Sephardim or Ashkenazim), (2) historical period when they were written (in the 19th century or in the 20th century) and (3) ethnic / time. For each classification task the dataset was divided into equal-sized suitable categories. That is, the number of documents in the positive and in the negative sets is the same in each sub-experiment. Table 2 presents the main results of all three sub-experiments.

Table 2. Classification results of Dataset # 2

Kind of Classif-ication	Classification by each set alone						Classification by the best combination of sets	Std-dev
	1 Lexical	2 Ortho-graphic	3 Topo-graphic	4 Quan-titative	5 Function	6 Vocab-ulary richness		
ethnic	82.16	52.25	60.34	**98.10**	72.62	51.02	{1,3,4,5,6} 98.85	0.18
time	81.30	61.56	58.24	**96.39**	75.62	57.59	{1,2,3,4,5,6} 99.06	0.16
ethnic / time	61.98	34.88	39.30	**91.63**	61.57	30.86	{1,3,4,5} 95.54	0.36

In general, the explanations to Table 2 and the results presented in it are very similar to the explanations and results described for Table 1. The same general conclusions that were drawn there can be also drawn from Table2.

In both experiments, in the first two classifications (ethnic only and time only) it seems that the classification performance when the quantitative feature set is applied is highly successful and almost the same as the classification performance when all (or almost all) feature sets are used. This is not really surprising - the SVMs are known to be robust with respect to irrelevant features. Only in the last sub-experiment, the most complex one (both ethnic and time), there is a meaningful contribution of other sets to the quantitative set.

Due to limitations of space, we cannot detail the differences in writing-style between writers who belong to different ethnic groups and / or different historical periods. However, it is interesting to mention that among the most successful features we can find the following quantitative features:

(1) Average number of punctuation-signs in a file -
    Sephardim use much more punctuation-marks than Ashkenazim and
    New authors use much more punctuation-marks than Old authors

(2) Average number of characters/words in a sentence -
   Ashkenazim use much more characters/words in a sentence than Sephardim
   Old authors use much more characters/words in a sentence than New authors

Such findings can yield results of great use to scholars in the humanities who want to identify the differences in writing-style between writers who belong to different ethnic groups and / or different historical periods.

7   Conclusions and Future Work

In this paper, identifying historical period and ethnic origin of documents using stylistic feature sets is investigated. To the best of our knowledge, identifying the ethnic origin of documents' authors using stylistic feature sets is the first proposed.

The application domain is Jewish Law articles written in Hebrew-Aramaic. Such documents present various interesting problems for stylistic classification. Firstly, these documents include words from both languages. Secondly, Hebrew and Aramaic are richer than English in their morphology forms.

All three classification tasks were highly successful. The quantitative feature set was superior to all other sets and in the first two experiments it was enough for an excellent classification. Only in the last sub-experiment, the most complex one (both ethnic and time), there is a meaningful contribution of other sets to the quantitative set. The more complex is the classification task, the lowest classification results we achieve. The more relevant features we use the higher results we achieve. The orthographic, topographic and the vocabulary richness sets were rather poor, especially the last one.

Many features proposed in this research in general and the quantitative features in particular are language-independent and domain-independent.

It will be interesting to compare these results to the results of the same classification tasks based on language-dependent and domain-dependent feature sets.

It will be also interesting to apply these stylistic feature sets in general and the quantitative set in particular into other domains as well as into other languages.

Another relevant future research would be to apply a co-training-like algorithm [2] based "views" created from various types of features. This may both increase accuracy and reduce the amount of documents needed for training.

Other general research proposals are: (1) Investigating whether the accuracy can be improved if the feature selection process will be applied to every single feature (2) Investigating other features and other machine learning techniques that might improve classification.

Concerning feature sets there are various potential research directions. For example: (1) Which feature sets are good for which classification tasks? (2) What are the specific reasons for sets to perform better or worse on different classification tasks? (3) What are the guidelines to choose the correct sets for a certain classification task?

Acknowledgements. The authors thank two anonymous reviewers for their fruitful comments.

References

1. Argamon-Engelson, S., Koppel, M., Avneri, G.: Style-based text categorization: What newspaper am I reading? Proceedings of the AAAI Workshop on Learning for Text Categorization (1998) 1-4
2. Blum, A., Mitchell, T.: Combining Labeled and Unlabeled Data with Co-Training Proceedings of the Conference on Computational Learning Theory (COLT), (1998) 92–100
3. Choueka, Y., Conley, E. S., Dagan, I.: A comprehensive bilingual word alignment system: application to disparate languages - Hebrew, English, in J. Veronis (Ed.), Parallel Text Processing, Kluwer Academic Publishers, (2000) 69–96
4. Cortes, C., Vapnik, V.: Support-Vector Networks. Machine Learning, 20 (1995) 273-297
5. Díaz, I., Ranilla, J., Montañés, E., Fernández, J., Combarro, E. F.: Improving performance of text categorization by combining filtering, supportvector machines. JASIST 55(7) (2004) 579-592
6. Dumais, S., Platt, J., Heckerman, D., Sahami, M.: Inductive Learning Algorithms, Representations for Text Categorization. in Proceedings of the 7th ACM International Conference on Information, Knowledge Management (CIKM), Bethesda, MD (1998) 148-155
7. Friedman, S.: The Manuscripts of the Babylonian Talmud: A Typology Based Upon Orthographic and Linguistic Features. In: Bar-Asher, M. (ed.) Studies in Hebrew and Jewish Languages Presented to Shelomo Morag (in Hebrew), Jerusalem (1996) 163-190
8. Gabrilovich, E., Markovitch, S.: Text categorization with many redundant features: using aggressive feature selection to make SVMs competitive with C4.5. Proceedings of the 21 Int. Conference on Machine Learning ICML (2004) 321-328
9. HaCohen-Kerner, Y., Kass, A., Peretz, A.: Baseline Methods for Automatic Disambiguation of Abbreviations in Jewish Law Documents, Proceedings of the 4th International Conference on Advances in Natural Language Processing, EsTal 2004, Lecture Notes in Artificial Intelligence 3230, Berlin: Springer-Verlag, (2004) 58-69
10. Joachims, T.: Text Categorization with Support Vector Machines: Learning with Many Relevant Features. In Proceedings of the 10th European Conference on Machine Learning (ECML), Chemnitz, Germany, (1998) 137-142
11. Joachims, T.: Learning to Classify Text using Support Vector Machines. Kluwer (2002)
12. Karlgren, J., Cutting, D.: Recognizing Text Genres with Simple Metrics Using Discriminant Analysis". In Proceedings of the 15th International Conference on Computational Linguistics, Kyoto, Japan, 2, (1994) 1071-1075
13. Knight, K.: Mining online text. Commun. ACM 42, 11 (1999) 58–61
14. Koppel, M., Argamon, S., Shimony A. R.: Automatically categorizing written texts by author gender, Literary, Linguistic Computing 17, 4 (2002) 401-412
15. Koppel, M., Mughaz D., Schler J.: Text categorization for authorship verification in Proc. 8th Symposium on Artificial Intelligence, Mathematics, Fort Lauderdale, FL (2004)
16. Koppel, M., Mughaz D., Akiva N.: New Methods for Attribution of Rabbinic Literature, Hebrew Linguistics: A Journal for Hebrew Descriptive, Computational, Applied Linguistics, Bar-Ilan University Press (2006) 57: v-xviii
17. Lim, C. S., Lee K. J., Kim G-C.: Multiple sets of features for automatic genre classification of web documents. Inf. Process. Manage. 41(5) (2005) 1263-1276
18. Melamed, E. Z.: Aramaic-Hebrew-English Dictionary, Feldheim (2005)
19. Meretakis, D., Wuthrich, B.: Extending Naive Bayes Classifiers Using Long Itemsets. In Proc. 5th ACM-SIGKDD Int. Conf. Knowledge Discovery, Data Mining (KDD'99), San Diego, USA, (1999) 165-174

20. Mughaz, D. Classification Of Hebrew Texts according to Style, M.Sc. Thesis (in Hebrew), Bar-Ilan University, Ramat-Gan, Israel (2003)
21. Mosteller, F., Wallace, D. L.: Inference and Disputed Authorship: The Federalist. Reading, Mass. : Addison Wesley (1964)
22. Pazienza, M. T., ed. Information Extraction. Lecture Notes in Computer Science, Vol. 1299. Springer, Heidelberg, Germany (1997)
23. Platt, J. C.: Fast training of support vector machines using sequential minimal optimization, in Advances in Kernel Methods - Support Vector Learning, (Eds) B. Scholkopf, C. Burges,, A. J. Smola, MIT Press, Cambridge, Massachusetts, chapter 12, (1999) 185-208
24. Radai, Y.: Hamikra haMemuchshav: Hesegim Bikoret uMishalot (in Hebrew), Balshanut Ivrit, 13 (1978) 92-99
25. Radai, Y.: Od al Hamikra haMemuchshav (in Hebrew), Balshanut Ivrit 15 (1979) 58-59
26. Radai, Y.: Mikra uMachshev: Divrei Idkun (in Hebrew), Balshanut Ivrit 19 (1982) 47-52
27. Rosenthal, F.: Aramaic Studies During the Past Thirty Years, The Journal of Near Eastern Studies, Chicago (1978) 81-82
28. Schneider, K. M.: Techniques for Improving the Performance of Naive Bayes for Text Classification. Proceedings of the 6th International Conference, CICLing 2005, Lecture Notes in Computer Science 3406 Springer, ISBN 3-540-24523-5 (2005) 682-693
29. Sebastiani, F.: Machine learning in automated text categorization, ACM Computing Surveys 34 (1) (2002) 1-47
30. Stamatatos, E., Fakotakis N., Kokkinakis, G.: Computer-based authorship attribution without lexical measures, Computers and the Humanities, 35 (2001) 193-214
31. Vapnik, V. N. The Nature of Statistical Learning Theory. Springer-Verlag, NY, USA, ISBN 0-387-94559-8 (1995)
32. Witten, I. H., Frank, E.: Weka 3: Machine Learning Software in Java: http://www.cs.waikato.ac.nz/~ml/weka (1999)
33. Yang, Y., Liu, X.: A Re-examination of Text Categorization Methods. in Proceedings of the 22nd ACM International Conference on Research, Development in Information Retrieval (SIGIR), Berkeley, CA (1999) 42-49
34. Yule, G.U.: On Sentence Length as a Statistical Characteristic of Style in Prose with Application to Two Cases of Disputed Authorship, Biometrika, 30 (1938) 363-390

A New Family of String Classifiers
Based on Local Relatedness

Yasuto Higa[1], Shunsuke Inenaga[2,1], Hideo Bannai[1], and Masayuki Takeda[1,3]

[1] Department of Informatics, Kyushu University, Japan
{y-higa, shunsuke.inenaga, bannai, takeda}@i.kyushu-u.ac.jp
[2] Japan Society for the Promotion of Science
[3] SORST, Japan Science and Technology Agency (JST)

Abstract. This paper introduces a new family of *string classifiers* based on local relatedness. We use three types of local relatedness measurements, namely, *longest common substrings* (*LCStr's*), *longest common subsequences* (*LCSeq's*), and *window-accumulated longest common subsequences* (*wLCSeq's*). We show that finding the optimal classier for given two sets of strings (the positive set and the negative set), is NP-hard for all of the above measurements. In order to achieve practically efficient algorithms for finding the best classifier, we investigate pruning heuristics and fast string matching techniques based on the properties of the local relatedness measurements.

1 Introduction

In recent years, we have witnessed a massive increase in the amount of *string* data available in many different domains, such as text data on the Internet, and biological sequence data. Finding meaningful knowledge in the form of string *patterns* from such data sets has become a very important research topic. In this light, we and others have been studying the problem of finding the optimal pattern which distinguishes between a positive set of strings and a negative set of strings [1,2,3,4,5,6,7,8,9,10,11]. The optimal pattern discovery problem aims to find the highest scoring pattern with respect to a certain scoring function, usually preferring patterns which are contained in all or most strings in the positive data set, but not contained in all or most strings of the negative data set.

In previous work, the classification models are mainly based on *distance*, where a string is classified to be consistent with a pattern when the string contains a substring that is within a certain fixed distance of the pattern, and not consistent otherwise. In this paper, we consider a classification model based on *relatedness*, where a string is classified to be consistent with the string classifier when it is locally more similar to the classifier than a certain fixed relatedness, and not consistent otherwise. Although the two seem to be equivalent concepts that are simply worded differently, the latter gives rise to a new, richer family of string classifiers.

We study and show the subtle differences between them from an algorithmic perspective. We consider three types of relatedness measurements, namely,

N. Lavrač, L. Todorovski, and K.P. Jantke (Eds.): DS 2006, LNAI 4265, pp. 114–124, 2006.

longest common substrings (*LCStr's*), *longest common subsequences* (*LCSeq's*), and *window-accumulated longest common subsequences* (*wLCSeq's*). We show that finding the optimal classier for given two sets of strings is NP-hard for all of the above measurements. In order to achieve practically efficient algorithms for finding the best classifier, we investigate pruning heuristics and fast string matching techniques based on the properties of the local relatedness measurements. Our preliminary experiments on DNA sequence data showed that the algorithm for the wLCSeq measurement runs in acceptable amount of time.

2 Preliminaries

2.1 Notations

Let $\mathcal{N}$ be the set of non-negative integers. Let Σ be a finite *alphabet*, and let Σ^* be the set of all *strings* over Σ. The *length* of string s is denoted by $|w|$. The empty string is denoted by ε, that is, $|\varepsilon| = 0$. For set $S \subseteq \Sigma^*$ of strings, we denote by $|S|$ the number of strings in S, and by $||S||$ the total length of strings in S.

Strings x, y and z are said to be a *prefix*, *substring*, *suffix* of string $s = xyz$, respectively. Let $Substr(s)$ be the set of the substrings of string s. When string x is a substring of string s, then s is said to be a *superstring* of x. The i-th character of string s is denoted by $s[i]$ for $1 \leq i \leq |s|$, and the substring that begins at position i and ends at position j is denoted by $s[i..j]$ for $1 \leq i \leq j \leq |s|$. We say that string q is a *subsequence* of string s if $q = s[i_1] \cdots s[i_{|q|}]$ for some $1 \leq i_1 < \cdots < i_{|q|} \leq |s|$. Let $Subseq(s)$ be the set of the subsequences of string s. When string x is a subsequence of string s, then s is said to be a *supersequence* of x.

For two strings $p, s \in \Sigma^*$, if $q \in Substr(p) \cap Substr(s)$, then q is called a *common substring* of p and s. When no common substrings of p and s are longer than q, q is called a *longest common substring* (*LCStr* for short) of p and s. Then we denote $lcstr(p, s) = |q|$. It is easy to see that $lcstr(p, s) \leq \min\{|p|, |s|\}$.

Similarly, if $q \in Subseq(p) \cap Subseq(s)$, then q is called a *common subsequence* of p and s. When no common subsequences of p and s are longer than q, q is called a *longest common subsequence* (*LCSeq* for short) of p and s. Then we denote $lcseq(p, s) = |q|$. It is easy to see that $lcseq(p, s) \leq \min\{|p|, |s|\}$.

2.2 Relatives of a String

The functions *lcstr* and *lcseq* are relatedness measures that quantify the affinities between two strings. Further, we consider an extended version of *lcseq*, *window-accumulated LCSeq measure*, where we compute the *lcseq* values for p against all substrings of s of length $\leq d$ (d is a positive integer). The new measure $wlcseq^d$ is defined as follows.

$$wlcseq^d(p, s) = \max\{lcseq(p, t) \mid t \in Substr(s) \text{ and } |t| \leq d\}.$$

We note that when d is long enough then $wlcseq^d(p, s) = lcseq(p, s)$.

Let δ be one of the measures $lcstr$, $lcseq$, and $wlcseq^d$. A k-*relative* of a string p under δ is any string s with $\delta(s, p) \geq k$. The set of k-relatives of p is denoted by $Re^\delta(p; k)$. That is,

$$Re^\delta(p; k) = \{s \in \Sigma^* \mid \delta(p, s) \geq k\}.$$

Definition 1. *The language class w.r.t. $Re^\delta(p; k)$ for each δ is as follows.*

$$\mathsf{LCSTRL} = \{Re^{lcstr}(p; k) \mid p \in \Sigma^* \text{ and } k \in \mathcal{N}\}$$
$$\mathsf{LCSEQL} = \{Re^{lcseq}(p; k) \mid p \in \Sigma^* \text{ and } k \in \mathcal{N}\}$$
$$\mathsf{wLCSEQL}^d = \{Re^{wlcseq^d}(p; k) \mid p \in \Sigma^* \text{ and } k \in \mathcal{N}\}$$

Remark 1. In [3] the subsequence pattern discovery problem was discussed, where a pattern $p \in \Sigma^*$ matches any supersequence of p. We note that LCSEQL subsumes the languages of the subsequence patterns since $Re^{lcseq}(p; |p|)$ is exactly the set of supersequences of p.

Remark 2. In [2] the pattern discovery problem for the window-accumulated subsequence patterns (episode patterns) was addressed, where a window-accumulated subsequence pattern is formally an ordered pair $\langle p, d \rangle \in \Sigma^* \times \mathcal{N}$ and matches a string s if there is a substring t of s such that $|t| \leq d$ and $p \in Subseq(t)$. Then, one can see that $\mathsf{wLCSEQL}^d$ generalizes the languages of window-accumulated subsequence patterns since we have $Re^{wlcseq^d}(p; |p|)$ is identical to the language of the window-accumulated subsequence pattern $\langle p, d \rangle$.

3 Finding Best String Classifiers

3.1 Score Function

Suppose Π is a set of descriptions over some finite alphabet, and each $\pi \in \Pi$ defines a language $L(\pi) \subseteq \Sigma^*$. Let $good$ be a function from $\Pi \times 2^{\Sigma^*} \times 2^{\Sigma^*}$ to the real numbers. The problem we consider in this paper is: *Given two sets $S, T \subseteq \Sigma^*$ of strings, find a description $\pi \in \Pi$ that maximizes score $good(\pi, S, T)$.* Intuitively, score $good(\pi, S, T)$ expresses the 'goodness' of π as a classifier for S and T. The definition of $good$ varies with applications. For example, the χ^2 values, entropy information gain, and Gini index are often used. Essentially, these statistical measures are defined by the number of strings that satisfy the rule specified by π. Any of the above-mentioned measures can be expressed by the following form:

$$good(\pi, S, T) = f(x_\pi, y_\pi, |S|, |T|),$$
$$\text{where } x_\pi = |S \cap L(\pi)| \text{ and } y_\pi = |T \cap L(\pi)|.$$

When S and T are fixed, $x_{\max} = |S|$ and $y_{\max} = |T|$ are regarded as constants. On this assumption, we abbreviate the function to $f(x_\pi, y_\pi)$. In the sequel, we assume that f is *pseudo-convex* (also called *conic* in the previous work [3]) and can

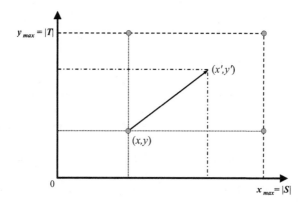

Fig. 1. Illustration of the domain of score function f. For pseudo-convex score functions, the highest score of an arbitrary point in the rectangle defined by the four highlighted points is the maximum score of the four points (Lemma 1).

be evaluated in constant time. We say that a function f from $[0, x_{max}] \times [0, y_{max}]$ to real numbers is pseudo-convex if

- for any $0 \leq y \leq y_{max}$, there exists an x_1 such that
  - $f(x, y) \geq f(x', y)$ for any $0 \leq x < x' \leq x_1$, and
  - $f(x, y) \leq f(x', y)$ for any $x_1 \leq x < x' \leq x_{max}$.
- for any $0 \leq x \leq x_{max}$, there exists a y_1 such that.
  - $f(x, y) \geq f(x, y')$ for any $0 \leq y < y' \leq y_1$, and
  - $f(x, y) \leq f(x, y')$ for any $y_1 \leq y < y' \leq y_{max}$.

The following is an important property of pseudo-convex functions.

Lemma 1 ([6]). *For any $0 \leq x < x' \leq x_{\max}$ and $0 \leq y < y' \leq y_{\max}$,*

$$f(x', y') \leq \max\{f(x, y), f(x, y_{\max}), f(x_{\max}, y), f(x_{\max}, y_{\max})\}.$$

3.2   Problem and Complexities

In this paper we consider the following problem for each relatedness measure δ.

Definition 2 (Finding best string classifier under δ according to f)

Input: *Two finite sets $S, T \subseteq \Sigma^*$ of strings, and a positive integer k.*
Output: *A string $p \in \Sigma^*$ that maximizes score $f(x_\pi, y_\pi)$, where $\pi = \langle p, k \rangle$, and
$x_\pi = |S \cap Re^\delta(p; k)|$ and $y_\pi = |T \cap Re^\delta(p; k)|$.*

Theorem 1. *The optimization problem of Definition 2 under lcstr is NP-hard.*

Proof. Reduction from MINIMUM SET COVER. (Given graph $G = (V, E)$, a set $C = \{C_1, \ldots, C_m\}$ where each $C_i \subseteq V$ for each $1 \leq i \leq m$, and integer c, does there exist a subset $C' \subseteq C$ such that $|C'| \leq c$ and any vertex in V is contained in at least one member of C'?) Consider the function

$$f(x,y) = \begin{cases} 0 & \text{if } x < x_{\max} \\ y_{\max} - y & \text{otherwise } (x = x_{\max}) \end{cases}$$

and create an instance of finding the best string classifier under *lcstr* as follows. Let $k = \lceil \log_2 m \rceil$, and let $\bar{i}$ denote the k-bit binary representation of integer $i \leq m$. S will consist of $|V|$ strings, each string corresponding to a node in V. For each $v \in V$, define $s_v = \sum_{i:v \in C_i} \$\bar{i}$. Let $x = \sum_{s \in X} \#s$, where $X = \{s_v[i..(i+k-1)] \mid v \in V, 1 \leq i \leq |s_v| - k + 1, \exists j \ (i \leq j \leq i+k-1), \text{ s.t. } s_v[j] = \$\}$, that is, x is the concatenation of all length k substrings of strings in S that contain the character $\$$, each preceded by $\#$. We define $T = \{\bar{i}x \mid 1 \leq i \leq m\}$, where each string corresponds to the member $C_i \in C$. Then, the existence of a set cover $C' = \{C_{i_1}, \ldots, C_{i_c}\}$ of size c implies the existence of a string p giving $f(x_\pi, y_\pi) = y_{\max} - c$ and vice versa: Suppose C' is a cover of V, and consider $p = \bar{i}_1 \text{¢} \bar{i}_2 \cdots \text{¢} \bar{i}_c$. Since each $v \in V$ is contained in some member $C_{i_j} \in C'$, each $s_v \in S$ will share the length k substring $\bar{i}_j$ with p. Noting that the character '¢' is not contained in any strings of S or T, p can only share length k substrings via each $\bar{i}_j$ ($1 \leq j \leq c$). Therefore, p will only be a k-relative to the c strings $\{\bar{i}_j x | C_{i_j} \in C'\} \subseteq T$. On the other hand, suppose p is a string that is a k-relative of all strings in S but for only c strings in T. Notice that p can only be a k-relative of strings in S or T by sharing length k substrings that correspond to some $\bar{i}$ ($1 \leq i \leq m$), since otherwise: (1) if the length k substring contained $\$$, then p would also be a k-relative of x and hence all strings in T, and (2) if the length k substring contained $\#$, then p would not be a k-relative of any string in S. For each different $\bar{i}$ that p contains, a unique string $\bar{i}x \in T$ also becomes a k-relative of p. Therefore, p will contain exactly c substrings $\bar{i}_1, \ldots \bar{i}_c$, where each $s_v \in S$ will contain at least one $\bar{i}_j$ ($1 \leq j \leq c$). This implies a set cover of size c consisting of C_{i_j} ($1 \leq j \leq c$). $\square$

Theorem 2. *The optimization problem of Definition 2 under lcseq is NP-hard.*

Proof. Reduction from MINIMUM SET COVER (See Theorem 1). Consider f as in Theorem 1. Consider the alphabet $\Sigma = \{\sigma_1, \ldots, \sigma_m\}$. Let $k = 1$, $S = \sum_{\sigma_i : v \in C_i}$, $T = \{\sigma_1, \ldots, \sigma_m\}$ ($\sigma_i \neq \sigma_j$ for $i \neq j$). Since $k = 1$, a string containing character σ_i will be a relative of a string s if and only if s also contains the character σ_i. Therefore, the existence of a string p giving $f(x_\pi, y_\pi) = y_{\max} - c$ implies the existence of a set cover $C' = \{C_{i_1}, \ldots, C_{i_c}\}$ of size c, and vice versa.
                $\square$

Recall that when d is long enough, we have $wlcseq^d(p, d) = lcseq(p, s)$. Thus, the optimization problem for the *lcseq* measure is a special case of that for the $wlcseq^d$ measure. Hence we get the following result.

Theorem 3. *The optimization problem of Definition 2 under $wlcseq^d$ is NP-hard.*

Since the optimization problem of finding the best π is NP-hard for all types of the similarity measures, we inherently face exponentially many candidates for the best string classifier. The two keys to a practically efficient computation of the best π are: *reducing the number of candidates (pruning)*, and *quickly counting x_π and y_π for each candidate π (fast string matching)*.

3.3 Branch-and-Bound Algorithms

Let δ be one of the measures *lcstr*, *lcseq*, and *wlcseq^d*. Consider given k. Note that, for any string p of length less than k, we always have $Re^\delta(p; k) = \emptyset$. On the other hand, we need to care about all the $|\Sigma|^k$ strings of length k, because all of those strings have a possibility of being a 'seed' of the best string with the highest score. Then, we examine longer candidates by appending new characters to the right of the $|\Sigma|^k$ strings. The next lemma states a useful property on $Re^\delta(p; k)$.

Lemma 2. *Let k be any positive integer. For any two strings p, p' in Σ^*, if p is a prefix of p', then $Re^\delta(p'; k) \supseteq Re^\delta(p; k)$.*

The following pruning lemma are derived from Lemmas 1 and 2.

Lemma 3. *Let k be any positive integer. For any two strings p, p' such that p is a prefix of p', let $\pi = \langle p, k \rangle$ and $\pi' = \langle p', k \rangle$. Then,*

$$f(x_{\pi'}, y_{\pi'}) \leq \max\{f(x_\pi, y_\pi), f(x_\pi, y_{\max}), f(x_{\max}, y_\pi), f(x_{\max}, y_{\max})\},$$

where $x_\pi = |S \cap Re^\delta(p; k)|$ and $y_\pi = |T \cap Re^\delta(p; k)|$.

What remains is how to compute the numbers x_π and y_π of strings s in S and T, respectively, such that $\delta(p, s) \geq k$, where $\pi = \langle p, k \rangle$.

Firstly, we consider the case of $\delta = lcstr$.

Definition 3 (Computing local relatedness under measure *lcstr*)

Given: *A finite set $S \subseteq \Sigma^*$ of strings, and a positive integer k.*
Query: *A string $p \in \Sigma^*$.*
Answer: *The cardinality of $Re^{lcstr}(p; k) \cap S$.*

Theorem 4. *The problem of computing local relatedness under the measure lcstr can be solved in $O(|p|)$ time using $O(|S|)$ extra space after $O(\|S\|)$ time and space preprocessing.*

Proof. We build the directed acyclic word graph (DAWG) [12] from the strings in S. We note that the nodes of DAWG represent the equivalence classes under some equivalence relation defined on $Substr(S) = \bigcup_{s \in S} Substr(s)$. Each equivalence class can be written as $\{x'y \mid x'$ is a suffix of $x\}$ for some strings x, y with $xy \in Substr(S)$, and therefore it can contain at most one string t of length k. For every node v containing such string t, we associate v with the list of ID's of strings in S that are superstrings of t. Such a data structure can be built only in $O(\|S\|)$ time and space. To compute the cardinality of $Re^{lcstr}(p;k)$ for every candidate p, we use this data structure as a finite-state sequential machine. The machine makes state-transitions scanning the characters of p one by one. Whenever the machine is in a state (node) such that the corresponding equivalence class contains no string of length $\leq k$, it makes "failure transition" navigated by the suffix links. If the current state has outputs, then it implies that all the strings s listed in the outputs satisfy $lcstr(p, s) \geq k$. The number of failure transitions executed is bounded by the number of ordinary state transitions executed, and is therefore at most $|p|$. The cardinality is thus computed in $O(|p|)$ time.   □

We can therefore compute x_π and y_π in $O(|p|)$ time using $O(|S| + |T|)$ extra space after $O(\|S\| + \|T\|)$ time and space preprocessing.

Secondly, we consider $\delta = lcseq$. We do not preprocess the input S and T, and simply compute $lcseq(p, s)$ against all strings s in $S \cup T$. Each $lcseq(p, s)$ is computable in $O(|p| \cdot |s|)$ time by a standard dynamic programming (DP) method. Section 4 explains the DP method. We can compute x_π and y_π in $O(|p| \cdot (\|S\| + \|T\|))$ time.

Lastly, we will devote the next full section to the case of $\delta = wlcseq^d$, as this case needs to be explained in details.

4   Computing Local Relatedness Under $wlcseq^d$

Given two strings p, s, a standard technique for computing $lcseq(p, s)$ is the *dynamic programming* method, where we compute the DP matrix of size $(|p| + 1) \times (|s| + 1)$ for which $DP[i, j] = lcseq(p[1..i], s[1..j])$ for $1 \leq i \leq |p|$ and $1 \leq j \leq |s|$. The recurrence for computing the DP matrix is the following:

$$DP[i, j] = \begin{cases} 0 & \text{if } i = 0 \text{ or } j = 0, \\ \max(DP[i-1, j], DP[i, j-1]) & \text{if } i, j > 0 \text{ and } p[i] \neq s[j], \\ DP[i-1, j-1] + 1 & \text{if } i, j > 0 \text{ and } p[i] = s[j]. \end{cases}$$

Therefore, to compute $lcseq(p, s) = DP[|p|, |s|]$, we need $O(|p| \cdot |s|)$ time and space. Pair (i, j) is said to be a *partition point* of DP, if $DP[i, j] = DP[i-1, j]+1$. P denotes the set of the partition points of DP. Remark that P is a compressed form of DP, and the size of P is $O(L|s|)$, where $L = lcseq(p, s)$. See Fig. 2 for an example of a DP matrix and its partition points. The partition point set P is implemented by double-linked lists as shown in Fig. 3, where the cells are vertically sorted by the $lcseq$ values, and the values in the cells represent the corresponding row indices.

The problem considered in this section is the following.

Definition 4 (Computing local relatedness under measure $wlcseq^d$)

Given: A finite set $S \subseteq \Sigma^*$ of strings, and a positive integer k.
Query: A string $p \in \Sigma^*$.
Answer: The cardinality of $Re^{wlcseq^d}(p; k) \cap S$.

		c	b	a	c	b	a	a	b	a
	0	0	0	0	0	0	0	0	0	0
b	0	0	1	1	1	1	1	1	1	1
c	0	1	1	1	2	2	2	2	2	2
d	0	1	1	1	2	2	2	2	2	2
a	0	1	1	2	2	2	3	3	3	3
b	0	1	2	2	2	3	3	3	4	4
a	0	1	2	3	3	3	4	4	4	5

Fig. 2. The DP matrix for $lcseq(p, s)$, where $p = $ **bcdaba** and $s = $ **cbacbaaba**. Note $lcseq(p, s) = DP[|p|, |s|] = DP[6, 9] = 5$. The colored entries represent the partition points of the DP matrix.

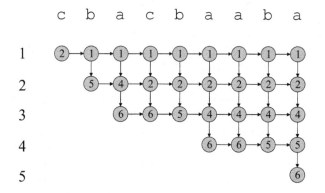

Fig. 3. Double-linked list implementation of the partition point set P for the DP matrix of Fig. 2. The cells are sorted vertically by the $lcseq$ values, and the values in the cells represent the corresponding row indices.

The above problem can be solved by computing the length of the LCSeq of p and every d-gram of $s \in S$, and by counting the number of strings s in S for which the maximum LCSeq length is not less than k. Here, we need to compute $lcseq(p, s[j..j + d - 1])$ for every position $1 \leq j \leq |s| - d + 1$ of each string $s \in S$. If naïvely computing the DP matrices for all pairs $p, s[j..j + d - 1]$, it takes total of $O(d|p| \cdot \|S\|)$ time and $O(d|p|)$ space. However, we can establish the following lemma that leads us to a more efficient solution.

Theorem 5. *The problem of computing local relatedness under the measure* $wlcseq^d$ *can be solved in* $O(d\|S\|)$ *time using* $O(\ell d)$ *extra space, where*

$$\ell = \max\{lcseq(p, s[j..j + d - 1]) \mid s \in S \text{ and } 1 \leq j \leq |s| - d + 1\}.$$

Proof. Let us concentrate on one string s in S. For any $1 \leq h \leq h' \leq |s|$, let $DP_{h'}^h$ and $P_{h'}^h$ denote the DP matrix and the corresponding partition point set for $lcseq(p, s[h..h'])$, respectively. For any $1 \leq j \leq |s| - d + 1$, let $r = j + d - 1$.

Now we are computing P_r^j for $j = |s| - d + 1, \ldots, 2, 1$, in the decreasing order of j. Namely, we compute the partition point set for a *sliding window* of width d (See Fig. 4). Computing P_r^j from P_{r+1}^{j+1} is done in two rounds. Firstly, compute P_{r+1}^j from P_{r+1}^{j+1}, then compute P_r^j from P_{r+1}^j.

Let $\ell_s = \max\{lcseq(p, s[j..r]) \mid 1 \le j \le |s| - d + 1\}$. The second step of computing P_r^j from P_{r+1}^j can be done in $O(\ell_s)$ time based on the following observations. It follows from the property of DP matrices that for any $1 \le i \le |p|$ and $j \le h \le r$, we have $DP^j[i, h] = DP^{j+1}[i, h]$. Therefore, we can compute P_r^j from P_{r+1}^j by simply deleting the $(r + 1)$-th column of P_{r+1}^j. The number of entries in that column is at most $lcseq(p, s[j..r + 1]) = lcseq(p, s[j..j + d]) \le lcseq(p, s[j..j + d - 1]) + 1 \le \ell_s + 1$. Thus it can be done in $O(\ell_s)$ time.

The first step of computing P_{r+1}^j from P_{r+1}^{j+1}, can be done efficiently based on the algorithm of Landau et al. [13]. They presented an algorithm which, given two strings p and s, computes $P_{|s|}^j$ representing $lcseq(p, s[j..|s|])$ for every $j = |s|, \ldots, 2, 1$. It runs in total of $O(L|s|)$ time and space, where $L = lcseq(p, s)$, by implementing the partition point with a double linked list (see Fig. 3). Combining this algorithm with the method mentioned in the above paragraph, we are able to compute P_{r+1}^j from P_{r+1}^{j+1}.

Now let us clarify the complexities for computing $lcseq(p, s[j..r])$ for all $j = |s| - d + 1, \ldots, 1$. The space consumption is clearly $O(\ell_s d)$, since the size of P_r^j is bounded by $O(\ell_s d)$. Regarding the time complexity, as mentioned above, deleting the last column takes $O(\ell_s)$ time. When adding a new column, in the worst case, a new partition point is inserted into each of the d columns of P_{r+1}^{j+1}. Insertion of these new partition points can be done in $O(d)$ time in aggregate [13], by the double-linked list implementation. Thus, the time cost is $O(\ell_s|s| + d|s|) = O(d|s|)$, since $\ell_s \le \min\{|p|, d\}$. In conclusion, the whole space requirement is $O(\max\{\ell_s \mid s \in S\} \cdot d) = O(\ell d)$, and the total time requirement is $O(\sum_{s \in S}(d|s|)) = O(d\|S\|)$. $\qquad\square$

Therefore, we can compute x_π and y_π for $wlcseq^d$ measure in $O(d(\|S\| + \|T\|))$ time using $O(\ell'd)$ space, where $\ell' = \max\{lcseq(p, s[j..j + d - 1]) \mid s \in S \cup T$ and $1 \le j \le |s| - d + 1\}$. We remark that $\ell' \le \min\{|p|, d\}$.

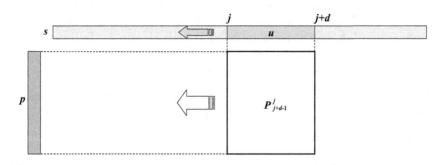

Fig. 4. Illustration for computing P_{j+d-1}^j representing $lcseq(p, s[j..j + d - 1])$ for every $1 \le j \le n - d + 1$, in a sliding window manner

5   Discussions

The problem of finding the optimal string classifier for two given sets S, T of strings, has been extensively studied in the recent years. The pursuit of better classification had been done mostly by enriching the 'pattern class' in which the best pattern is searched for. This paper suggested a new family of string classifiers based on local relatedness measures, $lcstr$, $lcseq$, and $wlcseq^d$. A big difference between the previous ones and our new family is that our string classifier does not necessarily appear as a pattern in the strings in S. Namely, the classifier can be composed of a superstring or supersequence of the strings in S. Therefore, when there do not exist good patterns which classify S and T, our new approach is expected to give us a good classifier that may lead to a more meaningful knowledge. Preliminary experiments conducted on DNA sequence data showed that the optimal string classifier discovery algorithm for the $wlcseq^d$ measure runs in acceptable amount of time for modest settings of k and d. Experiments supporting practical effectiveness of our method, remain as our future work.

References

1. Arimura, H., Wataki, A., Fujino, R., Arikawa, S.: A fast algorithm for discovering optimal string patterns in large text databases. In: International Workshop on Algorithmic Learning Theory (ALT'98). Volume 1501 of LNAI., Springer-Verlag (1998) 247–261

2. Hirao, M., Inenaga, S., Shinohara, A., Takeda, M., Arikawa, S.: A practical algorithm to find the best episode patterns. In: Proc. 4th International Conference on Discovery Science (DS2001). Volume 2226 of LNAI., Springer-Verlag (2001) 435–440

3. Hirao, M., Hoshino, H., Shinohara, A., Takeda, M., Arikawa, S.: A practical algorithm to find the best subsequence patterns. Theoretical Computer Science **292**(2) (2002) 465–479

4. Shinohara, A., Takeda, M., Arikawa, S., Hirao, M., Hoshino, H., Inenaga, S.: Finding best patterns practically. In: Progress in Discovery Science. Volume 2281 of LNAI., Springer-Verlag (2002) 307–317

5. Inenaga, S., Bannai, H., Shinohara, A., Takeda, M., Arikawa, S.: Discovering best variable-length-don't-care patterns. In: Proceedings of the 5th International Conference on Discovery Science. Volume 2534 of LNAI., Springer-Verlag (2002) 86–97

6. Shinozaki, D., Akutsu, T., Maruyama, O.: Finding optimal degenerate patterns in DNA sequences. Bioinformatics **19** (2003) ii206–ii214

7. Lanctot, J.K., Li, M., Ma, B., Wang, S., Zhang, L.: Distinguishing string selection problems. Information and Computation **185** (2003) 41–55

8. Takeda, M., Inenaga, S., Bannai, H., Shinohara, A., Arikawa, S.: Discovering most classificatory patterns for very expressive pattern classes. In: 6th International Conference on Discovery Science (DS 2003). Volume 2843 of LNCS., Springer-Verlag (2003) 486–493

9. Bannai, H., Hyyrö, H., Shinohara, A., Takeda, M., Nakai, K., Miyano, S.: Finding optimal pairs of patterns. In: 4th International Workshop on Algorithms in Bioinformatics (WABI 2004). Volume 3240 of LNBI., Springer-Verlag (2004) 450–462

10. Bannai, H., Hyyrö, H., Shinohara, A., Takeda, M., Nakai, K., Miyano, S.: An $O(N^2)$ algorithm for discovering optimal Boolean pattern pairs. IEEE/ACM Transactions on Computational Biology and Bioinformatics **1**(4) (2004) 159–170 (special issue for selected papers of WABI 2004).

11. Inenaga, S., Bannai, H., Hyyrö, H., Shinohara, A., Takeda, M., Nakai, K., Miyano, S.: Finding optimal pairs of cooperative and competing patterns with bounded distance. In: The 7th International Conference on Discovery Science (DS 2004). Volume 3245 of LNAI., Springer-Verlag (2004) 32–46

12. Crochemore, M., Rytter, W.: Text Algorithms. Oxford University Press (1994)

13. Landau, G.M., Myers, E., Ziv-Ukelson, M.: Two algorithms for LCS consecutive suffix alignment. In: Proc. Fifteenth Annual Combinatorial Pattern Matching Symposium (CPM2004). Volume 3109 of LNCS., Springer-Verlag (2004) 173–193

On Class Visualisation for High Dimensional Data: Exploring Scientific Data Sets

Ata Kabán[1], Jianyong Sun[1,2], Somak Raychaudhury[2], and Louisa Nolan[2]

[1] School of Computer Science
[2] School of Physics and Astronomy
The University of Birmingham, Birmingham, B15 2TT, UK
{axk, jxs}@cs.bham.ac.uk, {somak, lan}@star.sr.bham.ac.uk

Abstract. Parametric Embedding (PE) has recently been proposed as a general-purpose algorithm for class visualisation. It takes class posteriors produced by a mixture-based clustering algorithm and projects them in 2D for visualisation. However, although this fully modularised combination of objectives (clustering and projection) is attractive for its conceptual simplicity, in the case of high dimensional data, we show that a more optimal combination of these objectives can be achieved by integrating them both into a consistent probabilistic model. In this way, the projection step will fulfil a role of regularisation, guarding against the curse of dimensionality. As a result, the tradeoff between clustering and visualisation turns out to enhance the predictive abilities of the overall model. We present results on both synthetic data and two real-world high-dimensional data sets: observed spectra of early-type galaxies and gene expression arrays.

1 Introduction

Clustering and visualisation are two widespread unsupervised data analysis techniques, with applications in numerous fields of science and engineering. Two key strategies can be distinguished in the literature: (1) One is to produce a compression of the data first and then use that to visually detect distinct clusters. A wide range of linear and nonlinear dimensionality reduction techniques developed in machine learning follow this route, including PCA, GTM [4], etc. (2) The alternative strategy is to cluster the data first and visualise the resulting cluster assignments afterwards. This is more popular in the data mining community [13]. A recently proposed method, termed Parametric Embedding (PE) [7] proposes to take class posteriors produced by a mixture-based clustering algorithm and project them in 2D for visualisation.

Let us observe however, that for both of these data exploration strategies the two objectives – clustering and visualisation – are decoupled and essentially one is entirely subordinated to the other. This is worrying in that inevitably, the errors accumulated in the first stage cannot be corrected in a subsequent stage and may essentially compromise the process of understanding the data.

N. Lavrač, L. Todorovski, and K.P. Jantke (Eds.): DS 2006, LNAI 4265, pp. 125–136, 2006.

In this paper we consider the class visualisation problem, as in [7] and we identify a setting where a more fruitful coupling between clustering and visualisation can be achieved by integrating them both into a consistent probabilistic model. As we shall see, this is particularly useful in high-dimensional problems, where the number of data dimensions exceeds the number of observations. Such cases are encountered in modern scientific data analysis, e.g. in gene expression analysis, or the analysis of special objects in astronomy.

Our approach is based on a multi-objective formulation in a probabilistic formalism. As we shall see, our model can be decoupled and understood as a sum of two objectives: One term is responsible for clustering and one other for a PE-like projection of the estimated class assignments. These two steps are now interdependent, so that in high dimensional problems the projection step fulfils a regularisation role, guarding against the curse of dimensionality problem.

We use both synthetic data and two real-world high-dimensional data sets in our experiments: Observed spectra (of rare quality and coverage) of early-type galaxies and a benchmark gene expression array data set are used to demonstrate the working of the proposed approach. We find that in both cases we obtain not only a visualisation of the mixture posteriors, but a more predictive mixture model, as measured by out of sample test likelihood, as a result of appropriately combining the objectives of clustering and class projection.

The remainder of the paper is organised as follows: We begin with presenting our probabilistic model in Section 2. The interpretation by which this can be understood as a joint model for mixture based clustering and PE-like projection will become evident in Section 3, where the EM [3] methodology is used to derive a maximum a posteriori (MAP) estimation algorithm for our model. We then extend this work to take into account additional available knowledge on measurement uncertainties for real-world data analysis. Section 4 presents the application two very different experiments involving galaxy spectra and gene expression. The results are summarised in the concluding section.

2   The Model

Consider N independent, T-dimensional data points. The n-th point is denoted by $\boldsymbol{d}_n \in \mathcal{R}^{T \times 1}$ having features d_{tn}. We seek a 2D mapping of this data into points $\boldsymbol{x}_n \in \mathcal{R}^{2 \times 1}$ in the Euclidean space such as to reflect topological relationships based on some cluster structure that may be present in the data set. In building up our model, we begin by making the common assumption of conditional independence, in order to enforce the dependences among data features to be captured in the latent space.

$$p(\boldsymbol{d}_n|\boldsymbol{x}_n) = \prod_t p(d_{tn}|\boldsymbol{x}_n) \tag{1}$$

Further, in order to capture complicated density shapes, including possibly distinct clusters, we model the conditional probabilities of the data features as a mixtures of Gaussians. The mixing coefficients of these mixtures are instance-specific, so that each measurement that belongs to the same object will have the same mixing coefficient. This will ensure that the various features of an instance are likely to be allocated to the same (set of) mixture components.

$$p(d_{tn}|\boldsymbol{x}_n) = \sum_k p_{\theta_{tk}}(d_{tn}|k) P_{\boldsymbol{c}_k}(k|\boldsymbol{x}_n) \tag{2}$$

Observe we do not impose that each data point must belong to exactly one mixture component. This allows us to model the relationships between clusters.

Assuming that we work with real-valued observations, and $p(d_{tn}|k)$ is a Gaussian, then $\theta_{tk} = \{\mu_{tk}, v_{tk}\}$ are the mean and precision parameters respectively. Other choices are however possible as appropriate.

$$p(d_{tn}|k) = \mathcal{N}(d_{tn}|\mu_{tk}, v_{tk}) \tag{3}$$

The second factor in (2) is a nonlinear function that projects a point $\boldsymbol{x}_n$ from the Euclidean space onto a probability simplex. A parameterised softmax function can be used for this purpose.

$$P_{\boldsymbol{c}_k}(k|\boldsymbol{x}_n) = \frac{\exp\left\{-\frac{1}{2}(\boldsymbol{x}_n - \boldsymbol{c}_k)^2\right\}}{\sum_{k'} \exp\left\{-\frac{1}{2}(\boldsymbol{x}_n - \boldsymbol{c}_{k'})^2\right\}} \tag{4}$$

Our goal is then to determine $\boldsymbol{x}_n$ for each $\boldsymbol{d}_n$. In addition, we also need to estimate the parameters θ_{tk} and $\boldsymbol{c}_k$.

In order to somewhat narrow down the search space, we add smoothing priors, similarly to [7]:

$$\boldsymbol{x}_n \sim \mathcal{N}(\boldsymbol{x}_n|\boldsymbol{0}, \alpha\boldsymbol{I}); \ \boldsymbol{c}_k \sim \mathcal{N}(\boldsymbol{c}_k|\boldsymbol{0}, \beta\boldsymbol{I}); \tag{5}$$

In addition, the inverse variance parameters (precisions) are given exponential priors to prevent them produce singularities and encourage the extinction of unnecessary model complexity.

$$p(v_{tk}) = \gamma e^{-\gamma v_{tk}} \tag{6}$$

The hyperparameters α, β and γ must all have strictly positive values.

3   Parameter Estimation

Here we derive MAP estimates for our model specified in the previous section. The complete data log likelihood is proportional to the posterior over all hidden variables and this is the following.

$$
\begin{aligned}
\mathcal{L} &= \sum_{n,t} \log \sum_k p_{\theta_{tn}}(d_{tn}|k) P_{\boldsymbol{c}_k}(k|\boldsymbol{x}_n) + \sum_k \log P(\boldsymbol{c}_k) + \sum_n \log P(\boldsymbol{x}_n) + \sum_{t,k} \log P(v_{tk}) \\
&\geq \sum_n \sum_t \sum_k r_{ktn} \left\{\log p_{\theta_{tk}}(d_{tn}|k) + \log P_{\boldsymbol{c}_k}(k|\boldsymbol{x}_n) - \log r_{ktn}\right\} \\
&\quad + \sum_n \log P(\boldsymbol{x}_n) + \sum_k \log P(\boldsymbol{c}_k) + \sum_{t,k} \log P(v_{tk})
\end{aligned} \tag{7}
$$

where we used Jensen's inequality and $r_{ktn} \geq 0, \sum_k r_{ktn} = 1$ represent variational parameters that can be obtained from maximising (7):

$$r_{ktn} = \frac{p_{\theta_{tk}}(d_{tn}|k)P_{C_k}(k|\boldsymbol{x}_n)}{\sum_{k'} p_{\theta_{tk}}(d_{tn}|k)P_{C_k}(k|\boldsymbol{x}_n)} \tag{8}$$

We can also regard $k = 1, ..., K$ as the outcome of a hidden class variable and r_{ktn} are in fact true class posterior probabilities of this class variable, cf. Bayes' rule.

The re-writing (7-8) is convenient for deriving the estimation algorithm for the parameters of the model. Before proceeding, let us rearrange (7) in two main terms, so that the interpretation of our model as a combination of mixture-based clustering and a PE-like class projection becomes evident.

$$\text{Term}_1 = \sum_n \sum_t \sum_k r_{ktn} \log p_{\theta_{tk}}(d_{tn}|k) - \gamma \sum_{t,k} v_{tk} + const. \tag{9}$$

$$\text{Term}_2 = \sum_n \sum_t \sum_k r_{ktn} \{\log P_{C_k}(k|\boldsymbol{x}_n) - \log r_{ktn}\} + \sum_k \log P(c_k) + \sum_n \log P(\boldsymbol{x}_n)$$

$$= \sum_{n,t} -KL(r_{.,t,n}||P_{C_.}(.|\boldsymbol{x}_n)) - \alpha \sum_n ||\boldsymbol{x}||_n^2 - \beta \sum_k ||c||_k^2 + const. \tag{10}$$

Now, the first term can be recognised as a clustering model, essentially an instance of modular mixtures [2] or an aspect-mixture of Gaussians [6,12], which is known to be advantageous in high-dimensional clustering problems [2]. The second term, in turn, is a PE-like objective [7], which minimises the Kullback-Leibler (KL) divergence between the class-posteriors and their projections. Evidently, these two objectives are now interdependent. It remains to be seen in which cases their coupling is advantageous.

Carrying out the optimisation of (7) yields the following maximum likelihood estimates for the means and maximum a posteriori estimates for the precisions.

$$\mu_{tk} = \frac{\sum_n r_{ktn} d_{tn}}{\sum_n r_{ktn}}; \ v_{tk} = \frac{\sum_n r_{ktn}}{\sum_n r_{ktn}(d_{tn} - \mu_{tk})^2 + 2\gamma} \tag{11}$$

For the remaining parameters, there is no closed form solution, we employ numerical optimisation using the gradients (see Appendix):

$$\frac{\partial}{\partial \boldsymbol{x}_n} = \sum_k (c_k - \boldsymbol{x}_n) \sum_t (r_{ktn} - P_{C_k}(k|\boldsymbol{x}_n)) - \alpha \boldsymbol{x}_n \tag{12}$$

$$\frac{\partial}{\partial c_k} = \sum_n (\boldsymbol{x}_n - c_k) \sum_t (r_{ktn} - P_{C_k}(k|\boldsymbol{x}_n)) - \beta c_k \tag{13}$$

As expected, the form of parameter updates also reflects the interdependence of our two objectives: (11) is formally identical with the updates in [2,6,12] (up to the variation implied by the use of the prior for precisions[1]). The gradients (12)-(13) are in turn, as expected, very similar to the updates in PE [7].

---

[1] In [12], the authors propose an improper prior for the variances. Incidentally, our MAP estimate for the precision parameter (11) is formally identical to the inverse of their variance estimates – so we now see the expression can be derived with the use of a proper exponential prior on the precisions.

3.1   Empirical Bayesian Test Likelihood

Having estimated the model, the empirical Bayesian estimate [5] of the goodness of fit for new points is given by integrating over the empirical distribution $\frac{1}{N} \sum_n \delta(\boldsymbol{x} - \boldsymbol{x}_n)$. This is the following.

$$p(\boldsymbol{d}_{test}) = \frac{1}{N} \sum_n p(\boldsymbol{d}_{test}|\boldsymbol{x}_n) \tag{14}$$

3.2   Visualisation

The 2D coordinates $\boldsymbol{x}_n, n = 1 : N$ provide a visual summary of the data density. In addition, label markers (or colouring information), to aid the visual analysis, are obtained directly from $P_{\boldsymbol{C}_k}(k|\boldsymbol{x})$. This is a handy feature of our method, as opposed to techniques based on dimensionality reduction methods (such as e.g. PCA), where detecting meaningful clusters from the projection plot is not straightforward. The class labels may also serve as an interface to the domain expert, who may wish to specify and fix the labels of certain points in order to explore the structure of the data interactively.

For accommodating new data points on a visualisation produced from a training set, a fast 'folding in' [6] procedure can be used. This is to compute $\operatorname*{argmax}_{\boldsymbol{x}} p(\boldsymbol{d}_{test}|\boldsymbol{x})$ with all model parameters kept fixed to their estimated values. Conveniently, this optimisation task is convex, i.e. with all other parameters kept fixed, the Hessian w.r.t. $\boldsymbol{x}$ is positive semidefinite

$$\frac{\partial^2 \mathcal{L}}{\partial \boldsymbol{x}_n \partial \boldsymbol{x}_n^T} = \sum_k \sum_t r_{ktn} \boldsymbol{c}_k \boldsymbol{c}_k^T - \left\{ \sum_k \sum_t r_{ktn} \boldsymbol{c}_k \right\} \left\{ \sum_k \sum_t r_{ktn} \boldsymbol{c}_k \right\}^T \tag{15}$$

for the same reasons as in the case of PE [7]. Therefore the projection of test points is unique. However, PE [7] makes no mention of how to accommodate new points on an existing visualisation plot.

3.3   Taking into Account Estimates of Observational Error

It is often the case that data from science domains (e.g. astronomy) come with known observational errors. In this section we modify our algorithm to take these into account. Let σ_{tn} be the standard deviation of the known measurement error of the t-th feature of instance n. We handle this by considering d_{tn} as a hidden variable which stands for the clean data, and in addition we have $\mathcal{N}(y_{tn}|d_{tn}, 1/\sigma_{tn}^2)$.

Assuming that we are dealing with real valued data, $p(d_{tn}|\boldsymbol{x}_n)$ was defined as a mixture of Gaussians, and so the integration over the unseen clean data variable d_{tn} gives the following likelihood term for component k of feature t:

$$p(y_{tn}|k) = \int dd_{tn} \mathcal{N}(y_{tn}|d_{tn}, 1/\sigma_{tn}^2) \mathcal{N}(d_{tn}|\mu_{tk}, v_{tk}) = \mathcal{N}(y_{tn}|\mu_{tk}, (\sigma_{tn}^2 + 1/v_{tk})^{-1}) \tag{16}$$

In other words, the variance of the data likelihood now has two terms, one coming from the measurement error and one other coming from the modelling error. The latter needs to be estimated.

The estimation equations in this case modify as follows:

$$r_{ktn} = \frac{\mathcal{N}(y_{tn}|\mu_{tk},(\sigma_{tn}^2+1/v_{tk})^{-1})P_{c_k}(k|\boldsymbol{x}_n)}{\sum_{k'}\mathcal{N}(y_{tn}|\mu_{tk},\sigma_{tn}^2+1/v_{tk})P_{c_k}(k|\boldsymbol{x}_n)} \tag{17}$$

The update equation of μ_{tk} becomes

$$\mu_{tk} = \frac{\sum_n y_{tn}r_{ktn}/(\sigma_{tn}^2+1/v_{tk})}{\sum_n r_{ktn}/(\sigma_{tn}^2+1/v_{tk})} \tag{18}$$

and the updates of $\boldsymbol{x}_n$ and $\boldsymbol{c}_k$ remain unchanged.

For the precision parameters v_{tk} there is no closed form solution and so numerical optimisation may be employed, e.g. a conjugate gradient w.r.t. $\log v_{tk}$, since then the optimisation is unconstrained.

$$\frac{\partial}{\partial \log v_{tk}} = \frac{1}{v_{tk}}\sum_n r_{ktn}\left\{\frac{1}{2(\sigma_{tn}^2+1/v_{tk})} - \frac{(d_{tn}-\mu_{tk})^2}{2(\sigma_{tn}^2+1/v_{tk})^2}\right\} - \gamma v_{tk} = 0 \tag{19}$$

Observe that when $\sigma_{tn} = 0$, all equations of this subsection reduce to those presented for the noise-free case in Sec. 3. Yet another alternative is to treat d_{tn} as hidden variables and take a hierarchical EM approach.

It should be highlighted, that although many non-probabilistic methods simply ignore the measurement errors even when these are known, due to our probabilistic framework a principled treatment is possible. This prevents finding 'interesting' patterns in the visualisation plot as a result of measurement errors, at least in the cases when such errors are known. Furthermore, there are cases when further refinement of the noise model will be needed, e.g. in many cases the recorded error values are uncertain or known to be optimistic.

4   Experiments and Applications

4.1   Numerical Simulations

The first set of experiments is meant to demonstrate the working of our method and to highlight in which situations it is advantageous over the fully disjoint and sequential application of a mixture-based clustering and subsequent visualisation strategy. Illustrative cases are shown and these are important for knowing in what kind of problems is the method appropriate to use.

Throughout, we used smoothness hyperparameters $\alpha = \beta = 1$ and γ was determined by cross-validation under an initial assumption of a large (K=10) number of clusters. The priors on the precision parameters favour the extinction of unnecessary components and even if there are remaining redundant components, a good-enough γ parameter can be located. The typical value obtained was of the order of 10^{-3}. Then γ is fixed and a further cross-validation is run to determine the optimal number of clusters K (less or equal to the number of non-empty clusters found in the previous step). We noted the optimisation is very sensitive to initialisation and starting μ_k from K-means is beneficial. To alleviate problems with local optima, each run was repeated 20 times and the model that found better maximum of the model likelihood was retained for further testing.

Two sets of generated data were created. For the first set, 300 points were drawn from a 6-dimensional mixture of 5 independent Gaussians (60 points in each class). The second set was sampled from a 300-dimensional mixture of 5 independent Gaussians (again, 60 points per class). Fig. 1.a) shows the test likelihood averaged over 20 repeated runs, having generated the test data from the same model as the training data. The test likelihood obtained with a mixture of Gaussians (MoG) is superimposed for comparison. We see that for the relatively low dimensional data set the proposed joint model has little (no) advantage. This is simply because in this case there is enough data to reliably estimate a MoG. The obtained mixture posteriors could then safely be fed into e.g. a PE [7] for class visualisation.

For the case of high dimensional data, however the situation is different. The MoG overfits badly and is therefore unable to identify the clusters or to provide reliable input to a subsequent visualisation method. This is the situation when our proposed model is of use. The projection part of the objective guards against overfitting – as we can see from the test likelihood on Fig 1.b.

Fig 1.c shows the visualisation of the 300-dimensional data set. Each point is the 2D representation $(\boldsymbol{x}_n)$ of the corresponding 300-D datum point. The markers correspond to the maximum argument of the softmax outputs $P(k|\boldsymbol{x}_n)$, so they represent labels automatically assigned by the model. In this case, the

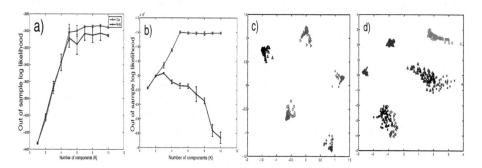

Fig. 1. Experiments on synthetic data: a) Out of sample test likelihood (higher is better) for the 6-dimensional data — our approach vs MoG. b) Same for the 300-dimensional data. Our approach is significantly less prone to overfitting than MoG in this case. c) The final visualisation plot for the 300-dimensional data set. The markers are automatically assigned by the model and in this case they are identical with the true generator classes. d) Illustration of the visualisation process when the number of assumed clusters (ten) is larger than the number of true clusters (five). Notice the true clusters remain compact in the visualisation space.

estimated labels are identical with the true labels. We also see that the true number of classes has been correctly recovered. In addition, we noted that in experiments where the number of classes was deliberately chosen larger than the true number of clusters, some of the unnecessary clusters progressively become empty indeed, due to the employed prior on the precision parameters, while others split a true cluster. However, notably, the visual image of the true cluster split by several components tends to remain compact. Such an example is seen on Fig 1.d.

4.2   Visualisation of Observed Spectra of Early-Type Galaxies

We apply the method developed above to a sample of measured spectra (in the ultraviolet to optical range of radiation) of 21 well-studied early-type (elliptical or lenticular) galaxies. In a previous work [8,9], we had studied this data set using various factor analysis techniques. Here, we seek to obtain a visual analysis of the data. Each of these spectra represent flux measurements at 348 values of wavelength in the range 2000-8000Å, in equal bins, for all spectra. Observational errors are associated with each value, which we take into account as described in Section 3.3. Thus, the clustering and class visualisation of these spectra is a high-dimensional problem.

This represents a pilot data set for an important study in the evolution of galaxies. It is generally believed that all early-type galaxies formed all their stars in the early universe, and then have evolved more-or-less passively until the present day- so one expects to find their spectrum to correspond to a collection of stars all of the same age. However, detailed observations in the last decade indicate a wealth of complex detail in a significant fraction of such galaxies, including evidence of a sub-population of very young stars in many cases. How common this effect is largely unknown, and can only be addressed through data mining of large spectral archives. Even though many $\times 10^5$ galaxy spectra are being assembled in large public archives (e.g. www.sdss.org), a sample as detailed as ours is rare and difficult to assemble, particularly with such wide a coverage in wavelength, which requires combining observations from both ground and space based observatories (see details in [9]). From this small sample, we would attempt to isolate those galaxies which have young stars from those that don't.

Needless to say, the fluxes are all positive values. In order to be interpretable, our method needs to ensure the estimated parameters (cluster prototypes) are also positive. In our previous work, we built in explicit constraints to ensure this [9]. Here, since each μ_{tk} in (11) is just a weighted average of positive data, its positivity is automatically satisfied.

The leave-one-out test likelihood of our model is shown on the left plot of Fig. 2. The peak at $K = 2$ indicates that two clusters have been identified. A mixture of Gaussians test likelihood is superimposed for comparison, showing the overfitting problem due to the high dimensionality. The MoG is therefore unable to identify any clusters in the data. Hence, a class visualisation of the data based on mixture posteriors would be clearly compromised in this case. The right hand plot shows the grouping of the actual data around the two identified

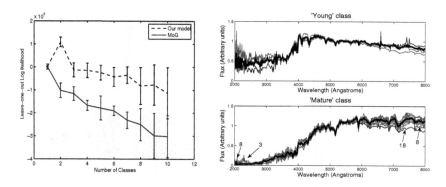

Fig. 2. Left: The leave-one-out test likelihood versus the number of clusters. The peak at $K = 2$ produced by our method indicates two clusters whereas a mixture of Gaussians overfits and therefore fails to identify clusters in the data. Right: The actual data, clustered around the two prototypes identified by our model. The parameters μ_k for $k = 1, 2$ are superimposed with thick lines. They are interpretable as a 'young' and an 'old' prototypical spectrum respectively. The identification number that marks some of the spectra correspond to those on Fig. 3. The marked spectra are the instances that apart from their overall shape present some features of the 'young' category too.

prototypes $\mu_k, k = 1, 2$ of our model. The latter are superimposed with thick lines. These can be recognised and interpreted as the prototype of the spectrum of a 'young' and 'mature' stellar population respectively. Thus, in this case, the clusters have a physical interpretation in astrophysical terms.

Identification numbers mark some of the spectra clustered in the 'mature' category on the left lower plot of Fig. 2. These are the galaxies that have a significantly non-zero class membership for either cluster, and they indeed include some morphological aspects of the 'young' category as well (the emission lines at < 2000Å and the slope of the spectral continuum in the range 6000-8000Å). Physically, this indicates the presence of a significant population of young (< 2 Gyr old) stars, whereas the rest of the stars are > 10 Gyr old.

The identification numbers are the same as those on Fig. 3, where we see the 2D visualisation of the sample on the left. For each spectrum y_n, the 2D latent representation x_n is plotted. The markers represent cluster labels automatically assigned by the model, as detailed in the right hand plot. We see the two clusters are well separated on the image and the 'hybrid' galaxies are indeed placed in between those that are clearly cluster members. Of these, the one marked as 18 represents a galaxy (NGC 3605) for which recent detailed physical analyses have been made [10]. It turns out that although more than 85% of its stellar mass is associated with an old (9–12 Gyr) stellar population, it does contain a younger stellar population too, at $\simeq 1$ Gyr [10].

We therefore conclude that it is possible to have an intuitive visual summary of a few hundreds of measurements per galaxy in just two coordinates with the application of our method.

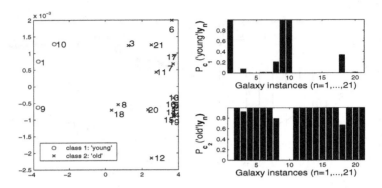

Fig. 3. Left: Visualisation of the data set of the spectra of 21 early-type galaxies. For each spectrum y_n, the 2D latent representation x_n is plotted. The markers are those assigned by the model, as shown on the right.

4.3   Visual Analysis of Gene Expressions

In a brief final experiment we show the potential use of our approach for the visual analysis of high dimensional gene expression arrays. Oligonucleotide arrays can provide a means of studying the state of a cell, by monitoring the expression level of thousands of genes at the same time [1] and have been the focus of extensive research. The main difficulty is that the number of examples is typically of the order of tens while the number of genes is of the order of thousands. Even after eliminating genes that have little variation, we are still left with at least hundreds of data dimensions. Straightforward mixture based clustering runs into the well-know curse of dimensionality problem. Here we apply our method to the ColonCancer data set, having 40 tumour and 22 normal colon tissue samples [1].

This is a benchmark data set, used in many previous classification and clustering studies. Our input matrix consisted of the 500 genes with highest overall variation × the 62 samples.

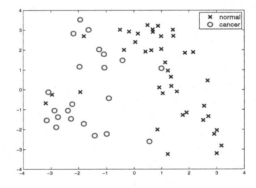

Fig. 4. Unsupervised class visualisation of the ColonCancer data set. The markers correspond to the true class for the ease of visual evaluation.

Fig. 4 shows the visualisation obtained in a purely unsupervised manner. The markers now correspond to the true labels (not used by the algorithm), and are given for the ease of visual evaluation of the representation produced. The separation of cancerous from noncancerous tissues is most apparent on the plot. The potential of such a visualisation tool lies mainly in that it would allow a domain expert to interactively explore the structure of the sample. Additionally, the gene-specific posteriors r_{ktn} provide quantitative gene-level class information.

5 Conclusions

We proposed and investigated a model for class visualisation of explicitly high dimensional data. We have shown this model relates closely to PE [7] in that it represents a probabilistic integration of the clustering and visualisation objectives into a single model. We derived empirical Bayesian estimates for our model which make this multi-objective interpretation easy to follow. Although this work may potentially further be enhanced by a fuller Bayesian estimation scheme, the empirical Bayesian methodology has been appropriate for our purposes [11] and it allows us to estimate an empirical latent density from a given example set of data and to reason about previously unseen data relative to that. We demonstrated gains in terms of the predictive capabilities of the proposed model over the fully modular and sequential approach to clustering and class visualisation in the case of high dimensional data.

Acknowledgements. This research is funded by PPARC grant PP/C503138/1, 'Designer Algorithms for Astronomical Data Mining'. AK also acknowledges partial support from a Wellcome Trust VIP Award (Project 10835).

References

1. U Alon, N Barkai, D Notterman, K Gish, S Ybarra, D Mack, A Levine. Broad Patterns of Gene Expression Revealed by Clustering Analysis of Tumour and Normal Colon Cancer Tissues Probed by Oligonucleotide Arrays. Cell Biol. 96, 6745–6750.
2. H Attias. Learning in High Dimension: Modular mixture models. Proc. Artificial Intelligence and Statistics, 2001.
3. C.M Bishop. Neural Networks for Pattern Recognition. Oxford University Press, Inc., New York, NY, 1995.
4. C.M Bishop, M Svensen and C.K.I Williams. GTM: The Generative Topographic Mapping. Neural Computation, vol. 10(1), 1998.
5. B.P Carlin and T.A Louis. Bayes and Empirical Bayes Methods for Data Analysis. Chapman and Hall, 2000.
6. Th Hofmann. Gaussian Latent Semantic Models for Collaborative Filtering. 26th Annual International ACM SIGIR Conference, 2003.
7. T Iwata, K Saito, N Ueda, S Stromsten, T.L Griffiths, J.B Tenenbaum. Parameteric Embedding for Class Visualisation. Proc. Neur. Information Processing Systems 17, 2005.

8. A Kabán, L Nolan and S Raychaudhury. Finding Young Stellar Populations in Elliptical Galaxies from Independent Components of Optical Spectra. Proc. SIAM Int'l Conf on Data Mining (SDM05), pp. 183–194.

9. L Nolan, M Harva, A Kabán and S Raychaudhury. A data-driven Bayesian approach to finding young stellar populations in early-type galaxies from their ultraviolet-optical spectra, Mon. Not. of the Royal Astron. Soc. 366,321-338, 2006.

10. L Nolan, J.S Dunlop, B Panter, R Jimenez, A Heavens, G Smith. The star-formation histories of elliptical galaxies across the fundamental plane, submitted to MNRAS.

11. J Rice. Reflections on SCMA III. In: Statistical challenges in astronomy. Eds: E.C Feigelson and G.J Babu. Springer. 2003.

12. S Rogers, M Girolami, C Campbell, R Breitling. The latent process decomposition of cDNA microarray datasets. IEEE/ACM Transact. Comput. Biol. Bioinformatics. 2: 143–156.

13. T Soukup and I Davidson. Visual Data Mining: Techniques and Tools for Data Visualisation and Mining. Wiley, 2002.

Appendix. Estimation of x_n

The terms containing x_n are the following.

$$\sum_n \sum_t \sum_k r_{ktn} \left\{ -\frac{1}{2}(x_n - c_k)^2 - \log \sum_{k'} \exp(-\frac{1}{2}(x_n - c_k)^2) \right\} - \alpha x_n^2$$

$$= \sum_n \sum_t \left\{ -\sum_k r_{ktn} \frac{1}{2}(x_n - c_k)^2 - \log \sum_{k'} \exp(-\frac{1}{2}(x_n - c_k)^2) \right\} - \alpha x_n^2$$

The gradient is then:

$$\frac{\partial}{\partial x_n} = \sum_t \left\{ -\sum_k r_{ktn}(x_n - c_k) + \sum_{k''} \frac{\exp(-\frac{1}{2}(x_n - c_{k''})^2)}{\sum_{k'} \exp(-\frac{1}{2}(x_n - c_{k'})^2)}(x_n - c_{k''}) \right\} - \alpha x_n$$

Renaming k'' by k and replacing the expression of $P_{C_k}(k|x_n)$ the following is obtained.

$$\frac{\partial}{\partial x_n} = \sum_t \sum_k \{-r_{ktn}(x_n - c_k) + P_{C_k}(k|x_n)(x_n - c_k)\} - \alpha x_n$$

$$= \sum_k (c_k - x_n) \sum_t (r_{ktn} - P_{C_k}(k|x_n)) - \alpha x_n$$

Mining Sectorial Episodes from Event Sequences[*]

Takashi Katoh[1], Kouichi Hirata[2], and Masateru Harao[2]

[1] Graduate School of Computer Science and Systems Engineering
[2] Department of Artificial Intelligence
Kyushu Institute of Technology
Kawazu 680-4, Iizuka 820-8502, Japan
t_katou@dumbo.ai.kyutech.ac.jp,
{hirata, harao}@ai.kyutech.ac.jp

Abstract. In this paper, we introduce a *sectorial episode* of the form $C \mapsto r$, where C is a set of events and r is an event. The sectorial episode $C \mapsto r$ means that every event of C is followed by an event r. Then, by formulating the *support* and the *confidence* of sectorial episodes, in this paper, we design the algorithm SECT to extract all of the *sectorial episodes that are frequent and accurate* from a given event sequence by traversing it just once. Finally, by applying the algorithm SECT to bacterial culture data, we extract sectorial episodes representing *drug-resistant change*.

1 Introduction

The *sequential pattern mining* [3,6,7,9,10] is one of the data mining methods from time-related data. The purpose of sequential pattern mining is to discover frequent *subsequences* as patterns in a sequential database. On the other hand, the *episode mining* [4,5] introduced by Mannila *et al.* [5] is known as another approach to discover frequent patterns from time-related data. The purpose of episode mining is to discover *frequent episodes*, not subsequences, that are a collection of events occurring frequently together in event sequences.

In episode mining, the frequency is formulated as the number of occurrences of episodes in every *window* that is a subsequence of event sequences under a fixed time span called the *width* of windows. Then, Mannila *et al.* [5] have introduced a *parallel episode* as a set of events and a *serial episode* as a sequence of events. By combining the above episodes, they have extended the forms of episodes as *directed acyclic graphs* of events of which edges specify the temporal precedent-subsequent relationship.

Concerned with episode mining, in this paper, we introduce a *sectorial episode* of the form $C \mapsto r$, where C is a parallel episode and r is an event. The sectorial episode $C \mapsto r$ means that every event of C is followed by an event r, so we can regard every event in C as a *causation* of r. Note that, since a sectorial episode

---

[*] This work is partially supported by Grand-in-Aid for Scientific Research 17200011 from the Ministry of Education, Culture, Sports, Science and Technology, Japan.

N. Lavrač, L. Todorovski, and K.P. Jantke (Eds.): DS 2006, LNAI 4265, pp. 137–148, 2006.

is captured as the direct precedent-subsequent relationship between events, it is just a candidate of *causality* in database (*cf.* [11]).

We formulate the *support* $supp(C \mapsto r)$ of a sectorial episode $C \mapsto r$ as the ratio of the number of k-windows (i.e., a window with width k) in which $C \mapsto r$ occurs for the number of all k-windows. For the *minimum support* σ such that $0 < \sigma < 1$, we say that $C \mapsto r$ is *frequent* if $supp(C \mapsto r) \geq \sigma$. Also we can show that the sectorial episode preserves *anti-monotonicity*, that is, for $C_1 \subseteq C_2$, if $C_2 \mapsto r$ is frequent then so is $C_1 \mapsto r$. Furthermore, by regarding a sectorial episode as an association rule [1,2], we introduce the *confidence* $conf(C \mapsto r)$ of a sectorial episode $C \mapsto r$ as the ratio of the number of k-windows in which $C \mapsto r$ occurs for the number of all k-windows in which C occurs. For the *minimum confidence* γ such that $0 < \gamma < 1$, we say that $C \mapsto r$ is *accurate* if $conf(C \mapsto r) \geq \gamma$.

The purpose of this paper is to design the algorithm to extract all of the *sectorial episodes that are frequent and accurate* from a given event sequence. Since our sectorial episode is a combination of parallel and serial episodes, it is possible to extract all of the sectorial episodes that are frequent and accurate by combining the algorithms designed in [5]. On the other hand, in this paper, we design another algorithm SECT to extract them efficiently and appropriate to sectorial episodes.

The algorithm SECT consists of the following two procedures. The first procedure SCAN is to extract all of the frequent sectorial episodes of the form $C \mapsto r$ such that $|C| = 1$ and then store the information of windows in which $C \mapsto r$ occurs. The second procedure is to extract all of the frequent sectorial episodes of the form $C \mapsto r$ such that $|C| \geq 2$ for every event r stored by SCAN. Here, the frequency of the constructed sectorial episodes is computed by using the anti-monotonicity, as similar as the well-known algorithm APRIORITID [1,2].

Then, we show that the algorithm SCAN runs in $O((l+k)|\mathcal{E}|^2)$ time and space, where l is the time span between the starting time and the ending time in a given event sequence, k is the width of windows and $\mathcal{E}$ is a set of all events, by traversing the event sequence just once. Hence, we show that the algorithm SECT extracts all of the sectorial episodes that are frequent and accurate in $O(M(l+k)|\mathcal{E}|^{2M+1})$ time and $O((l+k)|\mathcal{E}|^2 + |\mathcal{E}|^{M+1})$ space without traversing the event sequence, where $M = \max\{|C| \mid C \mapsto r \text{ is frequent}\}$. If it is not necessary to store all of the extracted sectorial episodes, then we can reduce the space complexity of the algorithm SECT to $O((l+k)|\mathcal{E}|^2)$, by outputting the episode whenever it is found.

Finally, we apply the algorithm SECT to bacterial culture data. Note that, from the medical viewpoint, in order to extract sectorial episodes concerned with *drug-resistant change*, it is necessary to extract them based on the same detected bacterium and the same sample. Hence, in this paper, we divide the database into *pages* for the detected bacterium and the sample in whole 44 attributes, and then extract sectorial episodes representing drug-resistant changes from them.

2   Sectorial Episodes

As similar as [5], we assume that an event has an associated time of occurrence as a natural number. Formally, let $\mathcal{E}$ be a set of *event types*. Then, a pair (e, t) is called an *event*, where $e \in \mathcal{E}$ and t is a natural number which is the *(occurrence)* *time* of the event. In the following, for a set $E \subseteq \mathcal{E}$ of event types, we denote $\{(e, t) \mid e \in \mathcal{E}\}$ by (E, t), and also call it by an *event* again. Furthermore, we denote a set $\{e_1, \ldots, e_m\} \subseteq \mathcal{E}$ of event types by a string $e_1 \cdots e_m$.

An *event sequence* $\mathcal{S}$ on $\mathcal{E}$ is a triple (S, T_s, T_e), where

$$S = \langle (E_1, t_1), \ldots, (E_n, t_n) \rangle$$

is an ordered sequence of events satisfying the following conditions.

1. $E_i \subseteq \mathcal{E}$ $(1 \le i \le n)$,
2. $t_i < t_{i+1}$ $(1 \le i \le n - 1)$, and
3. $T_s \le t_i < T_e$ $(1 \le i \le n)$.

In particular, T_s and T_e are called the *starting* time and the *ending* time of $\mathcal{S}$. We denote $T_e - T_s$ by $l_{\mathcal{S}}$. For an event sequence $\mathcal{S} = (S, T_s, T_e)$, we denote the set of all event types of $\mathcal{S}$ at t, that is, $\{E \subseteq \mathcal{E} \mid (E, t) \in S\}$ by $evtyp(\mathcal{S}, t)$.

A *window* in an event sequence $\mathcal{S} = (S, T_s, T_e)$ is an event sequence $W = (w, t_s, t_e)$ such that $t_s < T_e$, $t_e > T_s$ and w consists of all of the events (E, t) in S where $t_s \le t < t_e$. The time span $t_e - t_s$ is called the *width* of the window. For a window $W = (w, t_s, t_e)$, we denote the starting time t_s of W by $st(W)$ and the ending time t_e of W by $et(W)$.

We call a window with width k in $\mathcal{S}$ a k-*window*, and denote the k-window $(w, t, t + k)$ of $\mathcal{S}$ starting from t by $w(\mathcal{S}, t, k)$. For a k-window $W = w(\mathcal{S}, t, k)$, it is obvious that $st(W) = t$ and $et(W) = t + k$.

Note that we can regard a set $E = e_1 \cdots e_m$ of event types as a *parallel episode* [5]. Here, we call the above m the *size* of E. Furthermore, in this paper, we newly introduce the following *sectorial episode*.

Definition 1. Let $c_1 \cdots c_m \subseteq \mathcal{E}$ be a parallel episode and r an event type. Then, a *sectorial episode* is of the following form.

$$X = c_1 \cdots c_m \mapsto r.$$

In this paper, we regard every event in $c_1 \cdots c_m$ in a sectorial episode X as a *causation* of r, so we call every c_i and r a *causal type* and a *resulting type* of X, respectively. For a set $C = c_1 \cdots c_m$, we also denote $c_1 \cdots c_m \mapsto r$ by $C \mapsto r$. In particular, we call a sectorial episode $C \mapsto r$ such that $r \in C$ *trivial*.

We call both a parallel episode and a sectorial episode *episodes* simply.

Definition 2. Let $\mathcal{S}$ be an event sequence $\mathcal{S} = (S, T_s, T_e)$ and e an event type. Then, we say that e *occurs* in $\mathcal{S}$ if there exists an event $(E, t) \in S$ such that $e \in E$. We denote $\{t \mid (E, t) \in S \wedge e \in E\}$ by $T(e, \mathcal{S})$. Also we denote $st(e, \mathcal{S}) = \min\{t \mid t \in T(e, \mathcal{S})\}$ and $et(e, \mathcal{S}) = \max\{t \mid t \in T(e, \mathcal{S})\}$.

We say that a parallel episode $e_1 \cdots e_m$ *occurs* in $\mathcal{S}$ if every e_i occurs in $\mathcal{S}$. Also we say that a sectorial episode $c_1 \cdots c_m \mapsto r$ *occurs* in $\mathcal{S}$ if r occurs in $\mathcal{S}$, and, for every i $(1 \le i \le m)$, c_i occurs in $\mathcal{S}$ and $st(c_i, \mathcal{S}) < et(r, \mathcal{S})$.

Let S be an event sequence and k a natural number. Then, we denote the set of all k-windows by $W(S, k)$. Also, for an episode X, we denote the set of all k-windows such that X occurs in S by $W(X, S, k)$.

Note that we can number all k-windows in $W(S, k)$ from $T_s - k$ to T_e. We call such a number i $(T_s - k < i < T_e)$ the *label* of the i-th k-window. For an event sequence S and an episode X, we identify $W(X, S, k)$ with the set of all labels of k-windows in which X occurs in S.

0	1	2	3	4	5	6	7	8	9
a		a				a	a		
b	b		b		b	b		b	
c		c		c			c		
d	d		d			d			d
	e			e			e	e	

Fig. 1. An event sequence S in Example 1

Example 1. Let $\mathcal{E} = \{a, b, c, d, e\}$. Then, Figure 1 describes an event sequence $S = (S, 0, 10)$ on $\mathcal{E}$ where:

$$S = \langle (abcd, 0), (bde, 1), (ac, 2), (bd, 3), (ce, 4),$$
$$(b, 5), (abd, 6), (ace, 7), (be, 8), (d, 9) \rangle.$$

Furthermore, the event sequence $w = (\langle (ac, 2), (bd, 3), (ce, 4) \rangle, 2, 5)$ is a 3-window of S starting from 2, that is, $w = w(S, 2, 3) = (\langle (ac, 2), (bd, 3), (ce, 4) \rangle, 2, 5)$.

For the above event sequence S, it holds that $T(a, S) = \{1, 2, 6, 7\}$, $st(a, S) = 1$ and $et(a, S) = 7$. Also there exist 12 3-windows, of which starting time is from -2 to 9. Furthermore, for the above window w, sectorial episodes $ab \mapsto c$, $bc \mapsto e$ and $acd \mapsto e$ occur in w, for example.

Let S be an event sequence, X an episode and k a natural number. Then, the *frequency* $freq_{S,k}(X)$ and the *support* $supp_{S,k}(X)$ of X in S w.r.t. k are defined as follows.

$$freq_{S,k}(X) = |W(X, S, k)|, \; supp_{S,k}(X) = \frac{freq_{S,k}(X)}{|W(S, k)|}.$$

Definition 3. Let σ be the *minimum support* such that $0 < \sigma < 1$. Then, we say that an episode X is *frequent* if $supp(X) \geq \sigma$.

Lemma 1 (Anti-monotonicity for sectorial episodes). *Let C_1 and C_2 be parallel episodes such that $C_1 \subseteq C_2$. If $C_2 \mapsto r$ is frequent, then so is $C_1 \mapsto r$.*

Proof. It is sufficient to show that $W(C_2 \mapsto r, S, k) \subseteq W(C_1 \mapsto r, S, k)$. Suppose that $l \in W(C_2 \mapsto r, S, k)$ and let W_l be the l-th k-window in S. Then, it holds that $W_l = w(W(C_2 \mapsto r, S, k), l, k)$. For every $c \in C_2$, it holds that $st(c, W_i) < et(r, W_l)$. Since $C_1 \subseteq C_2$, it holds that $st(c', W_i) < et(r, W_l)$ for every $c' \in C_1$, so $C_1 \mapsto r$ occurs in W_l. Hence, it holds that $l \in W(C_1 \mapsto r, S, k)$. □

By regarding a sectorial episode $C \mapsto r$ as an association rule, we can introduce the *confidence* $conf_{\mathcal{S},k}(C \mapsto r)$ of $C \mapsto r$ in $\mathcal{S}$ w.r.t. k as follows.

$$conf_{\mathcal{S},k}(C \mapsto r) = \frac{|W(C \mapsto r, \mathcal{S}, k)|}{|W(C, \mathcal{S}, k)|}.$$

Definition 4. Let γ be the *minimum confidence* such that $0 < \gamma < 1$. Then, we say that a sectorial episode $C \mapsto r$ is *accurate* if $conf(C \mapsto r) \geq \gamma$.

In the following, the subscripts $\mathcal{S}$ and k in $freq_{\mathcal{S},k}(X)$, $supp_{\mathcal{S},k}(X)$ and $conf_{\mathcal{S},k}(X)$ are omitted, if they are clear by the context.

3   Algorithm to Extract Sectorial Episodes

In this section, we design the algorithm to extract all of the sectorial episodes that are frequent and accurate from an event sequence $\mathcal{S}$, the minimum support σ, the minimum confidence γ, and the width k of windows.

We assume the lexicographic order $\prec$ on $\mathcal{E}$. Also we extend $\prec$ to $2^{\mathcal{E}}$ as follows: For sets $Y = y_1 \cdots y_m$, $Z = z_1 \cdots z_n$ of event types, $Y \prec Z$ if there exists an i $(1 \leq i \leq n)$ such that $y_j = z_j$ $(1 \leq j \leq i-1)$ and $y_i \prec z_i$.

The algorithm SCAN described by Figure 2 extracts all of the frequent sectorial episodes of the form $c \mapsto r$, where c and r are events, and stores the set of labels of k-windows $W(c \mapsto r, \mathcal{S}, k)$ in which the sectorial episode $c \mapsto r$ occurs as $W[c][r]$ and the set of labels of k-windows $W(r, \mathcal{S}, k)$ for every $r \in \mathcal{E}$ as $V[r]$, by traversing $\mathcal{S}$ just once as similar as APRIORITID [1,2].

---

procedure SCAN($\mathcal{S}, \sigma, k$)
$C \leftarrow \emptyset$; $P \leftarrow \emptyset$; $size \leftarrow T_e - T_s + k - 1$;
foreach $e \in evtyp(\mathcal{S}, T_s - k)$ **do** $E \leftarrow E \cup \{e\}$; $V[e] \leftarrow V[e] \cup \{T_s - k\}$;
foreach $e \in evtyp(\mathcal{S}, T_e)$ **do** $E \leftarrow E \cup \{e\}$; $V[e] \leftarrow V[e] \cup \{T_e\}$;
for $i = T_s - k + 1$ **to** $T_e - 1$ **do begin**
  $R \leftarrow \emptyset$;
  foreach $r \in evtyp(w(\mathcal{S}, i, k), i + k - 1)$ **do begin**
    $R \leftarrow R \cup \{r\}$; $E \leftarrow E \cup \{r\}$;
    for $j = i$ **to** $i + k$ **do** $V[r] \leftarrow V[r] \cup \{j\}$;
    foreach $c \in C - \{r\}$ **do**
      $P \leftarrow P \cup \{(c, r)\}$; $t \leftarrow et(c, w(\mathcal{S}, i, k - 1))$;
      for $j = i$ **to** t **do** $W[c][r] \leftarrow W[c][r] \cup \{j\}$;
  end /* foreach */
  $C \leftarrow C - evtyp(w(\mathcal{S}, i, k), i)$; $C \leftarrow C \cup R$;
end /* for */
return $(\{(c, r, W[c][r]) \mid (c, r) \in P, |W[c][r]| \geq \sigma \cdot size\}, \{(e, V[e]) \mid e \in E\})$;

Fig. 2. The algorithm SCAN

Lemma 2. *Let S be an event sequence on $\mathcal{E}$ and k the width of windows. Then, the algorithm* SCAN *extracts all of the frequent sectorial episodes of the form $c \mapsto r$ in $O((l_S + k)|\mathcal{E}|^2)$ time and space, by traversing S just once.*

Proof. First, we show the correctness of the construction of R and C. Fix an index i ($T_s + k - 1 \leq i \leq T_e - 1$). Then, in the foreach-loop, the algorithm SCAN stores all elements in $evtyp(w(S, i, k), i + k - 1)$, where $et(w(S, i, k)) = i + k$, as R. Also it has stored all elements in $\bigcup\limits_{j=i}^{i+k-1} evtyp(w(S, i, k), j)$ as C.

Consider the case shifting i to $i+1$. Since every element of R becomes a causal type and every element of $evtyp(w(S, i, k), i) - \bigcup\limits_{j=i+1}^{i+k-1} evtyp(w(S, i, k), j)$ does not become a causal type, the algorithm SCAN updates C to $C - evtyp(w(S, i, k), i)$ and then adds R to C.

Next, for $r \in R$ and $c \in C - \{r\}$ on the i-th for-loop, consider the sectorial episode $c \mapsto r$. Let t be $et(c, w(S, i, k - 1))$. Then, the causal type c at t is the nearest one to the resulting type r in $w(S, i, k-1)$, so it holds that $c \mapsto r$ appears in the sequence $w(S, j, 1)$ of windows for $i \leq j \leq t$. Hence, by the inner for-loop, the algorithm SCAN adds the labels of such a sequence of windows to $W[c][r]$.

Hence, the algorithm SCAN stores all pairs (c, r) such that $c \mapsto r$ appears in some k-window in S as P, so SCAN outputs the set of all frequent sectorial episodes of the form $c \mapsto r$.

Since the number of the outer for-loop is $T_e - T_s + k$ and the running time in the for-loop is $O(|\mathcal{E}|^2)$, the time complexity of the algorithm SCAN is $O(2|\mathcal{E}| + (T_e - T_s + k)|\mathcal{E}|^2 + |\mathcal{E}|^2) = O((l_S + k)|\mathcal{E}|^2)$, where the first term $O(2|\mathcal{E}|)$ means the time of the first two foreach loops and the third term $O(|\mathcal{E}|^2)$ means the time checking frequency. Also the space of $W[c][r]$ and $V[e]$ are $O((l_S + k)|\mathcal{E}|^2)$ and $O((l_S + k)|\mathcal{E}|)$, respectively, so the space complexity of the algorithm SCAN is $O((l_S + k)|\mathcal{E}|^2)$. It is obvious that the algorithm SCAN traverses S just once. □

After applying the algorithm SCAN, we apply the algorithm SECT described by Figure 3 to extract all of the sectorial episodes that are frequent and accurate. Here, the algorithm SECT extracts all frequent sectorial episodes $C \mapsto r$ where $|C| \geq 2$, by designing an APRIORITID-like procedure for every resulting type $r \in R$ [1,2].

In order to reduce the space complexity, the algorithm SECT computes the support of $Xc \mapsto r$ for $X = c_1 \cdots c_n$ by using $W(c_i, S, k)$ and $W(c, S, k)$, without storing $W(X, S, k)$, and outputs episodes without storing them if they are frequent and accurate. The following lemma guarantees the correctness of it.

Lemma 3. *Let C and D be parallel episodes and r an event type such that $r \notin C \cup D$. Then, the following statement holds:*

1. $W(C \cap D, S, k) = W(C, S, k) \cap W(D, S, k)$.
2. $W(C \cap D \mapsto r, S, k) = W(C \mapsto r, S, k) \cap W(D \mapsto r, S, k)$.

procedure SECT$(\mathcal{S}, \sigma, \gamma, k)$ /* $\mathcal{S} = (S, T_s, T_e)$, $0 < \sigma, \gamma < 1$, $k > 0$ */
$(F, E) \leftarrow$ SCAN$(\mathcal{S}, \sigma, k)$;
$R \leftarrow \{r \mid (c, r, W) \in F\}$; $size \leftarrow T_e - T_s + k - 1$;
foreach $r \in R$ **do begin**
    $C_1 \leftarrow \{c \mid (c, r, W) \in F\}$; $W[c] \leftarrow \{W \mid (c, r, W) \in F\}$; $V[c] \leftarrow \{V \mid (c, V) \in E\}$;
    $D_1 \leftarrow \{c \mid (c, r, W) \in F$ and $|W[c]| \geq \gamma|V[c]|\}$; $m \leftarrow 1$;
    while $C_m \neq \emptyset$ **do begin**
        $C_{m+1} \leftarrow \emptyset$; $D_{m+1} \leftarrow \emptyset$;
        foreach $X \in C_m$ and $c \in C_1$ s.t. $X \prec c$ **do**
            if $|W[X] \cap W[c]| \geq \sigma \cdot size$ **then**
                $C_{m+1} \leftarrow C_{m+1} \cup \{Xc\}$; $W[Xc] \leftarrow W[X] \cap W[c]$; $V[Xc] \leftarrow V[X] \cap V[c]$;
                if $|W[Xc]| \geq \gamma|V[Xc]|$ **then** $D_{m+1} \leftarrow D_{m+1} \cup \{Xc\}$;
        $m \leftarrow m + 1$;
    end /* while */
    $C[r] \leftarrow \bigcup_{1 \leq i \leq m} C_i$; $D[r] \leftarrow \bigcup_{1 \leq i \leq m} D_i$;
end /* foreach */
return $\{D[r] \mapsto r \mid r \in R\}$;

Fig. 3. The algorithm SECT

Proof. The statement 1 is obvious. For the statement 2, suppose that $C = c_1 \cdots c_m$ and $D = d_1 \cdots d_n$. Also let l be a label such that $l \in W(C \cap D \mapsto r, \mathcal{S}, k)$. Then, $C \cap D \mapsto r$ occurs in the l-th k-window $w(\mathcal{S}, l, k)$, so it holds that $st(c_i, \mathcal{S}) < et(r, \mathcal{S})$ and $st(d_j, \mathcal{S}) < et(r, \mathcal{S})$ for every i and j ($1 \leq i \leq n$, $1 \leq j \leq m$), which implies that $l \in W(C \mapsto r, \mathcal{S}, k) \cap W(D \mapsto r, \mathcal{S}, k)$. The converse direction similarly holds.   □

In order to maintain the computation of the support, we use the bit vector a of labels of windows from $T_s - k + 1$ to $T_e - 1$, which is a familiar technique for implementing the algorithm APRIORITID [1,2]. For the bit vector a of the set A of labels of windows, $i \in A$ if and only if $a_i = 1$. Then, we can compute the intersection $A \cap B$ of two set of labels of windows as the bitwise logical product $a \odot b$ of two bit vectors a of A and b of B.

Theorem 1. *Let $\mathcal{S}$ be an event sequence on $\mathcal{E}$. Suppose that $M = \max\{|C| \mid C \in C[r], r \in R\}$. Then, the algorithm SECT extracts all of the sectorial episodes that are frequent and accurate in $O(M(l_\mathcal{S} + k)|\mathcal{E}|^{2M+1})$ time and $O((l_\mathcal{S} + k)|\mathcal{E}|^2 + |\mathcal{E}|^{M+1})$ space by traversing $\mathcal{S}$ just once.*

Proof. By Lemma 2, the algorithm SCAN returns the set F of triples $(c, r, W[c][r])$ such that $c \mapsto r$ is frequent and $W[c][r] = W(c \mapsto r, \mathcal{S}, k)$. Let R be a set $\{r \mid (c, r, W) \in F\}$. For every $r \in R$, it is sufficient to show that $C[r]$ in SECT is the set of all causal types X such that $X \mapsto r$ is frequent and accurate.

For $m \geq 1$, suppose that C_m is the set of all causal types X such that $|X| = m$ and $X \mapsto r$ is frequent. Consider the case $m + 1$, and also suppose that $Y \mapsto r$ is frequent, where $|Y| = m + 1$. Then, there exist X and c such that $Y = Xc$, where $|X| = m$ and $X \prec c$. Since $Xc \mapsto r$ is frequent and by Lemma 1, both $X \mapsto r$

and $c \mapsto r$ are frequent. By the induction hypothesis, it holds that $X \in C_m$ and $c \in C_1$. Since $W[X]$ in the algorithm SECT denotes $W(X \mapsto r, \mathcal{S}, k)$ and by Lemma 3.2, it holds that $W[X] \cap W[c] = W(X \mapsto r, \mathcal{S}, k) \cap W(c \mapsto r, \mathcal{S}, k) = W(Xc \mapsto r, \mathcal{S}, k)$. Then, the condition $|W[X] \cap W[c]| \geq \sigma \cdot size$ means that $Xc \mapsto r$ is frequent. Since $Xc \mapsto r$ is frequent, the algorithm SECT stores Xc in C_{m+1}, that is, $Y = Xc \in C_{m+1}$.

Since $C[r]$ is the set $\bigcup_{1 \leq i \leq m} C_i$ such that $C_i \neq \emptyset$ $(1 \leq i \leq m)$ and $C_{m+1} = \emptyset$, $C[r]$ in SECT is the set of all causal types X such that $X \mapsto r$ is frequent.

Finally, since $V[X]$ in SECT denotes $W(X, \mathcal{S}, k)$ and by Lemma 3.1, the condition $|W[X]| \geq \gamma|V[X]|$ means that $X \mapsto r$ is accurate. Hence, every D_i is the set of all causal types X such that $|X| = i$ and $X \mapsto r$ is frequent and accurate, so $D[r]$ in SECT is the set of all causal types X such that $X \mapsto r$ is frequent and accurate.

Next, consider the time and space complexity of SECT. For the inner foreach-loop, it holds that $|C_m| \leq |\mathcal{E}|^M$. Since both $W[X]$ and $W[c]$ are regarded as bit vectors with length $l_{\mathcal{S}} + k - 1$, for $X = c_1 \cdots c_m$, the time to check whether or not $|W[X] \cap W[c]| \geq \sigma \cdot size$ is $O(m(l_{\mathcal{S}} + k)) \leq O(M(l_{\mathcal{S}} + k))$ by computing $W[Xc] = W[X] \cap W[c] = W[c_1] \cap \cdots \cap W[c_m] \cap W[c]$. Since $V[e]$ is a bit vector with length $l_{\mathcal{S}} + k - 1$ and by using $W[Xc]$, the time to check whether or not $|W[Xc]| \geq \gamma|V[Xc]|$ is $O(l_{\mathcal{S}} + k)$. Then, the time complexity of the inner foreach-loop is $O(l_{\mathcal{S}} + k + M(l_{\mathcal{S}} + k)|\mathcal{E}|^M + l_{\mathcal{S}} + k) = O(M(l_{\mathcal{S}} + k)|\mathcal{E}|^M)$. Since the number of the outer and inner foreach-loop is at most $|\mathcal{E}|$ and $|\mathcal{E}|^M$, respectively, and by Lemma 2, the time complexity of SECT is $O((l_{\mathcal{S}} + k)|\mathcal{E}|^2 + (M(l_{\mathcal{S}} + k)|\mathcal{E}|^M)|\mathcal{E}|^M|\mathcal{E}|) = O(M(l_{\mathcal{S}} + k)|\mathcal{E}|^{2M+1})$. The space complexity is the sum of the space $O((l_{\mathcal{S}} + k)|\mathcal{E}|^2)$ of the bit vectors of $W[c][r]$ and $V[e]$ in SCAN, and the space $O(|\mathcal{E}|^{M+1})$ to store the obtained episodes, that is, $O((l_{\mathcal{S}} + k)|\mathcal{E}|^2 + |\mathcal{E}|^{M+1})$.

Finally, since the algorithm SECT does not traverse $\mathcal{S}$ and by Lemma 2, the algorithm SECT traverses $\mathcal{S}$ just once. □

If it is not necessary to store all of the extracted sectorial episodes in $D[r]$, then we can reduce the space complexity of the algorithm SECT to $O((l + k)|\mathcal{E}|^2)$, by outputting the episode whenever it is found, instead of using $D[r]$.

Example 2. Consider the event sequence $\mathcal{S}$ given in Figure 1. We give an running example of SECT$(\mathcal{S}, 0.5, 0.6, 3)$. Here, $W(\mathcal{S}, 3) = 12$.

Consider SCAN$(\mathcal{S}, 0.5, 3)$. Figure 4 (left) describes the sets C, $evtyp(w(S, i, 3), i)$ and R of event types at the end of foreach-loop for every i $(-1 \leq i \leq 8)$ in SCAN. Then, Figure 4 (right) describes the output of SCAN$(\mathcal{S}, 0.5, 3)$. Here, the column *freq* denotes the number of elements of $W[c][r]$ and the symbol •, which is a label that $c \mapsto r$ is frequent.

Hence, SCAN$(\mathcal{S}, 0.5, 3)$ returns (F, E), where F and E are the sets of triples and of pairs, respectively, as follows.

$$F = \left\{ \begin{array}{l} (a, b, 1111000110), (c, b, 1111110110), (b, d, 1110011011), \\ (c, d, 1111010010), (a, e, 1101001110), (b, e, 1101101100), \\ (d, e, 1101101100), (b, c, 0111101100), (d, c, 0111101100) \end{array} \right\},$$

$$E = \left\{ \begin{array}{l} (a, 111101111100), (b, 111111111110), (c, 111111111100), \\ (d, 111111111111), (e, 011111111110) \end{array} \right\}.$$

i	C	$evtyp(w(S,i,3),i)$	R
-2	$\emptyset$	$\emptyset$	$abcd$
-1	$abcd$	$\emptyset$	bde
0	$abced$	acd	ac
1	$abce$	be	bd
2	$abcd$	ac	ce
3	$bede$	bd	b
4	bce	ce	ad
5	abd	b	ac
6	acd	ad	be
7	bce	ac	d
8	abe	be	$\emptyset$

c	r	$W[c][r]$										$freq$
		-1	0	1	2	3	4	5	6	7	8	
a	b	1	1	1	1	0	0	0	1	1	0	6 •
c	b	1	1	1	1	1	1	0	1	1	0	8 •
d	b	1	1	1	0	1	0	0	1	0	0	5
a	d	1	1	1	1	0	0	0	0	1	0	5
b	d	1	1	1	0	0	1	1	0	1	1	7 •
c	d	1	1	1	1	0	1	0	0	1	0	6 •
a	e	1	1	0	1	0	0	1	1	1	0	6 •
b	e	1	1	0	1	1	0	1	1	0	0	6 •
c	e	1	1	0	1	0	0	0	1	1	0	5
d	e	1	1	0	1	1	0	1	1	0	0	6 •
b	a	0	1	1	0	0	1	1	1	0	0	5
c	a	0	1	0	0	0	1	0	0	0	0	2
d	a	0	1	1	0	0	0	1	1	0	0	4
e	a	0	1	1	0	0	1	0	0	0	0	3
a	c	0	1	0	1	0	0	1	1	0	0	4
b	c	0	1	1	1	1	0	1	1	0	0	6 •
d	c	0	1	1	1	1	0	1	1	0	0	6 •
e	c	0	1	1	0	0	0	0	0	0	0	2
e	b	0	0	1	0	1	1	0	1	1	0	5
e	d	0	0	1	0	0	1	0	0	1	1	4

Fig. 4. C, $evtyp(w(S,i,3),i)$ and R at the end of foreach-loop (left) and $W[c][r]$ in SCAN (right)

Next, consider $\text{SECT}(\mathcal{S}, 0.5, 0.6, 3)$. From the set F, we obtain R as $\{b, c, d, e\}$. In the following, we consider the constructions of $C[r]$ and $D[r]$, respectively, for every $r \in R$.

For $b \in R$, it holds that $C_1 = \{a, c\}$. Since $W[a] \cap W[c] = 1111000110 \odot 1111110110 = 1111000110$, it holds that $|W[a] \cap W[c]| = 6$, so it holds that $C_2 = \{ac\}$. Hence, it holds that $C[b] = \{a, c, ac\}$.

On the other hand, for $C_1 = \{a, c\}$, it holds that $|W[a]| = 6$ and $|V[a]| = 10$, while $|W[c]| = 8$ and $|V[c]| = 9$. Then, it holds that $D_1 = \{a, c\}$. Also since $V[a] \cap V[c] = 111101111100 \odot 111111111100 = 111101111100$, it holds that $|V[ac]| = 9$. Since $|W[ac]| = 6$, it holds that $D_2 = \{ac\}$, so $D[b] = \{a, c, ac\}$.

For $c \in R$, it holds that $C_1 = \{b, d\}$. Since $W[b] \cap W[d] = 0111101100 \odot 0111101100 = 0111101100$, it holds that $|W[b] \cap W[d]| = 6$, so it holds that $C_2 = \{bd\}$. Hence, it holds that $C[c] = \{b, d, bd\}$.

On the other hand, for $C_1 = \{b, d\}$, it holds that $|W[b]| = 6$ and $|V[b]| = 11$, while $|W[d]| = 6$ and $|V[d]| = 12$. Then, it holds that $D_1 = \emptyset$. Also since $V[b] \cap V[d] = 111111111110 \odot 111111111111 = 111101111110$, it holds that $|V[bd]| = 11$. Since $|W[bd]| = 6$, it holds that $D_2 = \emptyset$, so $D[b] = \emptyset$.

For $d \in R$, it holds that $C_1 = \{b, c\}$. Since $W[b] \cap W[c] = 1110011011 \odot 1111010010 = 1110010010$, it holds that $|W[b] \cap W[d]| = 5$, so it holds that $C_2 = \emptyset$. Hence, it holds that $C[d] = \{b, c\}$.

On the other hand, for $C_1 = \{b, c\}$, it holds that $|W[b]| = 7$ and $|V[b]| = 11$, while $|W[c]| = 6$ and $|V[c]| = 10$. Then, it holds that $D_1 = \{b, c\}$. Also since $V[b] \cap V[c] = 111111111110 \odot 111111111100 = 111101111100$, it holds that $|V[bc]| = 10$. Since $|W[bc]| = 5$, it holds that $D_2 = \emptyset$, so $D[d] = \{b, c\}$.

For $e \in R$, it holds that $C_1 = \{a, b, d\}$. Then, the following statement holds.

$$W[a] \cap W[b] = 1101001110 \odot 1101101100 = 1101001100,$$
$$W[a] \cap W[d] = 1101001110 \odot 1101101100 = 1101001100,$$
$$W[b] \cap W[d] = 1101101100 \odot 1101101100 = 1101101100.$$

Then, it holds that $|W[a] \cap W[b]| = 5$, $|W[a] \cap W[d]| = 5$ and $|W[b] \cap W[d]| = 6$. Hence, it holds that $C_2 = \{bd\}$, so it holds that $C[e] = \{a, b, d, bd\}$.

On the other hand, for $C_1 = \{a, b, d\}$, it holds that $|W[a]| = 6$, $|V[a]| = 9$, $|W[b]| = 6$, $|V[b]| = 11$, $|W[d]| = 6$ and $|V[d]| = 12$. Then, it holds that $D_1 = \{a\}$. Also since $|V[a] \cap V[b]| = 9$, $|V[a] \cap V[d]| = 9$ and $|V[b] \cap V[d]| = 11$, it holds that $D_2 = \emptyset$, so $D[e] = \{a\}$.

As the result, the algorithm $\textsc{Sect}(\mathcal{S}, 0.5, 0.6, 3)$ returns the following frequent sectorial episodes, where ones with bold faces are frequent and accurate.

$$
\begin{array}{lll}
\mathbf{a \mapsto b} & \mathbf{c \mapsto b} & \mathbf{ac \mapsto b} \\
b \mapsto c & d \mapsto c & bd \mapsto c \\
\mathbf{b \mapsto d} & \mathbf{c \mapsto d} & \\
a \mapsto e & b \mapsto e & d \mapsto e \quad bd \mapsto e
\end{array}
$$

4   Empirical Results

In this section, by applying the algorithm $\textsc{Sect}$ to bacterial culture data, which are complete data in [8] from 1995 to 1998, we extract sectorial episodes concerned with *drug-resistant change*. Here, we regard a pair of "attribute=value" as an event type.

First, we fix the width of windows as 30 days. Since the database contains the patient information not related to drug-resistant change such as date, gender, ward and engineer [8], it is necessary to focus the specified attributes. Then, we select the attributes age, department, sample, fever, catheter, tracheo, intubation, drainage, WBC (white blood cell) count, medication, Urea-WBC, Urea-Nitocide, Urea-Occultblood, Urea-Protein, the total amount of bacteria, the detected bacterium, and the sensitivity of antibiotics as the causal types, and the sensitivity of antibiotics as the resulting type in whole 44 attributes.

From the medical viewpoint, in order to extract sectorial episodes concerned with drug-resistant change, it is necessary to extract them based on the same detected bacterium and the same sample. Hence, in this paper, we divide the database into *pages* for the detected bacterium and the sample described as Figure 5. Here, the detected bacteria are Staphylococci (bac1), Enteric bacteria (bac7), glucose-nonfermentative gram-negative bacteria (bac8) and Anaerobes (bac11), and the samples are catheter/others (spl1), urinary and genital organs (spl3) and respiratory organs (spl5). Also the column "patients" in Figure 5 denotes the number of different patients consisting of more than two records in a page and the column "max." denotes the maximum number of records for patients in a page.

bacterium	sample	patients	max.	episodes	antibiotics
bac1	spl1	296	34	2668	RFPFOM(8), TC(1)
	spl5	319	19	27174	CBP(225), RFPFOM(148), ML(1), TC(1)
bac7	spl3	131	10	21957	Cep3(340), AG(323), CepAP(79), TC(16), Cep2(5), Cep1(3), Aug(2)
	spl5	301	21	76951	CBP(1334), CepAP(1158), Cep3(184), PcAP(128), TC(98), Aug(2), Cep1(2), Cep2(2)
bac8	spl1	81	7	8127	AG(1), CepAP(1), PcAP(1)
	spl5	296	21	37938	CepAP(40), AG(8), CBP(1)
bac11	spl1	208	22	16720	CBP(304), Cep2(81), Cep3(81), Cep1(1), ML(1), PcAP(1), PcB(1)

Fig. 5. The pages for the detected bacterium and the sample from bacterial culture data, the number of extracted sectorial episodes under $\sigma = 0.15$ and $\gamma = 0.8$, and the antibiotics, where $Ant(n)$ denotes that n is the number of extracted sectorial episodes of the form $C \mapsto (Ant=R)$ such that $(Ant=S) \in C$

Finally, since the drug-resistant change over different patients is also meaningless, we collect all records in a page for every patient, and construct one event sequence by connecting them such that the span between all records for one patient and ones for other patient is at least 30 days (the width of windows). Then, we apply the algorithm SECT to the event sequence.

For the minimum support $\sigma = 0.15$ and the minimum confidence $\gamma = 0.8$, the column "episodes" in Figure 5 describes the number of the extracted sectorial episodes. Furthermore, we focus on the extracted sectorial episode $C \mapsto r$ such that, for antibiotics Ant, C contains "$Ant=S$ (susceptibility)" and r is "$Ant=R$ (resistant)." In the column "antibiotics" in Figure 5, $Ant(n)$ denotes that n is the number of extracted sectorial episodes of the form $C \mapsto (Ant=R)$ such that $(Ant=S) \in C$. Here, antibiotics are benzilpenicillin (PcB), augmentin (Aug), anti-pseudomonas penicillin (PcAP), 1st generation cephems (Cep1), 2nd generation cephems (Cep2), 3rd generation cephems (Cep3), anti-pseudomonas cephems (CepAP), aminoglycosides (AG), macrolides (ML), tetracyclines (TC), carbapenems (CBP), and RFP/FOM (RFPFOM).

Figure 5 means that different sectorial episodes representing drug-resistant change are extracted for every detected bacterium and sample. For (bac1,spl5), (bac7,spl5) and (bac11,spl1), the drug-resistant change for CBP occurs in the extracted episodes. In particular, for (bac7,spl5), the drug-resistant change CepAP also occurs in the extracted episodes. On the other hand, for (bac7,spl3), the drug-resistant change Cep3 and AG occurs in the extracted episodes.

5   Conclusion

In this paper, we have newly introduced the *sectorial episode* together with the *parallel episode* [5]. Then, we have designed the algorithm SECT to extract all of the sectorial episodes that are frequent and accurate. Finally, we have applied the algorithm SECT to bacterial culture data, and extracted sectorial episodes representing drug-resistant change.

Since the number of extracted sectorial episodes in Section 4 is large, it is a future work to introduce the concept of *closed* sectorial episodes like as closed sequential patterns [9,10], in order to reduce the number of extracted episodes.

In Section 4, we have treated a page for the same detected bacterium and the same sample as one event sequence. Then, it is a future work to introduce another frequency measure like as the frequency for a patient in a page. Also it is a future work to apply our algorithm to another time-related data and extract sectorial episodes from them.

As stated in Section 1, since a sectorial episode has been captured as the direct precedent-subsequent relationship of events, it is just a candidate of *causality* in database (*cf.* [11]). Hence, it is an important future work to incorporate such causality with our sectorial episodes and to design the algorithm to extract sectorial episodes concerned with causality.

References

1. R. Agrawal, H. Mannila, R. Srikant, H. Toivonen, A. I. Verkamo: *Fast discovery of association rules*, in U. M. Fayyed, G. Piatetsky-Shapiro, P. Smyth, R. Uthurusamy (eds.): *Advances in Knowledge Discovery and Data Mining*, AAAI/MIT Press, 307–328, 1996.
2. R. Agrawal, R. Srikant: *Fast algorithms for mining association rules in large databases*, Proc. 20th VLDB, 487–499, 1994.
3. R. Agrawal, R. Srikant: *Mining sequential patterns*, Proc. 11th ICDE, 3–14, 1995.
4. C. Bettini, S. Wang, S. Jajodia, J.-L. Lin: *Discovering frequent event patterns with multiple granularities in time sequences*, IEEE Trans. Knowledge and Data Engineering **10**, 222–237, 1998.
5. H. Mannila, H. Toivonen, A. I. Verkamo: *Discovery of frequent episodes in event sequences*, Data Mining and Knowledge Discovery **1**, 259–289, 1997.
6. J. Pei, J. Han, B. Mortazavi-Asi, J. Wang, H. Pinto, Q. Chen, U. Dayal, M.-C. Hsu: *Mining sequential patterns by pattern-growth: The PrefixSpan approach*, IEEE Trans. Knowledge and Data Engineering **16**, 1–17, 2004.
7. R. Srikant, R. Agrawal: *Mining sequential patterns: Generalizations and performance improvements*, Proc. 5th EDBT, 3–17, 1996.
8. S. Tsumoto: *Guide to the bacteriological examination data set*, in E. Suzuki (ed.): Proc. International Workshop of KDD Challenge on Real-World Data (KDD Challenge 2000), 8–12, 2000.
9. J. Wang, J. Han: *BIDE: Efficient mining of frequent closed sequences*, Proc. 20th ICDE, 2004.
10. X. Yan, J. Han, R. Afshar: *CloSpan: Mining closed sequential patterns in large datasets*, Proc. 3rd SDM, 2003.
11. C. Zhang, S. Zhang: *Association rule mining*, Springer-Verlag, 2002.

A Voronoi Diagram Approach
to Autonomous Clustering

Heidi Koivistoinen, Minna Ruuska, and Tapio Elomaa

Institute of Software Systems, Tampere University of Technology
P. O. Box 553, FI-33101 Tampere, Finland
`firstname.lastname@tut.fi`

Abstract. Clustering is a basic tool in unsupervised machine learning and data mining. Distance-based clustering algorithms rarely have the means to autonomously come up with the correct number of clusters from the data. A recent approach to identifying the natural clusters is to compare the point densities in different parts of the sample space.

In this paper we put forward an agglomerative clustering algorithm which accesses density information by constructing a Voronoi diagram for the input sample. The volumes of the point cells directly reflect the point density in the respective parts of the instance space. Scanning through the input points and their Voronoi cells once, we combine the densest parts of the instance space into clusters.

Our empirical experiments demonstrate the proposed algorithm is able to come up with a high-accuracy clustering for many different types of data. The Voronoi approach clearly outperforms k-means algorithm on data conforming to its underlying assumptions.

1 Introduction

Clustering is the fundamental task of grouping together similar unlabeled objects in order to obtain some generalization through categorization. This need is encountered in a vast array of application. As similarity of objects varies from application to application, there cannot be an universally superior method of clustering instances. Thus, it should not come as a surprise that the number of different unsupervised clustering algorithms put forward in the literature is huge (see e.g., [1]).

There are two basic approaches to obtaining a clustering: one can use either the bottom-up *agglomerative* or the top-down *divisive* construction. The former begins with each instance in a singleton cluster and successively merges clusters together until a stopping criterion is met. In the latter approach one begins with all instances in a single cluster and partitions the clusters until satisfaction of a stopping criterion. Usually a *distance measure* is required to determine which clusters to merge together next due to their adjacency or which cluster to divide because of the sparseness of the points in it.

Of course, it is also possible to choose an intermediate value for the number of clusters and start manipulating the clusters from there. A simple and widely used

N. Lavrač, L. Todorovski, and K.P. Jantke (Eds.): DS 2006, LNAI 4265, pp. 149–160, 2006.

clustering algorithm is k-means clustering [2,3,4]: Given k initial (e.g., random) cluster centers $C = \{c_1, \ldots, c_k\}$ for the metric observations $S = \{s_1, \ldots, s_n\}$, iterate the following until the cluster centers do not change.

1. For each observation point $s_i \in S$ determine the center $c_j \in C$ that is closest to it and associate s_i with the corresponding cluster.
2. Recompute the center locations (as the center of the mass of points associated with them).

The iterative k-means minimizes the squared error — the squared Euclidean distance between the cluster centers and the observations associated with them.

The obvious shortcomings of the basic k-means clustering are that the number of clusters needs to determined in advance and its computational cost with respect to the number of observations, clusters, and iterations. Though, k-means is efficient in the sense that only the distance between a point and the k cluster centers — not all other points — needs to be (re)computed in each iteration. Typically the number of observations n is much larger than k.

There have been many approaches trying to alleviate both of these problems. Variations of k-means clustering that are supposed to cope without prior knowledge of the number of cluster centers have been presented [5,6]. Several proposals for scaling up clustering and, in particular, k-means for massive data sets have been proposed [7,8,9,10]. Quite often these studies assume that a particular distance measure is applied. Moore [10] and Elkan [11] have shown how the triangle inequality can be used for any distance function to reduce the number of distance calculations required during the execution of k-means by carrying information over from iteration to iteration.

While the proposed solutions to k-means clustering certainly improve its practical behavior, they do not overcome the fundamental problems associated with the algorithm. For example, the G-means algorithm of Hamerly and Elkan [6] makes k-means autonomously find approximately the right number of clusters at the expense of increased running time. However, the algorithm still errs consistently even on simple cases [12]. On the other hand, the speed of clustering alone, without exactness of the end result, is not of much use. Rather than stubbornly try to make distance-based clustering algorithms cope with the problem of identifying the right number of clusters, one needs to change the point of view and find an approach that is better suited to the task at hand. We propose density-based algorithms as a better alternative to identifying the correct number of clusters in the data.

In this paper we propose an agglomerative clustering algorithm that begins from the situation in which each instance makes up a cluster on its own. For singleton cluster selection we use *Voronoi diagrams* [13]. The initial clusters are merged together with neighboring cells as long as the cell volumes are below a user defined threshold. Observe that we do not need a distance measure, since only neighboring cells are candidates for merging and the Voronoi diagram readily provides us with the neighborhood information. On the other hand, we need to have the threshold volume to limit merging of cells. Our approach is efficient in the sense that there is no need for iterative processing.

To the best of our knowledge the only prior clustering algorithm taking directly advantage of Voronoi diagrams is that of Schreiber [14]. However, his algorithm only adjusts k-means clustering with the help of the Voronoi diagram. The Voronoi diagram is built over the clusters, not for the data points. Clusters with the largest error are halved in two. Voronoi diagrams and their duals Delaunay tessellations are, though, widely used in nearest neighbor methods [4]. Also clustering of graphs can use the Delaunay triangulation as the starting point of processing and then apply, e.g., the minimum spanning tree clustering to the graph at hand [1].

This paper is organized as follows. Section 2 motivates the algorithm introduced in this paper and discusses related work. Basics of Voronoi diagrams are recapitulated and the agglomerative clustering algorithm is introduced in Section 3. Finer points of the algorithm are further discussed in Section 4. Our empirical experiments in Section 5 show that clustering based on Voronoi diagrams significantly outperforms the k-means algorithm already on normally distributed spherical data, which suits the latter clustering algorithm well. The difference between the two is even higher on more complex (real-world) data. Section 6 contains the concluding remarks of this paper and outlines future work.

2   Motivation and Related Work

Clustering based on the regular-shaped classes easily runs into trouble when faced with data in which the existing clusters do not conform to the allowed shape of clusters. This is e.g. the case with k-means clustering and its spherical clusters. For example, Figure 2 demonstrates a data set with different geometric shapes that cannot be captured or approximated well by using only one shape of clusters.

Hence, a clear requirement for a general-purpose clustering algorithm is that it be able to discover clusters of arbitrary shape. Visually (non-overlapping) clusters are usually quite easy to detect because the density of data points within a cluster differs (significantly) from that of points surrounding it. There have been many approaches trying to utilize this intuition. For example, the DBSCAN algorithm of Ester, Kriegel, and Xu [15] searches for clusters that are maximal sets of density-connected points, that is, points that have at least a minimum number of neighbors in their immediate neighborhood form a cluster.

The subsequent algorithm DBCLASD by Xu et al. [16] explicitly approximates the cluster density by comparing the approximate subspace and overall volume of the point set. This tells us whether the subspace is denser than average in the given sample. Because the clusters may have arbitrary shape, it is not immediately clear how to compute the volume of the part of the instance space that a point set occupies. Xu et al. use a (hyper)rectangular grid division of the instance space in volume estimation. This makes volume approximation simple but, at the same time, highly dependent on the chosen granularity.

The elegant density-based clustering algorithm of Hinneburg and Keim [17] computes an influence function over the whole data set. *Density attractors* — local maxima of the overall density — then determine the clusters. For a continuous and differentiable density function, a hill-climbing algorithm can be used to discover the clusters efficiently. On the downside, two (unintuitive) parameter values that heavily influence the quality of the obtained clustering need to be determined.

In addition to the practical clustering algorithms studied in data mining and machine learning literature, a lot of theoretical work has been devoted to this problem. Theoretically what we have at hand is an intractable optimization problem that can be efficiently solved only approximately. For example, k-means clustering is NP-hard already when $k = 2$. Recently the first linear-time $(1 + \varepsilon)$-approximation algorithm for fixed k and ε based on sampling was devised by Kumar, Sabharwal, and Sen [18].

Not only the different shapes of clusters, but also the different sizes of clusters can cause problems to some algorithms. For example, the k-means algorithm cannot cope well with different-sized clusters; i.e., clusters of different radius or with different numbers of points in them. Consider, e.g., two cluster centers distance d apart from each other. If one of the clusters has radius $2d/3$ and the other one radius $d/3$, then k-means will unavoidably place part of the points of the first cluster to the second cluster simply because its center is nearer to the points. Similarly, consider two clusters one containing 100 and the other one 1,000 points. Minimizing the squared distance in the denser cluster by placing both centers within it may be more profitable than letting both clusters contain one center.

The starting point for our work is that one ought to be able to discover arbitrary-shaped clusters, like in the algorithms mentioned above. Density within and without a cluster is the measure that actually defines the clusters. Therefore, we change the point of view from distance-based to density-based clustering. Optimizing the correct measure is expected to lead to much better results than fixing algorithms attempting to optimize a measure that is only indirectly related to the task at hand. A lesson learned from prior work is that taking global characteristics of the data into account leads to better results than relying on local properties alone. Another design principle is that we do not want to restrict the number of clusters in advance like is required, e.g., in k-means. Moreover, we aim at as autonomous clustering as possible. That is, there should not be many parameters whose value the user needs to set. Any parameters that the user needs to set have to be intuitive, though not necessarily concrete. For practical success the parameters need to be easy to learn.

3   Agglomerative Voronoi Clustering

3.1   Voronoi Diagrams and Delaunay Tessellations

A multidimensional Voronoi diagram [13] is a partitioning of the instance space $\mathbb{R}^d$ in regions R_j with the properties: Each center c_j lies exactly in one region R_j, which consists of all points $x \in \mathbb{R}^d$ that are closer to c_j than any other center c_k, $j \neq k$: $R_j = \left\{ x \in \mathbb{R}^d : \|x - c_j\| < \|x - c_k\|, \forall j \neq k \right\}$. The centers c_j are

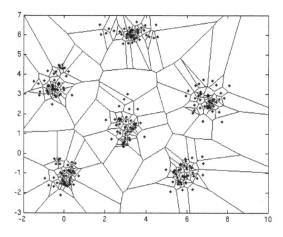

Fig. 1. A Voronoi diagram for data coming from six clusters

called *Voronoi points*. In the setting of unsupervised learning, where a sample $S = \{s_1, \ldots, s_n\}$ is given, the sample points s_i, naturally, become the Voronoi points. Observe that the end result of the iterative k-means algorithm is also a Voronoi diagram with the k centers as Voronoi points.

Let us call the region associated with a Voronoi point c_j its *cell*. Because a cell contains all those points that are closer to its center than any other center, the cell borders lie exactly in the middle of two centers. Figure 1 demonstrates the Voronoi diagram for a data set that clearly has six separate clusters.

The dual of the Voronoi diagram for a point set is its *Delaunay tessellation* (also Delaunay triangulation). It is a graph in which the vertices are the centers of the Voronoi cells (the initial data points) and edges connect any two vertices that have a common boundary in the Voronoi diagram. The same set of edges is obtained by connecting any two points p and q for which there exists a ball B that passes through p and q and does not contain any other point of S in its interior or boundary. The name Delaunay triangulation stems from the fact that in the plane the *triangulation* of a point set S is a planar graph with vertices S and a maximal set of straight line edges. Adding any further straight line edge would lead to crossing other edges. A subset of a edges of a triangulation is called a *tessellation* of S if it contains the edges of the convex hull and if each point of S has at least two adjacent edges.

Voronoi diagram is straightforward to calculate for a point set in the plane, but becomes more complicated in higher dimensions. Therefore, the uses of Voronoi diagrams are mostly limited to $\mathbb{R}^2$. However, the qhull algorithm of Barber, Dobkin, and Huhdanpaa [19] works for general dimension. The algorithm, among other things, is able to compute the Voronoi diagram for an arbitrary point set. We use qhull to compute the Voronoi diagrams that the clustering algorithm operates on.

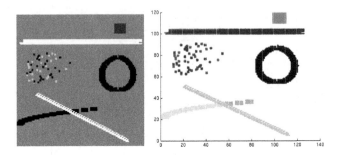

Fig. 2. An example data set with different geometric shapes (left) and the result of applying Voronoi clustering to it (right). The colors denote clusters. Intuitively, there are classification errors only in the cloud of points in the middle left and in the fringe points of the horizontal line on top.

3.2   The Method

Given a data set of n points $S = \{s_1, \ldots, s_n\} \subseteq \mathbb{R}^d$ to be clustered we begin by constructing the Voronoi diagram for S. The computational complexity of Voronoi diagram construction in the general case is $\Theta(n \log n)$ [13], but requires only linear time in some restricted cases [20]. For the ease of illustration, let us consider the two-dimensional plane.

The algorithm is given as input parameter a threshold value max indicating the maximum volume allowed to a cell that still can be combined into an evolving cluster. We approximate local instance space density by cell volume, so only it matters in deciding whether the density is high enough for further combination into a cluster. Hence, individual clusters may grow to any size as long as the local density is sufficient. After all cells of at most volume max have been taken care of, we only have relatively large cells remaining.

In the Voronoi diagram for the data each point makes up its own cell. The volumes (areas in case of the plane) of the cells are computed (or approximated) next. The cell volumes are needed, because we go through the data points in the increasing order of the cell volume associated with the point. Our current implementation computes the exact cell area in the plane as the area of a polygon, but in higher dimensions d we simply approximate the cell to be a hyperball with volume $(\pi^{\lfloor n/2 \rfloor} r^d)/\Gamma(d/2 + 1)$, where $\Gamma(z + 1) = z!$ for integer values and the radius r is the average distance of cell corners from its center. This approximation is not very accurate in small dimensions, but it improves as the number of dimensions grows. Each cell is associated with a class label. We just assign increasing integer values as class labels.

The point having the smallest Voronoi cell is handled first. It has no known neighboring cells, so we just assign the class label 1 to this cell. The cell considered next might be a neighbor of the first cell, which would be detected by the two sharing a corner point. In that case, if the volume of the latter cell is below the threshold value max, they get combined together and assume the class label of the first cell. Otherwise, the second cell may turn out not to be a neighbor

Table 1. The Voronoi clustering algorithm

```
Algorithm Voronoi_Clustering( set S, real max )
```

1. Construct the Voronoi diagram for the sample $S = \{s_1, \ldots, s_n\}$.
2. Approximate the Voronoi cell volumes and order the points accordingly. Without loss of generality, let the obtained order be $s_1, \ldots, s_n$.
3. For $i = 1$ to n do
   If the volume of the cell R_i associated with s_i is at most max then
   (a) For $j = 1$ to $i - 1$ do
       If cluster C_j and cell R_i are adjacent then
       − Merge them together;
       − If R_i has no class number yet then it assumes that of C_j else both assume the minimum class label among those of C_j and R_i;
   (b) If cell R_i has no class number then assign a new one to it;
   else assign R_i to the closest neighboring cluster.

of the first one, in which case it gets a class label of its own. However, it is still possible that the two points belong to the same cluster and later get combined into the same cell due to having a common neighboring cell.

In the general case, a cell has many neighboring cells with different class labels and some that have not yet been labeled (those that are larger in volume than the cell at hand). The known neighbors are processed in the order of their size (i.e., in order of their class values). The cell under consideration is merged to its neighbor with the smallest class label, and its labeled neighbors also assume the same class. Cells are combined as long as their volume stays under the max value. If the max value is ever reached, there is no need to go through the remaining neighbor cells.

When all points have been considered once, all the cluster combinations have been executed. There is no need to iterate the process. However, we still need to do a simple post processing of the clustering obtained to make it sensible. Consider for instance the example of Figure 1. It is easy to see that the cell combining procedure described above will detect the six clusters with ease for a wide range of values for the threshold max. However, it is as clear that the outermost points of the clusters cannot get combined, because then the clusters would grow together.

In this and all other imaginable applications the following heuristics takes care of the points at the fringe of a cluster (of any shape). Center points with a cell of volume larger than the threshold max simply get combined to the closest neighboring class. Table 1 represents the algorithm in more compact form.

4 Discussion

After initialization the algorithm goes through the data points once doing only relatively simple calculations. Hence, the approach is efficient in practice. There are two $O(n \log n)$ phases in the algorithm: Voronoi diagram construction and sorting of the Voronoi points according to their associated cell volumes. In our current implementation neither of these tasks, however, constitutes the most significant phase in time consumption. The most time intensive task in the algorithm is adjacency information processing because of the heavy data structures chosen. Our current implementation uses corner point representation to identify neighboring Voronoi cells. The number of corner points, unfortunately, grows with increasing number of instance space dimensions. Thus, this implementation becomes inefficient for domains with many attributes. It is our intention to turn to using Delaunay tessellations instead to record neighborhood information. The graph directly provides cell adjacency information as more efficient to handle.

In the Voronoi diagram of a point set the cell boundaries are by definition equally distant from the relevant points. Thus, the Voronoi diagram naturally gives raise to kind of *maximum margin separation* of the data. Our algorithm combines existing cells instead of creating artificial center points — like, e.g., k-means does. Hence, the maximal margins are maintained also in the clustered data. Since there are no reasons for placing the cluster border closer to one or other cluster, maximal margin clustering seems a natural choice. Moreover, results on boosting and kernel methods have shown maximal margin separation to be a good strategy [21].

One can view Voronoi clustering as a change of paradigm from the k-means, whose Euclidean distance minimization can be made autonomous with respect to the number of clusters only by optimizing a parameter that is too hard to learn [6,12]. In the future we will study whether the one input parameter in our clustering algorithm can be deduced without user intervention.

5   Empirical Evaluation

To evaluate the proposed algorithm empirically, we test it on generated and real-world data. As the comparison algorithm we use the basic k-means algorithm randomly choosing the initial cluster centers (as implemented by Matlab). For k-means, the correct number of clusters is given as a parameter value. Moreover, to favor it more, we mostly use generated data that ought to, by default, suit k-means well. Let us describe the data sets:

2DEqualDiscs: In the first experiment 10 cluster centers were drawn randomly from the instance space and 150 points were uniformly drawn to each circular cluster of equal radius. The instance space height and width is 80 units and the radius of a cluster is 3 units. Because of the randomly drawn cluster centers, the clusters do sometimes partially overlap. To better account for the effects of overlapping clusters, we report the results of 5 independent runs in this setting, each consisting of 10 repetitions.

2DEqualDiscs — long distances: To see what is the effect of cluster overlapping, we changed the above situation by drawing the cluster centers from an instance space of height and width 150 units. Thus, the clusters are not as probable to overlap as in the previous experiment.

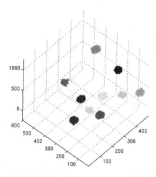

Fig. 3. An example of the 3DEqualBalls data: 10 randomly drawn clusters of equal size

Table 2. Average accuracies over 10 repetitions in the test domains

DATA SET		k-MEANS	VORONOI
2DEQUALDISCS	A	71.6 ±8.8	76.0 ±8.2
	B	77.4 ±8.8	79.1 ±14.1
	C	78.0 ±9.7	81.6 ±15.3
	D	74.6 ±13.4	81.7 ±12.4
	E	74.4 ±13.0	84.6 ±18.0
2DEQUALDISCS — LONG DISTANCES	A	77.0 ±7.3	100.0 ±0.0
	B	71.6 ±12.5	98.9 ±3.3
	C	76.7 ±7.7	93.8 ±11.1
2DUNEQUALDISCS	A	76.8 ±9.2	90.7 ±5.8
	B	76.8 ±5.9	92.9 ±6.6
	C	75.0 ±11.0	90.0 ±8.6
	D	76.8 ±8.8	92.3 ±6.7
	E	83.3 ±10.5	93.1 ±8.4
3DEQUALBALLS	A	71.1 ±16.0	91.0 ±28.5
	B	70.6 ±12.7	99.0 ±3.1
	C	71.6 ±13.0	99.0 ±3.2
	D	75.1 ±9.8	95.0 ±10.8
	E	74.9 ±12.8	100.0 ±0.0
3DUNEQUALBALLS	A	73.6 ±7.3	95.6 ±8.5
	B	79.6 ±7.4	94.9 ±9.5
	C	72.0 ±12.7	96.0 ±9.2
3DOBJECTS	A	74.4 ±11.8	80.2 ±10.6
	B	78.1 ±16.1	81.6 ±7.9
	C	69.9 ±15.0	73.5 ±7.7

2DUnequalDiscs: This experiment is the same as the first one, except that this time the cluster sizes vary from one hundred to one thousand points (random selection, possible sizes multiples of one hundred).

3DEqualBalls: This experiment corresponds to the first one, except that the instance space now has three dimensions. See Figure 3 for an example situation.

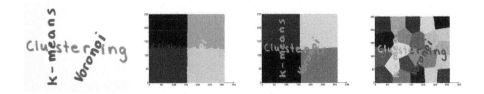

Fig. 4. The original text data and results by k-means with 4, 5, and 25 clusters

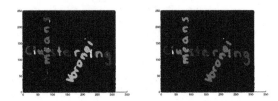

Fig. 5. Text recognition with Voronoi clustering using parameter value 50 and 70

3DUnequalBalls: As the second experiment, but in three dimensions.
3DObjects: Classification of different 3D objects which are shown as a 2D projection in Figure 2.

The results of our empirical comparison are given in Table 2. A general remark that can be made concerning these results is the fact that standard deviation is quite high in all the test domains. That partially accounts for the fact that in all the experiments Voronoi clustering was found to be statistically significantly more accurate than k-means clustering (as determined by t-test at 95% significance level).

k-means fares quite well in comparison when 2D spherical data is processed, however Voronoi clustering is clearly more accurate on even this data conforming to the underlying assumptions of k-means. When the cluster distances are increased, the relative advantage of Voronoi clustering becomes even clearer. The same effect can be observed when the clusters have unequal numbers of points in them. On 3D balls the superiority of Voronoi clustering is further enhanced. The fact that it is not more clearly better than k-means on the 3D object data is somewhat surprising and will need further analysis.

Our final experiment applies the two clustering algorithms to a text recognition task. Figure 4 depicts the original picture and results of k-means clustering into 4, 5, and 25 clusters. The obvious problem is that squared distance is minimized by dividing the input space into the requested number of more or less equal-sized clusters. Figure 5 shows the result of Voronoi clustering using two different max values. This time the hand-written words stand out better or worse, because there is no need to partition the instance space into equal-sized clusters.

6   Conclusion and Further Work

Clustering algorithms are widely used and useful tools in unsupervised learning. The user is usually required to provide the value for at least one parameter of the algorithm. Ideally, she should be liberated from this task and the algorithm be able to autonomously adapt to the characteristic properties of the given data. This being too idealistic, the parameter values requested from the user should be as easy to determine as possible.

In this paper we proposed a density-based clustering algorithm that initially builds the Voronoi diagram for the data supplied. It only requires the user to provide one, quite intuitive, parameter. Our empirical experiments demonstrated that Voronoi clustering is an effective algorithm over a wide range of data. It outperforms k-means clustering significantly on all our test domains.

There are many details and implementation issues that still need our attention. For instance, the current implementation suffers from being dependent on the number of corner points, which limits its efficiency in high-dimensional domains. Delaunay tessellation is expected to relieve us from this inconvenience.

Acknowledgements

Work supported by Academy of Finland project "ALEA: Approximation and Learning Algorithms". In addition the work of the first author is financially supported by the graduate school of Tampere University of Technology.

References

1. Jain, A.K., Murty, M.N., Flynn, P.J.: Data clustering: a review. ACM Computing Surveys **31** (1999) 264–323
2. MacQueen, J.B.: On convergence of k-means and partitions with minimum average variance (abstract). Annals of Mathematical Statistics **36** (1965) 1084
3. Forgy, E.: Cluster analysis of multivariate data: Efficiency vs. interpretability of classifications. Biometrics **21** (1965) 768
4. Duda, R.O., Hart, P.E.: Pattern Classification and Scene Analysis. John Wiley & Sons, New York, NY (1973)
5. Pelleg, D., Moore, A.: X-means: Extending k-means with efficient estimation of the number of clusters. In Langley, P., ed.: Proc. 17th International Conference on Machine Learning, San Francisco, CA, Morgan Kaufmann (2000) 727–734
6. Hamerly, G., Elkan, C.: Learning the k in k-means. In Thrun, S., Saul, L.K., Schölkopf, B., eds.: Advances in Neural Information Processing Systems 16. MIT Press, Cambridge, MA (2004) 281–288
7. Ng, R.T., Han, J.: Efficient and effective clustering methods for spatial data mining. In Bocca, J.B., Jarke, M., Zaniolo, C., eds.: Proc. 20th International Conference on Very Large Data Bases, San Francisco, CA, Morgan Kaufmann (1994) 144–155
8. Zhang, T., Ramakrishnan, R., Livny, M.: Birch: An efficient data clustering method for very large databases. In: Proc. ACM SIGMOD International Conference on Management of Data, New York, NY, ACM Press (1995) 103–114

9. Guha, S., Rastogi, R., Shim, K.: Cure: An efficient clustering algorithm for large datasets. In: Proc. ACM SIGMOD International Conference on Management of Data, New York, NY, ACM Press (1998) 73–84

10. Moore, A.W.: The anchors hierarchy: Using the triangle inequality to survive high dimensional data. In Boutilier, C., Goldszmidt, M., eds.: Proc. 16th Conference on Uncertainty in Artificial Intelligence, San Francisco, CA, Morgan Kaufmann (2000) 397–405

11. Elkan, C.: Using the triangle inequality to accelerate k-means. In Fawcett, T., Mishra, N., eds.: Proc. 20th International Conference on Machine Learning, Menlo Park, AAAI Press (2003) 147–153

12. Elomaa, T., Koivistoinen, H.: On autonomous k-means clustering. In Hacid, M.S., Murray, N., Raś, Z.W., Tsumoto, S., eds.: Foundations of Intelligent Systems, Proc. 15th International Symposium, ISMIS'05. Volume 3488 of Lecture Notes in Artificial Intelligence., Berlin Heidelberg New York, Springer (2005) 228–236

13. Aurenhammer, F., Klein, R.: Voronoi diagrams. In Sack, J., Urrutia, G., eds.: Handbook of Computational Geometry. North-Holland, Amsterdam, The Netherlands (2000) 201–290

14. Schreiber, T.: A Voronoi diagram based adaptive k-means-type clustering algorithm for multidimensional weighted data. In Bieri, H., Noltemeier, H., eds.: Computational Geometry — Methods, Algorithms and Applications. Volume 553 of Lecture Notes in Computer Science., Berlin Heidelberg New York, Springer (1991) 265–275

15. Ester, M., Kriegel, H.P., Sander, J., Xu, X.: A density-based algorithm for discovering clusters in large spatial databases with noise. In: Proc. 2nd International Conference on Knowledge Discovery and Data Mining, Menlo Park, CA, AAAI Press (1996) 226–231

16. Xu, X., Ester, M., Kriegel, H.P., Sander, J.: A distribution-based clustering algorithm for mining in large spatial databases. In: Proc. 14th International Conference on Data Engineering, Los Alamitos, CA, IEEE Computer Society Press (1998) 324–331

17. Hinneburg, A., Keim, D.A.: A general approach to clustering in large databases with noise. Knowledge and Information Systems **5** (2003) 387–415

18. Kumar, A., Sabharwal, Y., Sen, S.: A simple linear time $(1 + \varepsilon)$-approximation algorithm for k-means clustering in any dimensions. In: Proc. 45th Annual IEEE Symposium on Foundations on Computer Science, Los Alamitos, CA, IEEE Press (2004) 454–462

19. Barber, C.B., Dobkin, D.P., Huhdanpaa, H.T.: The Quickhull algorithm for convex hulls. ACM Transactions on Mathematical Software **22** (1996) 469–483

20. Aggarwal, A., Guibas, L.J., Saxe, J.B., Shor, P.W.: A linear-time algorithm for computing the voronoi diagram of a convex polygon. Discrete & Computational Geometry **4** (1989) 591–604

21. Shawe-Taylor, J., Cristianini, N.: Kernel Methods for Pattern Analysis. Cambridge University Press, Cambridge, UK (2004)

Itemset Support Queries Using Frequent Itemsets and Their Condensed Representations

Taneli Mielikäinen[1], Pance Panov[2], and Sašo Džeroski[2]

[1] HIIT BRU, Department of Computer Science, University of Helsinki, Finland
[2] Department of Knowledge Technologies, Jožef Stefan Institute, Ljubljana, Slovenia

Abstract. The purpose of this paper is two-fold: First, we give efficient algorithms for answering itemset support queries for collections of itemsets from various representations of the frequency information. As index structures we use itemset tries of transaction databases, frequent itemsets and their condensed representations. Second, we evaluate the usefulness of condensed representations of frequent itemsets to answer itemset support queries using the proposed query algorithms and index structures. We study analytically the worst-case time complexities of querying condensed representations and evaluate experimentally the query efficiency with random itemset queries to several benchmark transaction databases.

1 Introduction

Discovery of frequent itemsets aims to find all itemsets occurring sufficiently many times in a given transaction database [1,2]. An example of a transaction database is a collection of documents represented as sets of words occurring in them. There, an itemset is a set of words and the frequency of an itemset is the fraction of the documents containing all the words in the itemset. Some other examples of frequent itemsets include sets of products often bought together from a supermarket and sets of web pages often referred together by other web pages. A variety of methods enabling the discovery of large numbers of frequent itemsets have been developed; see [3,4] for a representative collection of the state of the art methods for frequent itemset mining.

Frequent itemsets can be used to summarize data directly or as an intermediate step to construct association rules [1,5]. Frequent itemset collections describing the data in detail tend to be very large and hence they are rather problematic as summaries of the data. Due to such problems major efforts have been done to exclude redundant itemsets from the output, i.e., to obtain a condensed representation of frequent itemsets; see [6,7].

Condensed representations of frequent itemsets are considered also as promising building blocks in inductive databases [8,9,10,11]. One important issue in (inductive) databases is query answering. An inductive database should be able to answer efficiently data mining queries, for example itemset support queries. An itemset support query asks how large fraction of transactions in the given transaction database contain a given itemset.

N. Lavrač, L. Todorovski, and K.P. Jantke (Eds.): DS 2006, LNAI 4265, pp. 161–172, 2006.

In this paper we study the task of answering itemset support queries and examine how the condensed representations of frequent itemsets could be used for such queries. Efficient answering of itemset support queries can be used in many data mining algorithms such as rule induction [12], decision tree construction [13] and learning Naïve Bayes classifiers [14], see [15]. Also, they have a close connection to statistical queries studied in learning theory: an infrastructure for answering itemset support queries can be considered as an efficient implementation of the statistical query oracle for the concept class of conjunctions of positive literals [16]. Furthermore, assessing the applicability of different condensed representations for itemset support queries would be helpful also in focusing the research and use of different condensed representations.

Related work. There has been some work on approximate frequency estimates for and using frequent itemsets. In [17] frequent itemsets are used to induce probabilistic models describing the joint probability distributions of binary datasets. [5,18] study the approximations of the frequencies of boolean formulas in transaction data. Techniques for obtaining frequency approximations based on random subsets of transactions are described in [19,20]. Also the problem of determining exact frequencies of itemsets have been studied. [21] gives algorithms for answering several types queries to a transaction database represented as a tree in a main memory. [22] proposes FP-trees that are trie structures for transaction data developed for frequent itemset mining. A data structure called AD-tree is proposed in [23]. AD-tree represents for all attribute-value combinations occurring in the projections of a relational database their counts in a trie. [24] studies the use of automata to represent itemset collections and their condensed representations. Protocols for private itemset support queries are given in [25]. To the best of our knowledge this is the first systematic study of using condensed representations of frequent itemsets to facilitate the answering of itemset support queries.

Roadmap. Section 2 gives the central concepts in frequent itemset mining. Section 3 defines the itemset support query problem, and describes data structures and algorithms for answering itemset support queries using different condensed representations. The proposed approaches for itemset support query answering are evaluated experimentally in Section 4. Section 5 concludes the paper.

2   Preliminaries

In this section we briefly define the central concepts in frequent itemset mining and their condensed representations. See [2,3,4,6,7] for more details.

Two most important ingredients in frequent itemset mining are transaction databases and itemsets. Transaction databases comprise the data. Itemsets are in dual role as building blocks of transaction databases and representation of the discovered knowledge:

Definition 1. *[Transaction databases and itemsets] A transaction database $\mathcal{D}$ is a set of transactions. A transaction t is a pair $\langle i, X \rangle$ consisting of a transaction* *identifier tid*(*t*) = *i and an itemset is*(*t*) = *X. A transaction identifier is a natural number occurring in at most one transaction in a transaction database. An itemset X is a finite subset of* $\mathcal{I}$*, the set of possible items. The set* $\mathcal{D}_u$ *consists of all different itemsets in* $\mathcal{D}$*, i.e.,* $\mathcal{D}_u = \{X : \langle i, X \rangle \in \mathcal{D}\}$*.*

Example 1. $\mathcal{D} = \{\langle 1, \{1, 2, 3, 4\} \rangle, \langle 2, \{3, 4, 5, 6\} \rangle, \langle 3, \{1, 2, 3, 4, 5, 6\} \rangle\}$ *is an example of a transaction database.*

As examples of itemsets, the itemsets occurring in that database, i.e., the sets in $\mathcal{D}_u$ are $\{1, 2, 3, 4\}$, $\{3, 4, 5, 6\}$, and $\{1, 2, 3, 4, 5, 6\}$.

Also certain statistics computed from the transaction database to itemsets are central for frequent itemset mining:

Definition 2 (Occurrences, counts, covers, supports and frequencies). *The occurrence set of an itemset X in a transaction database* $\mathcal{D}$ *is* $occ(X, \mathcal{D}) = \{i : \langle i, Y \rangle \in \mathcal{D}, X = Y\}$*. The count of X in* $\mathcal{D}$ *is* $count(X, \mathcal{D}) = |occ(X, \mathcal{D})|$*. The cover of X in* $\mathcal{D}$ *is* $cover(X, \mathcal{D}) = \{i : \langle i, Y \rangle, X \subseteq Y\}$*, the support of X in* $\mathcal{D}$ *is* $supp(X, \mathcal{D}) = |cover(X, \mathcal{D})|$ *and the frequency of X in* $\mathcal{D}$ *is* $fr(X, \mathcal{D}) = supp(X, \mathcal{D}) / |\mathcal{D}|$*.*

Using the concepts defined above we can define frequent itemsets:

Definition 3 (Frequent itemsets). *An itemset X is σ-frequent in a transaction database* $\mathcal{D}$ *if* $fr(X, \mathcal{D}) \geq \sigma$*. The collection of all σ-frequent itemsets in* $\mathcal{D}$ *is* $\mathcal{F}(\sigma, \mathcal{D})$*.*

Frequent itemset mining has been a central issue in data mining since its introduction almost 15 years ago [1], many efficient algorithms for frequent itemset mining have been proposed [2,3,4], and similar techniques have been applied to a variety of other types of patterns than itemsets and quality functions than support.

The frequent itemset collections representing the data well are often very large—even larger than the underlying transaction database—and contain much redundant information. To cope with this problem many techniques for removing redundant itemsets from the frequent itemset collection have been proposed, see [6,7].

For example, in many transaction databases there are several frequent itemsets with exactly the same cover. This can be used to partition the collection of frequent itemsets to equivalence classes. From each equivalence class we can select, e.g., all maximal or minimal itemsets (called closed and free itemsets):

Definition 4 (Closed and free frequent itemsets). $X \in \mathcal{F}(\sigma, \mathcal{D})$ *is closed σ-frequent itemset in* $\mathcal{D}$*, if* $fr(X, \mathcal{D}) > fr(Y, \mathcal{D})$ *for all* $Y \supsetneq X$*. The collection of closed σ-frequent itemsets is* $\mathcal{C}(\sigma, \mathcal{D})$*.*

$X \in \mathcal{F}(\sigma, \mathcal{D})$ *is free σ-frequent itemset in* $\mathcal{D}$*, if* $fr(X, \mathcal{D}) < fr(Y, \mathcal{D})$ *for all* $Y \subsetneq X$*. The collection of free σ-frequent itemsets is* $\mathcal{G}(\sigma, \mathcal{D})$*. Free itemsets are known also as generators.*

Example 2. The frequent itemsets in the transaction database given in Example 1 with minimum frequency threshold $2/3$ are the itemsets in $2^{\{1,2,3,4\}} \cup 2^{\{3,4,5,6\}}$, in total 27 itemsets. The closed frequent itemsets comprise in the itemsets $\{1,2,3,4\}$, $\{3,4\}$, and $\{3,4,5,6\}$. The free frequent itemsets are $\emptyset$, $\{1\}$, $\{2\}$, $\{5\}$, and $\{6\}$.

Although there is a many-to-one mapping between free and closed frequent itemsets in $\mathcal{D}$, i.e., that there are always at least as many free frequent itemsets as there are closed frequent itemsets, the collections of free frequent itemsets have the advantage of being downward closed. (An itemset collection $\mathcal{S}$ is downward closed if $X \in \mathcal{S}, Y \subseteq X \Rightarrow Y \in \mathcal{S}$.) This enables the re-use of virtually all known frequent itemset mining algorithms with minor modifications to discover only free frequent itemsets.

The deduction rules for closed and free itemsets to purge the frequent itemset collection are quite simple. Hence, they do not purge the collection very much. Smaller condensed representations can be obtained using more powerful deduction rules, such as Bonferroni inequalities [6,7]:

Definition 5 (Non-derivable frequent itemsets). $X \in \mathcal{F}(\sigma, \mathcal{D})$ *is non-derivable σ-frequent itemset in $\mathcal{D}$, if $\underline{supp}(X, \mathcal{D}) < \overline{supp}(X, \mathcal{D})$ where*

$$\underline{supp}(X, \mathcal{D}) = \max_{Y \subsetneq X, |X \setminus Y| \ odd} \sum_{Y \subseteq Z \subsetneq X} (-1)^{|X \setminus Z|+1} supp(Z, \mathcal{D})$$

$$\overline{supp}(X, \mathcal{D}) = \min_{Y \subsetneq X, |X \setminus Y| \ even} \sum_{Y \subseteq Z \subsetneq X} (-1)^{|X \setminus Z|+1} supp(Z, \mathcal{D})$$

are the greatest lower bound $\underline{supp}(X, \mathcal{D})$ and smallest upper bound $\overline{supp}(X, \mathcal{D})$ for the support of an itemset X given the supports of all subsets of X. The collection of non-derivable σ-frequent itemsets is $\mathcal{N}(\sigma, \mathcal{D})$.

3   Itemset Support Queries

The itemset support query problem asks for the supports of a collection of itemsets w.r.t. some support function $supp|_{\mathcal{S}}$. The problem can be formulated as follows:

Problem 1 (Itemset support query). Given a representation of supports of itemsets in a collection $\mathcal{S}$, and a collection $\mathcal{Q}$ of itemsets, find the supports for all itemsets in $\mathcal{Q} \cap \mathcal{S}$.

In this paper, we focus on the case where the collection $\mathcal{S}$ is equal to $\mathcal{F}(\sigma, \mathcal{D})$ in some transaction database $\mathcal{D}$ and $\mathcal{Q}$ being a single itemset, as the main goal of the current paper is study how well condensed representations of itemset collections are suited to answer itemset support queries.

The representations we consider are $supp|_{\mathcal{F}(\sigma, \mathcal{D})}$, $supp|_{\mathcal{C}(\sigma, \mathcal{D})}$, $supp|_{\mathcal{G}(\sigma, \mathcal{D})}$ $supp|_{\mathcal{N}(\sigma, \mathcal{D})}$ and $count|_{\mathcal{D}_u}$. That is, we use index structures containing supports for all, closed, free and non-derivable frequent itemsets, and the occurrence counts of different itemsets in the database $\mathcal{D}$, respectively, to answer the itemset support queries. We represent itemset collections $\mathcal{S}$ that we use in the itemset support queries as tries, similarly to [26].

Definition 6 (Itemset tries). *An itemset trie for an itemset collection $\mathcal{S} \subseteq 2^{\mathcal{I}}$ is a rooted labeled tree $T(\mathcal{S}) = T = \langle V, E, l : E \to \mathcal{I} \rangle$, such that for each itemset $X \in \mathcal{S}, |X| = k$, there is an unique node $v_X \in V$ such that the labels $label(\langle v_{i-1}, v_i \rangle)$ of the path $\langle root(T), v_1 \rangle, \ldots, \langle v_{k-1}, v_X \rangle$ form an increasing sequence corresponding to the itemset $X = \{label(\langle root(T), v_1 \rangle), \ldots, label(\langle v_{k-1}, v_X \rangle)\}$. The itemset corresponding to a node v is denoted by X_v.*

Example 3. The tries $T(\mathcal{D}_u)$, $T(\mathcal{C}(\sigma, \mathcal{D}))$, $T(\mathcal{G}(\sigma, \mathcal{D}))$ and $T(\mathcal{F}(\sigma, \mathcal{D}))$ of Examples 1 and 2 are as follows:

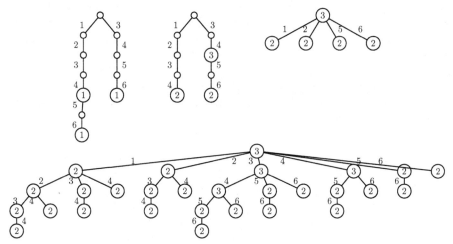

The edges are labeled by the items in the itemsets and the labeled nodes correspond to the itemsets in the collection: the itemset associated to a label node is determined by the edge labels of the path from root to that node, and the node label corresponds to the value of the node.

Itemset tries provide simple representations of itemset collections $\mathcal{S}$ for deciding very efficiently whether a given itemset is in the collection or not. The tries are adapted straightforwardly to answer also many other kinds of queries by adding to each node v in $T(\mathcal{S})$ a value $val(v)$. The exact use of the values of the nodes depend on what kind of function is represented by the trie. For example, $val(v) = count(X_v, \mathcal{D}), X_v \in \mathcal{D}_u$, and $val(v) = supp(X_v, \mathcal{D}), X_v \in \mathcal{S}$ for all $supp|_{\mathcal{S}}, \mathcal{S} \in \{\mathcal{F}(\sigma, \mathcal{D}), \mathcal{C}(\sigma, \mathcal{D}), \mathcal{G}(\sigma, \mathcal{D}), \mathcal{N}(\sigma, \mathcal{D})\}$.

The algorithm for querying the value of a given itemset Q from an itemset trie $T(\mathcal{S})$ is given in Algorithm 1. The algorithm retrieves the value of an itemset Q (represented as an ascending sequence $Q_1, \ldots, Q_{|Q|}$ of items) in time $\mathcal{O}(|Q|)$. We assume that $val(v) = 0$ for all $X_v \notin \mathcal{S}$. Hence, the the value returned for an itemset not in $\mathcal{S}$ is always 0.

Algorithm 1 provides the worst-case optimal solution for querying the values associated to itemsets from an itemset trie representing the collection explicitly.

Algorithm 1 The algorithm for obtaining the value of an itemset Q from $T(\mathcal{S})$ if $Q \in \mathcal{S}$

1: **function** IVQ-EQ($T(\mathcal{S}), Q$)
2:      $u \leftarrow root(T(\mathcal{S}))$
3:      **for all** $j = 1, \ldots, |Q|$ **do**
4:          **if** $child(u, Q_j)$ exists **then**
5:              $u \leftarrow child(u, Q_j)$
6:          **else**
7:              **return** 0
8:      **return** $val(u)$

As the condensed representations of frequent itemsets are proper subcollections of frequent itemsets, Algorithm 1 can be used to query only the supports of the itemsets that appear explicitly in the condensed representations. For example, Algorithm 1 returns non-zero supports only for closed frequent itemsets from $T(\mathcal{C}(\sigma, \mathcal{D}))$ and only for free frequent itemsets from $T(\mathcal{G}(\sigma, \mathcal{D}))$.

The supports of frequent itemsets can be retrieved from the supports of closed and free frequent itemsets, and the counts in $\mathcal{D}_u$ using the formula of form

$$val(X) = \circ\{val(Y) : Y \in \mathcal{S}, Y \supseteq X\} \tag{1}$$

where the operation $\circ$ forms a commutative monoid with some value set M. ($M = \mathbb{N}$ for supports and counts.) Namely, the support of $X \in \mathcal{F}(\sigma, \mathcal{D})$ can be obtained from those representations as follows:

$$supp(X, \mathcal{D}) = \sum_{Y \in \mathcal{D}_u, Y \supseteq X} count(Y, \mathcal{D})$$

$$= \max_{Y \in \mathcal{C}(\sigma, \mathcal{D}), Y \supseteq X} supp(Y, \mathcal{D})$$

$$= \min_{Y \in \mathcal{G}(\sigma, \mathcal{D}), \mathcal{I} \setminus Y \supseteq \mathcal{I} \setminus X} supp(Y, \mathcal{D})$$

Hence, the operations $\circ$ are in these cases $+$, max and $\min^{>0}$, respectively. (These formulas to deduce the supports follow immediately from the definitions of $\mathcal{D}_u$, $\mathcal{C}(\sigma, \mathcal{D})$, and $\mathcal{G}(\sigma, \mathcal{D})$.)

To answer the itemset value queries using Equation 1, the itemset tries $T(\mathcal{S})$ are preprocessed for these queries in such a way that the values $val(v)$ of the nodes v in $T(\mathcal{S})$ are replaced by $\circ\{val(u) : u \in subtrie(v, T(\mathcal{S}))\}$. Such preprocessing can be done in time $\mathcal{O}(|V|)$ for an itemset trie $T = (V, E)$.

Example 4. After the preprocessing, the tries $T(\mathcal{D}_u)$ and $T(\mathcal{C}(\sigma, \mathcal{D}))$ of Example 3 look the following:

Preprocessing replaces the values $val(v)$ in $T(\mathcal{D}_u)$ by $\sum_{u \in subtrie(v, T(\mathcal{S}))} val(u)$ and the values $val(v)$ in $T(\mathcal{C}(\sigma, \mathcal{D}))$ by $\max\{val(u) : u \in subtrie(v, T(\mathcal{S}))\}$.

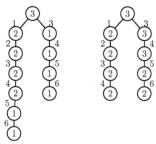

The evaluation of Equation 1 for (preprocessed) itemset tries $T(\mathcal{S})$ is described as Algorithm 2. (Note that the pseudocode omits many lower-level details of the algorithm for the sake of readability. For example, recursive function calls are not be used in practice as they cause some slowdown.)

Algorithm 2 The algorithm for obtaining the value itemset Q from $T(\mathcal{S})$ by Equation 1

1: **function** IVQ-SUPSET($T(\mathcal{S}), Q$)
2:　　**return** IVQ-SUPSET-NODE($T(\mathcal{S}), Q, root(T(\mathcal{S})), 1$)

1: **function** IVQ-SUPSET-NODE($T(\mathcal{S}), Q, v, j$)
2:　　$val \leftarrow 0$
3:　　**for all** $u \in children(v), label(v, u) \leq Q_j$ and $subtrie(u, T(\mathcal{S}))$ can contain the itemset $\{Q_j, \ldots, Q_{|Q|}\}$ **do**
4:　　　　**if** $label(v, u) < Q_j$ **then**
5:　　　　　　$val \leftarrow val \circ$ IVQ-SUPSET-NODE($T(\mathcal{S}), Q, u, j$)
6:　　　　**else if** $j < |Q|$ **then**
7:　　　　　　$val \leftarrow val \circ$ IVQ-SUPSET-NODE($T(\mathcal{S}), Q, u, j + 1$)
8:　　　　**else**
9:　　　　　　$val \leftarrow val \circ val(u)$
10:　　**return** val

The worst-case time complexity of Algorithm 2 is not as good as of Algorithm 1 Namely, in the worst case the whole itemset trie $T(\mathcal{S})$ have to be traversed. This is the case, for example, when $\mathcal{S} = \{\{i, n/2, \ldots, n-1\} : i = 1, \ldots, n/2 - 1\}$ and $Q = \{n\}$. In this example the ordering of the items in the trie makes a great difference in terms of time and space requirements. Even more severe example is the following: $\mathcal{S}\{X \cup \{2n+1\} : X \subseteq \{1, \ldots, 2n\}, |X| = n\}$ and $Q = \{2n + 1\}$.

The performance of the query evaluation can be improved by efficient and effective detection of subtries that cannot have paths containing the query itemset. A few examples of such criteria are $depth(subtrie(u, T(\mathcal{S}))) < |Q| - i$ (i.e., in the subtrie of u there should be a path of length at least $|Q| - i$) and $\max label(subtrie(u, T(\mathcal{S}))) < Q_{|Q|}$ (i.e., it should be possible that the subtrie contains the item $Q_{|Q|}$).

Algorithm 2 is not very appropriate for free frequent itemsets because then the complements of the itemsets must be used. This often leads to much larger itemsets. An alternative of using the complements of the free frequent itemsets is to use a simple modification of Equation 1:

$$val(X) = \circ \{val(Y) : Y \in \mathcal{S}, Y \subseteq X\} \tag{2}$$

The support of $X \in \mathcal{F}(\sigma, \mathcal{D})$ can be obtained from the representations $T(\mathcal{D}_u)$, $T(\mathcal{C}(\sigma, \mathcal{D}))$ and $T(\mathcal{G}(\sigma, \mathcal{D}))$ using Equation 2 by

$$supp(X, \mathcal{D}) = \sum_{Y \in \mathcal{D}_u, \mathcal{I} \setminus Y \subseteq \mathcal{I} \setminus X} count(Y, \mathcal{D})$$

$$= \max_{Y \in \mathcal{C}(\sigma, \mathcal{D}), \mathcal{I} \setminus Y \subseteq \mathcal{I} \setminus X} supp(Y, \mathcal{D})$$

$$= \min_{Y \in \mathcal{G}(\sigma, \mathcal{D}), Y \subseteq X} supp(Y, \mathcal{D})$$

The method implementing Equation 2 is given as Algorithm 3 For this algorithm the itemset tries $T(\mathcal{S})$ should represent the itemset collection and their values as they are, i.e., without any preprocessing.

Algorithm 3 The algorithm for obtaining the value itemset Q from $T(\mathcal{S})$ by Equation 2.

1: **function** IVQ-SUBSET($T(\mathcal{S}), Q$)
2:     **return** IVQ-SUBSET-NODE($T(\mathcal{S}), Q, root(T(\mathcal{S})), 1$)

1: **function** IVQ-SUBSET-NODE($T(\mathcal{S}), Q, v, j$)
2:     $val \leftarrow 0$
3:     **for all** $u \in children(v), label(v, u) \in Q$ **do**
4:         **if** $label(v, u) < Q_{|Q|}$ **then**
5:             $val \leftarrow val \circ$ IVQ-SUBSET-NODE($T(\mathcal{S}), Q, u, \min\{i : Q_i > label(v, u)\}$)
6:         **else**
7:             $val \leftarrow val \circ val(u)$
8:     **return** val

In this case, the time complexity can also be quite high: in the worst case the whole trie $T(\mathcal{S})$ has to be traversed even when no itemset in $\mathcal{S}$ is contained in any itemset in $\mathcal{Q}$. As an example, consider the collection $\mathcal{S} = \{X \cup \{n\} : X \subseteq \{1, \ldots, n - 2\}\}$ and the query $Q = \{1, \ldots, n - 1\}$.

In general case not much additional pruning of unpromising parts of the search trie can be done. In some special cases, however, that is possible. For example, in the case free frequent itemsets, maintaining the smallest support in the subtrie can be used to skip those subtries that cannot have smaller supports than the value obtained that far in the traversal.

Answering itemset support queries using the representation $T(\mathcal{N}(\sigma, \mathcal{D}))$ is slightly more complicated, but can be implemented as follows: (1) query $\mathcal{Q}$ is transformed into downward closed collection $\mathcal{Q}'$, (2) Algorithm 1 is applied to $T(\mathcal{Q}')$, (3) the still undetermined supports of the itemsets in $T(\mathcal{Q}')$ are deduced using Bonferroni-inequalities [27], and (4) Algorithm 1 is applied to retrieve supports of itemsets in $\mathcal{Q}$ from $T(\mathcal{Q}')$.

Unfortunately the proposed algorithm for itemset support queries from the representation based on non-derivable itemsets is not very practical, because the fastest known algorithm to obtain the support of an itemset of size k using Bonferroni-inequalities has the worst-case time complexity $\mathcal{O}(2^k)$ [27]. Hence, we omit non-derivable itemsets from our experiments.

4 Experiments

We examine the performance of itemset query algorithms using random queries to several benchmark databases in frequent itemset mining from the FIMI transaction database repository (http://fimi.cs.helsinki.fi). To see the potential differences in the performance of different representations, we selected 8 transaction databases where the number of all frequent itemsets is higher than the number of closed or free frequent itemsets. For each database $\mathcal{D}$ we used a minimum support threshold $\sigma |\mathcal{D}|$ that produces a large number of frequent itemsets, because the goal of using itemset collections as index structures for itemset support queries is to be able to answer as large fraction of the queries as possible. This differs somewhat from the classical use of frequent itemsets and association rules as a small number of nuggets of knowledge. The basic statistics of the selected databases are shown in Table 1.

Table 1. The transaction databases used in the experiments. The columns are the transaction database name, the minimum support threshold, the number frequent items, the number of transactions, the number of different transactions after removing infrequent items, the number of frequent itemsets, the number of free frequent itemsets, and the number of closed frequent itemsets.

$\mathcal{D}$	$|\mathcal{I}|$	$|\mathcal{D}|$	$\sigma |\mathcal{D}|$	$|\mathcal{I}^{\geq\sigma}|$	$|\mathcal{D}_u^{\geq\sigma}|$	$|\mathcal{F}(\sigma,\mathcal{D})|$	$|\mathcal{G}(\sigma,\mathcal{D})|$	$|\mathcal{C}(\sigma,\mathcal{D})|$
CONNECT	129	67 557	50 000	30	214	1 928 335	26 417	26 417
KOSARAK	41 270	990 002	900	1 384	387 603	1 598 294	1 143 594	1 124 819
MUSHROOM	119	8 124	500	67	7 032	1 442 503	14 925	9 864
PUMSB	2 113	49 046	35 000	34	3305	1897479	519 725	194 538
PUMSB*	2 088	49 046	13 000	63	25 404	1 293 828	57 172	32 115
RETAIL	16 470	88 162	5	10 988	83 119	1 506 775	532 342	504 142
WEBVIEW-1	497	59 601	35	369	18 184	1 177 607	118 696	76 260
WEBVIEW-2	3 340	77 511	15	2 643	48 117	1 599 210	397 283	343 818

We selected randomly 1 000 000 itemsets from $\mathcal{F}(\sigma, \mathcal{D})$ with uniform distribution over the collection to assess average-case efficiency of different representations of the frequent itemsets to answer itemset support queries. As a baseline representation we used the representation of $\mathcal{D}_u^{\geq\sigma}$ as list of itemsets together with their occurrence counts. The sizes of the trie representations and the average query answering performances for $1\,000\,000$ random itemset support queries from the collection $\mathcal{F}(\sigma, \mathcal{D})$ for the databases of Table 1 are shown in Table 2.

Table 2. The space requirements of different trie representations and their query answering time speedups w.r.t. a list representation of $\mathcal{D}_u^{\geq \sigma}$ for $1\,000\,000$ random itemset support queries selected from frequent itemsets.

| | the sizes of the tries in KBs | | | | the speed-ups in the query times | | | |
$\mathcal{D}$	$\mathcal{D}_u^{\geq \sigma}$	$\mathcal{F}(\sigma, \mathcal{D})$	$\mathcal{G}(\sigma, \mathcal{D})$	$\mathcal{C}(\sigma, \mathcal{D})$	$\mathcal{D}_u^{\geq \sigma}$	$\mathcal{F}(\sigma, \mathcal{D})$	$\mathcal{G}(\sigma, \mathcal{D})$	$\mathcal{C}(\sigma, \mathcal{D})$
CONNECT	36	75 326	1 032	1 032	14.4	7.0	1.1	1.4
KOSARAK	117 735	62 433	44 672	44 286	0.8	8 705.7	170.8	65.7
MUSHROOM	776	56 348	583	420	29.5	244.7	71.6	126.2
PUMSB	752	74 120	20 302	10 416	43.9	117.8	6.2	5.6
PUMSB*	6 695	50 540	2 233	1 395	41.1	1 505.3	248.1	302.9
RETAIL	26 024	58 858	20 795	20 240	1.5	1 759.3	35.0	6.4
WEBVIEW-1	2 166	46 000	4 637	3 308	5.2	320.5	38.0	11.4
WEBVIEW-2	8 692	62 469	15 519	13 950	22.5	820.8	24.8	11.7

The trie representations show often considerable speedup compared to the baseline approach. The experimental results show also that answering itemset support queries to $T(\mathcal{D}_u^{\geq \sigma})$, $T(\mathcal{C}(\sigma, \mathcal{D}))$ and $T(\mathcal{G}(\sigma, \mathcal{D}))$ are often much slower than answering them using $T(\mathcal{F}(\sigma, \mathcal{D}))$. Especially queries to $T(\mathcal{D}_u^{\geq \sigma})$ for KOSARAK were considerable slower than to the other representations, even slower than the baseline representation. One possible explanation is that the branching factor of the trie $T(\mathcal{D}_u^{\geq \sigma})$ is higher than in $T(\mathcal{F}(\sigma, \mathcal{D}))$, $T(\mathcal{C}(\sigma, \mathcal{D}))$ and $T(\mathcal{G}(\sigma, \mathcal{D}))$. Furthermore, the pruning conditions in Algorithm 2 are more expensive than in the baseline approach. If the pruning fails, then the pruning attempts result just slowdown.

In general, condensed representations of frequent itemsets seem to offer a reasonable alternative for itemset support queries using all frequent itemsets only when very few frequent itemsets are contained in the condensed representation. (Note that in the case of $T(\mathcal{G}(\sigma, \mathcal{D}))$ and $T(\mathcal{C}(\sigma, \mathcal{D}))$ the query optimization of first checking whether the itemset is in the collection could improve the performance considerably.) The preliminary experimental results are promising, but more comprehensive evaluation of itemset support queries is needed.

5   Conclusions

In this paper we studied the use of frequent itemsets, the condensed representations of the frequent itemsets, and the concise representations of transaction databases for answering itemset support queries. Answering such queries is an important task in inductive databases for performing interactive data mining queries and as building blocks of data mining algorithms. We proposed efficient trie structures for representing itemset collections and the answering itemset support queries. We evaluated experimentally the applicability of the major condensed representations of frequent itemsets to answer itemset support queries. Trie representations seem to offer a reasonable approach for facilitating itemset support querying.

As a future work we plan to work on efficient indices for querying approximate representations of frequent itemsets. Some representations, e.g., [19,28], fit readily to itemset tries, but it is not clear what would be the best representation for answering multiple itemset support queries. Another potential line of research is of hierarchical descriptions of data to facilitate the efficient query answering. For example, using the projections of the transaction database to the maximal frequent itemsets can result significant space saving compared other condensed representations [29]. Querying projections with only a very small number of different transactions is fast. Most likely there is no particular represenation that would be always the best for all itemset support queries, but different indices suit for different kinds of transaction databases. This suggests to develop techniques for finding most efficient representation for a particular database and distribution queries. Furthermore, combining different representations for a single transaction databases have some promise in both helping to comprehend the underlying itemset collections and speeding up the itemset support queries.

References

1. Agrawal, R., Imielinski, T., Swami, A.N.: Mining association rules between sets of items in large databases. In Buneman, P., Jajodia, S., eds.: SIGMOD Conference. (1993) 207–216
2. Goethals, B.: Frequent set mining. In Maimon, O., Rokach, L., eds.: The Data Mining and Knowledge Discovery Handbook. Springer (2005) 377–397
3. Goethals, B., Zaki, M.J., eds.: FIMI '03, Frequent Itemset Mining Implementations, Proceedings of the ICDM 2003 Workshop on Frequent Itemset Mining Implementations, 19 December 2003, Melbourne, Florida, USA. Volume 90 of CEUR Workshop Proceedings. (2003)
4. Bayardo Jr., R.J., Goethals, B., Zaki, M.J., eds.: FIMI '04, Proceedings of the IEEE ICDM Workshop on Frequent Itemset Mining Implementations, Brighton, UK, November 1, 2004. Volume 126 of CEUR Workshop Proceedings. (2004)
5. Mannila, H., Toivonen, H.: Multiple uses of frequent sets and condensed representations (extended abstract). In: KDD. (1996) 189–194
6. Calders, T., Rigotti, C., Boulicaut, J.F.: A survey on condensed representations for frequent sets. [30] 64–80
7. Mielikäinen, T.: Transaction databases, frequent itemsets, and their condensed representations. [31] 139–164
8. Boulicaut, J.F.: Inductive databases and multiple uses of frequent itemsets: The cInQ approach. In Boulicaut, J.F., Raedt, L.D., Mannila, H., eds.: Database Support for Data Mining Applications. Volume 3848 of LNCS. (2004) 1–23
9. Imielinski, T., Mannila, H.: A database perspective on knowledge discovery. Communications of the ACM **39** (1996) 58–64
10. Mannila, H.: Inductive databases and condensed representations for data mining. In: ILPS. (1997) 21–30

11. Siebes, A.: Data mining in inductive databases. [31] 1–23
12. Clark, P., Niblett, T.: The CN2 induction algorithm. Machine Learning **3** (1989) 261–283
13. Quinlan, J.R.: C4.5: Programs for Machine Learning. Morgan Kaufmann (1992)
14. Maron, M.E.: Automatic indexing: An experimental inquiry. J. ACM **8** (1961) 404–417
15. Panov, P., Džeroski, S., Blockeel, H., Loškovska, S.: Predictive data mining using itemset frequencies. In: Proceedings of the 8th International Multiconference Information Society. (2005) 224–227
16. Kearns, M.J.: Efficient noise-tolerant learning from statistical queries. J. ACM **45** (1998) 983–1006
17. Pavlov, D., Mannila, H., Smyth, P.: Beyond independence: Probabilistic models for query approximation on binary transaction data. IEEE Transactions on Knowledge and Data Engineering **15** (2003) 1409–1421
18. Seppänen, J.K., Mannila, H.: Boolean formulas and frequent sets. [30] 348–361
19. Mielikäinen, T.: Separating structure from interestingness. In Dai, H., Srikant, R., Zhang, C., eds.: PAKDD. Volume 3056 of LNCS. (2004) 476–485
20. Toivonen, H.: Sampling large databases for association rules. In Vijayaraman, T.M., Buchmann, A.P., Mohan, C., Sarda, N.L., eds.: VLDB. (1996) 134–145
21. Kubat, M., Hafez, A., Raghavan, V.V., Lekkala, J.R., Chen, W.K.: Itemset trees for targeted association querying. IEEE Transactions on Knowledge and Data Engineering **15** (2003) 1522–1534
22. Han, J., Pei, J., Yin, Y., Mao, R.: Mining frequent patterns without candidate generation: A frequent-pattern tree approach. Data Min. Knowl. Discov. **8** (2004) 53–87
23. Moore, A.W., Lee, M.S.: Cached sufficient statistics for efficient machine learning with large datasets. JAIR **8** (1998) 67–91
24. Mielikäinen, T.: Implicit enumeration of patterns. [32] 150–172
25. Laur, S., Lipmaa, H., Mielikäinen, T.: Private itemset support counting. In Qing, S., Mao, W., Lopez, J., Wang, G., eds.: ICICS. Volume 3783 of LNCS. (2005) 97–111
26. Mielikäinen, T.: An automata approach to pattern collections. [32] 130–149
27. Calders, T., Goethals, B.: Quick inclusion-exclusion. [31] 86–103
28. Geerts, F., Goethals, B., Mielikäinen, T.: What you store is what you get. [33] 60–69
29. Mielikäinen, T.: Finding all occurring patterns of interest. [33] 97–106
30. Boulicaut, J.F., Raedt, L.D., Mannila, H., eds.: Constraint-Based Mining and Inductive Databases, European Workshop on Inductive Databases and Constraint Based Mining, Hinterzarten, Germany, March 11-13, 2004, Revised Selected Papers. Volume 3848 of LNCS. (2005)
31. Bonchi, F., Boulicaut, J.F., eds.: Knowledge Discovery in Inductive Databases, 4th International Workshop, KDID 2005, Porto, Portugal, October 3, 2005, Revised Selected and Invited Papers. Volume 3933 of LNCS. (2006)
32. Goethals, B., Siebes, A., eds.: KDID 2004, Knowledge Discovery in Inductive Databases, Proceedings of the Third International Workshop on Knowledge Discovery in Inductive Databases, Pisa, Italy, September 20, 2004, Revised Selected and Invited Papers. Volume 3377 of LNCS. (2005)
33. Boulicaut, J.F., Dzeroski, S., eds.: Proceedings of the Second International Workshop on Inductive Databases, 22 September, Cavtat-Dubrovnik, Croatia. (2003)

Strategy Diagram for Identifying Play Strategies in Multi-view Soccer Video Data

Yukihiro Nakamura[1], Shin Ando[1], Kenji Aoki[1],
Hiroyuki Mano[1], and Einoshin Suzuki[2]

[1] Electrical and Computer Engineering, Yokohama National University, Japan
{nakamu, ando, aokikenji, mano}@slab.dnj.ynu.ac.jp
[2] Department of Informatics, Graduate School of Information Science and Electrical
Engineering, Kyushu University, Japan
suzuki@i.kyushu-u.ac.jp

Abstract. In this paper, we propose a strategy diagram to acquire knowledge of soccer for identifying play strategies in multi-view video data. Soccer, as the most popular team sport in the world, attracts attention of researchers in knowledge discovery and data mining and its related areas. Domain knowledge is mandatory in such applications but acquiring domain knowledge of soccer from experts is a laborious task. Moreover such domain knowledge is typically acquired and used in an ad-hoc manner. Diagrams in textbooks can be considered as a promising source of knowledge and are intuitive to humans. Our strategy diagram enables a systematic acquisition and use of such diagrams as domain knowledge for identifying play strategies in video data of a soccer game taken from multiple angles. The key idea is to transform multi-view video data to sequential coordinates then match the strategy diagram in terms of essential features. Experiments using video data of a national tournament for high school students show that the proposed method exhibits promising results and gives insightful lessons for further studies.

1 Introduction

Soccer is said to be the world's most popular team sport. For instance, the World Cup is known to be the biggest sport event as the number of TV viewers in the world is larger than that for the Olympic Games. Soccer has recently attracted attention of researchers in various fields related with machine learning [1,3,5,6,9,10,11,12,13]. Most of them perform multiple objects tracking thus provide fundamental facilities for analyzing a soccer game. Several papers, however, are devoted to concept learning from a soccer game thus open the possibility of adaptive analysis. [5] clusters pass patterns from manually inputted coordinates and [12] learns classifiers which predict a play break, a pass, and a goal from a video clip. We have proposed decision tree induction for play evaluation and defensive/attack role analysis from a set of image sequences [1]. [3] goes one step further, as it proposes a fully automatic method for analyzing and summarizing soccer videos thus covers many of the problems tackled by other studies.

N. Lavrač, L. Todorovski, and K.P. Jantke (Eds.): DS 2006, LNAI 4265, pp. 173–184, 2006.

From our experience in [1], the most difficult tasks are image processing and knowledge acquisition. For the latter task, concepts specific to soccer such as pressing and marking are mandatory in any non-trivial analysis of a soccer game. The process of acquiring and implementing such concepts is laborious as they have to be treated carefully and separately. The most difficult case is when the concept represents a play strategy i.e. it is related with tactic movements of several players. We anticipate that other researchers have met similar situations as no systematic method has been proposed in the references.

As shown in the research field of diagrammatic reasoning [4], it is widely believed that using diagrams instead of symbols reduces cognitive labor of the user considerably. In this paper, we propose a method with which we acquire play strategies of soccer from diagrams in a textbook of soccer and uses such strategies in a systematic manner. The effectiveness of the method is evaluated with a task of identifying play strategies from video data of a soccer game. We consider that our method reduces the burden of coding soccer strategies in various applications.

2   Generation of Sequential Coordinate Data from Video Data

2.1   Overall Flow

We filmed two semi-final games of Kanagawa Prefecture preliminaries in 2004 National High-School Championship Tournament in Japan under permission of the organizers. Seven fixed video camera recorders (VCRs) were used as shown in Figure 1 (a). The digits in the Figure represent the identifiers of VCRs. From the video data, we sampled, per second, 25 still bit-map images each of which size is 352×240 (pixel). We show in Figure 1 (b) examples of still images, where the digits represent identifiers of VCRs. Using the method in section 2.2, seven still images are transformed into a set of coordinates of players and the ball in the field as shown in Figure 1 (c).

In our spatio-temporal data, a player is represented as a point on the 2-dimensional field and so is the ball. Each player is allocated with an identifier, from which his/her team can be also identified. This representation neglects various information including the posture of a player but still contains information essential to a soccer game such as trajectories and space[1].

Formally, sequential coordinate data C represents a sequence $c_1, c_2, \ldots$ of coordinates c_t, where t represents a time tick. Coordinates c_t is described in terms of the players $\alpha_1, \alpha_2, \ldots, \alpha_{11}$ of a team, the players $\alpha_{12}, \alpha_{13}, \ldots, \alpha_{22}$ of the other team, and the ball β. Let the x coordinate and the y coordinate of an object γ at t be $x(\gamma, t)$ and $y(\gamma, t)$, respectively, then $c_t = (x(\alpha_1, t), y(\alpha_1, t), x(\alpha_2, t), y(\alpha_2, t), \ldots, x(\alpha_{22}, t), y(\alpha_{22}, t), x(\beta, t), y(\beta, t))$.

---

[1] One possible extension is to employ a 3-dimensional point for the ball to represent an air ball but we leave this for future work.

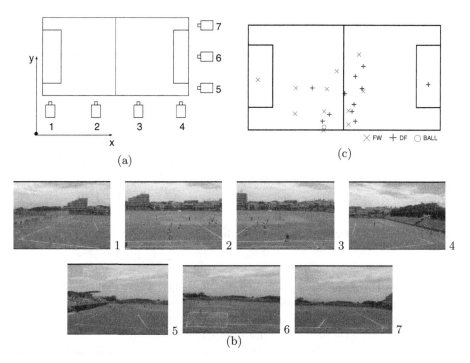

Fig. 1. (a) Positions of VCRs, (b) 7 images as input, (c) sequential coordinate data as output

2.2 Transforming Multi-view Video Clips to Sequential Coordinate Data

We have modified our semi-automatic method in [1] which processes each image to obtain coordinates of the object in the image then transforms the coordinates of all images into a sequence of coordinates on the field for each player and the ball. For image processing, we showed an automatic rule-based method which extracts the field, recognizes objects, and identifies teams and the ball in [1]. The method exhibits good performance but shows some errors as an object is sometimes as small as three pixels and the grass on the field is damaged as seen in Figure 1 (b). To obtain more reliable results, we asked a collaborator to manually mark players and the ball in each still image as objects and process the mark automatically to obtain coordinates on the image.

For coordinate transformation, the method in [1] agglomerates the seven sets of coordinates for the seven VCRs and corrects the set of coordinates. An obtained coordinate (X, Y) in an image filmed with a VCR is transformed to a coordinate (x, y) on the field. We assume that a filmed image is projected on a screen located between the field and a VCR as shown in Figure 2. This transformation represents a combination of Affine-transformation and perspective projection thus can be expressed as follows.

$$X = \frac{ax + by + c}{px + qy + r}, \qquad Y = \frac{dx + ey + f}{px + qy + r}$$

where $a, b, c, d, e, f, p, q, r$ represent coefficients which can be calculated with parameters in the transformation.

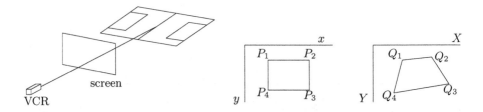

Fig. 2. Image as a projection on a screen

Fig. 3. Points for calculating parameters

In order to calculate these coefficients, we specified a rectangle $P_1 P_2 P_3 P_4$ on the xy plane and the corresponding square on the XY plane $Q_1 Q_2 Q_3 Q_4$ as shown in Figure 3. A simple calculation shows that we can assume $r \neq 0$ thus we have obtained formula for transforming (X, Y) to (x, y).

Next we combine coordinates on the xy plane obtained with the seven VCRs. In this procedure, we use a procedure agglo(c_{at}, c_{bt}) for agglomerating coordinates c_{at}, c_{bt} of frames s_{at}, s_{bt} filmed at time t with cameras a and b, respectively. Let $c_{it} = \{p_{it1}, p_{it2}, \ldots, p_{itm(it)}\}$, where p_{itj} represents a position, i.e. coordinates of a player j in c_{it} and $m(it)$ represents the number of players identified in c_{it}. We obtain a Euclidean distance for each pair $(p_{it\alpha}, p_{jt\beta})$ for $\alpha = 1, 2, \ldots, m(it)$ and $\beta = 1, 2, \ldots, m(jt)$ and consider that a pair represents different players if the value is no smaller than R. The correspondence which shows the minimum value for the add-sum of the distances is chosen with depth-first search. If a corresponding player is not found in a set, the coordinates in the other set is used.

For two VCRs located side by side in Figure 1 (a) such as VCRs 1 and 2 or VCRs 5 and 6, we identify each player and the ball with this method and use the average value if the values are different. More precisely, we use agglo(agglo(c_{1t}, c_{2t}), agglo(c_{3t}, c_{4t})) and agglo(agglo(c_{5t}, c_{6t}), c_{7t}) as agglomerated coordinates for VCRs 1 - 4 and VCR 5 - 7, respectively, where we use $R = 12$ pixel.

The agglomeration of the two sets of coordinates above needs to handle discrepancies and occlusions. We use $R = 16$ pixel and in case of discrepancies we use the x coordinate of VCRs 1-4 and the y coordinate of VCRs 5-7 to exploit data of better precision. In case of occlusions, we avoid conflicts in the agglomeration process by using the same x or y coordinate of an occluding player for the occluded player. We describe the set of coordinates obtained with the procedure by c'_t.

Since an object can be surrounded by other objects and the coordinate agglomeration can fail, the number of identified players varies during a game. To circumvent this problem, we use the following method to correct c'_t into c'''_t and keep the number to be always 22. Let t be the identifier of the starting frame.

1. Obtain the set c''_t of coordinates of all players at time t manually.
2. $c'''_t = \text{agglo}(c'_t, c''_t)$
3. $c'''_{t+1} = \text{agglo}(c'_{t+1}, c'''_t)$
4. Repeat the procedures 2 and 3 by incrementing t.

The transformation task is difficult even for sophisticated methods in computer vision due to various reasons such as occlusions and the size of the ball can be as small as three pixels. Thus we consider manual revision mandatory and have developed an interface which transforms the obtained coordinates $c'''_1, c'''_2, \ldots, c'''_T$ into a tgif[2] file. The final sequence $(c_1, c_2, \ldots, c_T)$ is obtained by this manual correction procedure.

3   Identifying Play Strategies with the Strategy Diagram

3.1   Strategy Diagram

A strategy diagram represents a play strategy in a soccer game. In a strategy diagram, a circle and an arrow represent a player and an action (pass or run), respectively. In this paper, we use [8] as a textbook of soccer for drawing strategy diagrams with TGIF. In Figure 4, we show, in a half field, an example of a strategy diagram which represents a play strategy of changing the field sides of the ball with two passes. We hereafter call the play strategy in the Figure a 2-step side change and distinguish from a 1-step side change in which players change the field sides of the ball with one pass. As we see from the Figure, a player of the attacking team without the ball, a player of the attacking team with the ball, and a player of the defending team are represented by a shaded circle (FW), a black circle (FW_BALL), and a white circle (DF), respectively. A dribble and a pass are represented by a short dotted arrow (RUN) and a long dotted arrow (PASS), respectively.

As a strategy diagram describes movements of the players and the ball, it represents their positions at time ticks $q = 1, 2, \ldots$. We define the coordinates of the players and the ball in a strategy diagram as we did for sequential coordinate data in section 2. It should be noted that the number $\nu(D)$ of players in a strategy diagram D is typically smaller than 22 because a play strategy can be defined in terms of a subset of the players. We describe coordinates of the players at time tick $q = 1$ unless the circle of the player is not directed by a short dotted arrow (RUN). A short dotted arrow (RUN) is drawn from $(x(\alpha_i, q), y(\alpha_i, q))$ to $(x(\alpha_i, q+1), y(\alpha_i, q+1))$.

In a strategy diagram, not all objects are essential and not all features are essential. For example, in Figure 4, essential players are the three players of the attacking team on the left-hand side while essential features are the field-side positions and the pass of the players. We specify essential circles and essential arrows with red and distinguish from other circles and arrows in black. A small pop-up window displays features for each essential player and the user specifies features which s/he

---

[2] http://bourbon.usc.edu:8001/tgif/

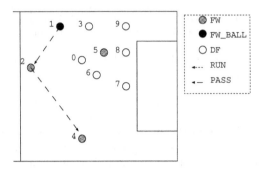

Fig. 4. Strategy diagram for the 2-step side change

considers essential with red. Each essential player has its own essential features at a specific time.

We propose seven essential features which describe plays in soccer such as "marking". We show in Figure 5 a half field and define the x, y axes and areas in the half field. We define, on the goal line, the coordinates of the lower corner, the coordinates of the upper corner, and the bottom corner of the goal mouse as $(x_r, y_r), (x_r, y_l)$ and (x_r, y_g), respectively.

A player α_j at time q can have at most seven values for seven essential features $p_1(\alpha_j, q), p_2(\alpha_j, q), \ldots, p_7(\alpha_j, q)$ which are shown in Table 1. We employ six thresholds $l_1, l_2, \ldots, l_6$ in defining the essential features. The first feature $p_1(\alpha_j, q)$ represents the position of α_j at q along the y axis. The second feature $p_2(\alpha_j, q)$ represents the direction of α_j at q and is defined by the intersection (x_r, y_d) of the goal line and the direction. The third feature $p_3(\alpha_j, q)$ represents a pass, where $D(\gamma, \gamma', q)$ represents the distance between a pair of objects γ and γ' at q. The fourth feature $p_4(\alpha_j, q)$ represents the marked player j by α_j at q. The meanings of the remaining features $p_5(\alpha_j, q), p_6(\alpha_j, q)$, and $p_7(\alpha_j, q)$ are clear from the Table, where $D'(\gamma, q, q+1)$ represents the distance of the positions of object γ at q and $q+1$. Each of the features $p_3(\alpha_j, q), p_4(\alpha_j, q), \ldots, p_7(\alpha_j, q)$ has no value if α_j does not satisfy the conditions in the Table at q.

3.2 Play Strategy Identification from Sequential Coordinate Data

To represent essential movements of objects in a play strategy, a strategy model $Z(D)$ is extracted from a strategy diagram D with essential features. An essential player α_j at time tick q has a set $P(\alpha_j, q)$ of values for essential attributes, where a missing value is not included in $P(\alpha_j, q)$. A strategy model $Z(D)$ consists of all $P(\alpha_j, q)$ for essential players in D.

For example, the strategy model extracted from the strategy diagram in Figure 4 is shown in Table 2. The Table represents the strategy model for the 2-step side change: FW1 in the right side with the ball issues a pass at $q = 0$, FW2 in the middle receives the pass and issues a pass at $q = 1$, then FW4 in the left side receives the pass at $q = 2$. This model is consistent with the explanation in the textbook [8].

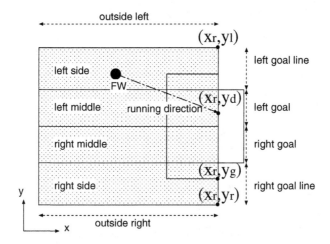

Fig. 5. Areas in a half field

A set $P(\alpha_i, t)$ of values for essential attributes of a player α_i at time tick t is obtained from sequential coordinate data. Therefore, sequential coordinate data are transformed into a set Y which consists of $P(\alpha_i, t)$ for all α_i at t. Intuitively, a play strategy is identified if $Z(D)$ is included in Y.

The procedure for identifying a play strategy performs a linear scans for time ticks of the sequential coordinate data to find matching scenes in the strategy model. Note that players who are essential for the strategy are also identified in the corresponding scene. The input are sequential coordinate data C, a strategy model $Z(D)$, and the maximum time ticks $q_{\max}$ in the search. If the elements of $Z(D)$ at $q = 1$ are all included in Y at time tick t, the procedure searches for t' at which Y includes the elements of $Z(D)$ at $q = 2$, where $t' = t+1$, $t+2$, ..., $t+5$. The procedure is iterated for the remaining time ticks q in $Z(D)$.

4   Experimental Evaluation

4.1   Experiments

In the experiments, we sampled a still image for each second from the video data in section 2. The six play strategies and the corresponding numbers of selected scenes are shown in Table 3. In the Table, "scenes" and "near misses" correspond to the scenes that match the play strategy and the scenes that are judged similar but do not match the play strategy, respectively. In the Table, the first three play strategies are 2-time tick length while the remaining three play strategies are 3-time tick length. We show in Figure 6 an example of sequential coordinate data of a scene of the 2-step side change, where we limit the field to the relevant areas and describe the time tick in the upper-left.

Table 1. Definitions of essential features

Feature	Value	Condition
p_1	"outside right"	$y(\alpha_j, q) < y_r$
	"right side"	$y_r \leq y(\alpha_j, q) < \frac{y_l + 3y_r}{4}$
	"right middle"	$\frac{y_l + 3y_r}{4} \leq y(\alpha_j, q) < \frac{y_l + y_r}{2}$
	"left middle"	$\frac{y_l + y_r}{2} \leq y(\alpha_j, q) < \frac{3y_l + y_r}{4}$
	"left side"	$\frac{3y_l + y_r}{4} \leq y(\alpha_j, q) < y_l$
	"outside left"	$y_l < y(\alpha_j, q)$
p_2	"outside right"	$y_d < y_r$
	"right goal line"	$y_r \leq y_d < \frac{y_l + 3y_r}{4}$
	"right goal"	$\frac{y_l + 3y_r}{4} \leq y_d < \frac{y_l + y_r}{2}$
	"left goal"	$\frac{y_l + y_r}{2} \leq y_d < \frac{3y_l + y_r}{4}$
	"left goal line"	$\frac{3y_l + y_r}{4} \leq y_d < y_l$
	"outside left"	$y_l < y_d$
p_3	"issue a pass"	$l_1 < D(\alpha_j, \beta, q)$ and α_j has the ball at q
	"receive a pass"	α_j has the ball at $q+1$ and α_j does not have the ball at q
p_4	"mark j'"	$D(\alpha_j, \alpha_{j'}, q) \leq l_2$ and α_j is in the triangle $(j$, both sides of the goal mouse$)$ at q or $D(\alpha_j, \alpha_{j'}, q) \leq l_3$
p_5	"exist in space"	$D(\alpha_j, \alpha_{j'}, q) \geq l_4$ for any DF j'
p_6	"move long"	$D'(\alpha_j, q, q+1) \geq l_5$
p_7	"intersect with j'"	$D(\alpha_j, \alpha_{j'}, q) \leq l_6$

Table 2. Strategy model of 2-step side change extracted from Figure 4

q	player	$P(\alpha_j, q)$
0	FW1 (possess the ball)	p_1="right side", p_3="issue a pass"
	FW2	
	FW4	
1	FW1	
	FW2 (possess the ball)	p_3="receive a pass", p_3="issue a pass", p_1="right middle"
	FW4	
2	FW1	
	FW2	
	FW4 (possess the ball)	p_1="left side", p_3="receive a pass"

For each play strategy of 2-time tick length, five non-specialists of soccer drew his own strategy diagram. The strategy diagrams drawn were identical to those in the textbook in most of the cases. Each subject could easily draw strategy diagrams in moderate time and reported that the process was easy. We noticed that several subjects drew diagrams inappropriately for the pressing strategy due to a careless misinterpretation of the textbook [8]. The diagram in the textbook shows three defenders for making a pressure on the player with the ball but it is explained in the text that plural defenders make a pressure. They should specify two defenders instead of three to make their diagrams consistent with the explanation of the textbook. We excluded the diagrams from the experiments because it suffices to notice the user of our method to specify two players for plural players. Throughout

Table 3. Numbers of employed scenes

Strategy	# of scenes	# of near misses
Overlap	3	2
Pressing	3	1
1-step side change	3	0
2-step side change	3	0
1-2 pass	1	1
Switch play	2	1

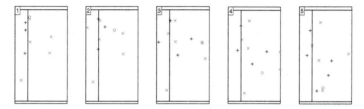

Fig. 6. Sequential coordinate data for the 2-step side change

the experiments, we used threshold values as follows: $l_1 = 10$ (3.3m), $l_2 = 30$ (9.9m), $l_3 = 15$ (4.95m), $l_4 = 30$ (9.9m), $l_5 = 40$ (13.2m), $l_6 = 10$ (3.3m)[3].

We show the experimental results in Table 4 and several examples of the output sequential coordinate data in Figure 7. In the Table, x/y represents that our method succeeded in detecting the play strategy with related players in x scenes out of y scenes. In the Figure, the coordinates of the essential players and the ball are shown using arrows while the coordinates of other players are for the time tick $q = 0$.

Table 4. Experimental result

Strategy	Scenes	Near misses	Other detected strategies
Overlap	3/3	2/2	Pressing 4/4, 1-2 pass 1/1
Pressing	3/3	0/1	
1-step side change	2/3		Pressing 2/3
2-step side change	3/3		Pressing 3/3
1-2 pass	1/1	1/2	Pressing 1/2
Switch play	2/2	1/1	Pressing 3/3

4.2   Discussions

We see from Table 4 that our method has succeeded in identifying the right play strategy accurately in most of the cases. Moreover, other strategies, most of which were unexpected to us, were identified by our method. This fact supports the usefulness of our method because these strategies, especially pressing, had been unnoticed in our inspection. We attribute the good performance to the adequacy of

---

[3] These values should be determined for each filming condition.

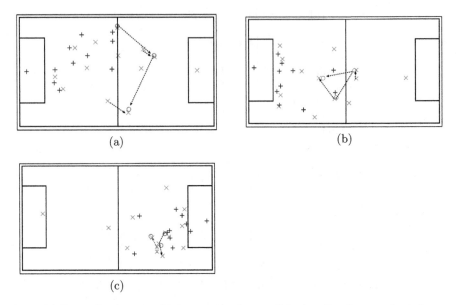

Fig. 7. (a) Example of output for 2-step side change, (b) example of output for 1-2 pass, (c) example of output for switch play

our strategy diagram in representing a play strategy and the adequacy of the play strategies to the actual movements of the players and the ball. Figure 7 supports our belief.

The Table shows, however, that our method is not free from errors. We consider that most of the problems comes from the subjective nature of the "space" and the "ball ownership" of a player. For instance, pressing is judged according to various factors especially the relative positions and the velocities of players and the ball. This analysis suggests that a machine learning approach is worth pursuing for this kind of problem.

Anyway, due to the complex nature of the scenes and the previous experience of hard-coding soccer concepts, we are satisfied with the results. It is difficult to compare the efforts spent for brute-force coding and those for our method accurately, but we feel that our method requires far less efforts. Annotating scenes with play strategies gives a non-trivial analysis program opportunities to enrich their analysis exploiting concepts of soccer as we humans do.

We have investigated the possibility of applying our method for other play strategies in the textbook [8]. It turned out that 41 out of 44 of the play strategies can be included in our method in the current framework by adding new essential attributes. The remaining three play strategies cannot be included because they use information which are not included in the sequential coordinate data such as gestures of players. We believe that our strategy diagram is an effective method for acquiring knowledge for play strategies of soccer in the level of sequential coordinate data.

5   Conclusions

Sport data have been an appealing target for AI programs e.g.[1-3,5-7,9-13]]. Soccer seems most appealing to us because it is the world's most popular team sport and can be naturally modeled as spatial-temporal data of coordinates due to the large size of the field. We have proposed a method for acquiring and employing play strategies in textbooks for identifying the play strategies used in multi-view soccer video data. The method will serve as a fundamental but essential method for various intelligent applications including knowledge discovery and data mining. We plan to make our spatio-temporal data public for academic and educational use in order to contribute to extending soccer analysis to a wide range of researchers.

Acknowledgments

This work was partially supported by the grants-in-aid for scientific research on fundamental research (B) 16300042 and 18300047 from the Japanese Ministry of Education, Culture, Sports, Science and Technology. We are grateful to Kanagawa Football Association for kindly letting us record the soccer games and permit academic and educational use. We appreciate Kazuhisa Sakakibara for negotiating with the association.

References

1. K. Aoki, H. Mano, Y. Nakamura, S. Ando, and E. Suzuki: "Mining Multiple Video Clips of a Soccer Game", *Proc. First International Workshop on Mining Complex Data (MCD)*, pp. 17–24, 2005.
2. I. S. Bhandari, E. Colet, J. Parker, Z. Pines, R. Pratap, and K. Ramanujam: "Advanced Scout: Data Mining and Knowledge Discovery in NBA Data", *Data Mining and Knowledge Discovery*, Vol. 1, No. 1, pp. 121–125, 1997.
3. A. Ekin, A. Tekalp, and R. Mehrotra: "Automatic Soccer Video Analysis and Summarization", *IEEE Trans. Image Processing*, Vol. 12, No. 7, pp. 796-807, 2003.
4. J. Glasgow, N. H. Narayanan, and B. Chanrasekaran: *Diagrammatic Reasoning*, AAAI Press, Menlo Park, 1995.
5. S. Hirano and S. Tsumoto: "Finding Interesting Pass Patterns from Soccer Game Records", *Principles and Practice of Knowledge Discovery in Databases*, LNAI 3202, pp. 209–218, Springer, 2004.
6. S. Iwase and H. Saito: "Tracking Soccer Players Based on Homography among Multiple Views", *Proc. Visual Communications and Image Processing (VCIP), SPIE*, Vol. 5150, pp. 283–292, 2003.
7. M. Lazarescu, S. Venkatesh, G. A. W. West, and T. Caelli: "On the Automated Interpretation and Indexing of American Football", *Proc. IEEE International Conference on Multimedia Computing and Systems (ICMCS)*, Vol. 1, pp. 802–806, 1999.
8. H. Maeda: *Strategies and Techniques of Soccer*, Sinsei Pub., Tokyo, 2003.   (in Japanese).

9. T. Misu, S. Gohshi, Y. Izumi, Y. Fujita, and M. Naemura: "Robust Tracking of Athletes using Multiple Features of Multiple Views", *Journal of the International Conference in Central Europe on Computer Graphics and Visualization (WSCG)*, Vol. 12, pp. 285–292, 2004.
10. T. Misu, M. Naemura, W. Zheng, Y. Izumi, and K. Fukui: "Robust Tracking of Soccer Players Based on Data Fusion", *Proc. Sixteenth International Conference on Pattern Recognition (ICPR)*, Vol. 1, pp. 556–561, 2002.
11. C. J. Needham and R. D. Boyle: "Tracking Multiple Sports Players through Occlusion, Congestion and Scale", *Proc. British Machine Vision Conference 2001 (BMVC)*, 2001.
12. L. Xie, P. Xu, S.-F. Chang, A. Divakaran, and H. Sun: "Structure Analysis of Soccer Video with Domain Knowledge and Hidden Markov Models", *Pattern Recognition Letters*, Vol. 25, pp. 767–775, 2004.
13. A. Yamada, Y. Shirai, and J. Miura: "Tracking Players and a Ball in Video Image Sequence and Estimating Camera Parameters for 3D Interpretation of Soccer Games", *Proc. Sixteenth International Conference on Pattern Recognition (ICPR)*, Vol. 1, pp. 303–306, 2002.

Prediction of Domain-Domain Interactions Using Inductive Logic Programming from Multiple Genome Databases

Thanh Phuong Nguyen and Tu Bao Ho

School of Knowledge Science
Japan Advanced Institute of Science and Technology
1-1 Asahidai, Nomi, Ishikawa 923-1292, Japan
{phuong, bao}@jaist.ac.jp

Abstract. Protein domains are the building blocks of proteins, and their interactions are crucial in forming stable protein-protein interactions (PPI) and take part in many cellular processes and biochemical events. Prediction of protein domain-domain interactions (DDI) is an emerging problem in computational biology. Different from early works on DDI prediction, which exploit only a single protein database, we introduce in this paper an integrative approach to DDI prediction that exploits multiple genome databases using inductive logic programming (ILP). The main contribution to biomedical knowledge discovery of this work are a newly generated database of more than 100,000 ground facts of the twenty predicates on protein domains, and various DDI findings that are evaluated to be significant. Experimental results show that ILP is more appropriate to this learning problem than several other methods. Also, many predictive rules associated with domain sites, conserved motifs, protein functions and biological pathways were found.

1 Introduction

Understanding functions of proteins is a main task in molecular biology. Early work in computational biology has focused on finding protein functions via prediction of protein structures, e.g., [13]. Recently, detecting protein functions via prediction of protein-protein interactions (PPI) has emerged as a new trend in computational biology, e.g., [2], [10], [24].

Within a protein, a domain is a fundamental structural unit that is self-stabilizing and often folds independently of the rest of the protein chain. Domains often are named and singled out because they figure prominently in the biological function of the protein they belong to; for example, the *calcium-bindingdomain* of *calmodulin*. The domains form the structural or functional units of proteins that partake in intermolecular interactions. Therefore, domain-domain interaction (DDI) problem has biological significance in understanding protein-protein interactions in deepth.

Concerning protein domains, a number of domain-based approaches to predict PPIs have recently been proposed. One of the pioneering works is an association

N. Lavrač, L. Todorovski, and K.P. Jantke (Eds.): DS 2006, LNAI 4265, pp. 185–196, 2006.

method developed by Sprinzak and Margalit [23]. Kim *et al.* improved the association method by considering the number of domains in each protein [10]. Han *et al.* proposed a domain combination-based method by considering the possibility of domain combinations appearing in both interacting and non-interacting sets of protein pairs [8]. A graph-oriented approach is proposed by Wojcik and Schachter called the 'interacting domain profile pairs' (IDPP) approach [26]. That method uses a combination of sequence similarity search and clustering based on interaction patterns. Therefore, the only purpose of the above mentioned work was to predict and/or to validate protein interactions. They all confirmed the biological role of DDIs in PPIs, however, they did not much take domain-domain interactions into account.

Recently, there are several works that not only use protein domains to predict protein interactions, but also attempt to discover DDIs. An integrative approach is proposed by Ng *et al.* to infer putative domain-domain interactions from three data sources, including experimentally derived protein interactions, protein complexes and Rosetta stone sequences [15]. The interaction scores for domain pairs in these data sources were obtained with a calculation scheme similar to the association method by considering frequency of each domain among the interacting protein pairs. The maximum likelihood estimation (MLE) is applied to infer domain interactions by maximizing the likelihood of the observed protein interaction data [5]. The probabilities of interaction between two domains (only single-domain pairs are considered) are optimized using the expectation maximization (EM) algorithm. Chen et al. used domain-based random forest framework to predict PPIs [2]. In fact, they used the PPI data from DIP and a random forest of decision trees to classify protein pairs into sets of interacting and non-interacting pairs. Following the branches of trees, they found a number of DDIs. Riley et al. proposed a domain pair exclusion analysis (DPEA) for inferring DDIs from databases of protein interactions [22]. DPEA features a log odds score, E_{ij}, reflecting confidence that domains i and j interact.

The above mentioned works mostly use protein interaction data to infer DDIs, and all of them have two limitations. First, they used only the protein information (particularly protein-protein interaction data) or the co-occurrence of domains in proteins, and ignored other domain-domain interaction information between the protein pairs. However, DDIs also depend on other features of proteins and domains as well–not only protein interactions [11], [25]. Second, each of them usually exploited only a single protein database and none of the single protein databases can provide all information needed to do better DDI prediction.

In this paper, we present an approach using ILP and multiple genome databases to predict domain-domain interactions. The key idea of our computational method of DDI prediction is to exploit as much as possible background knowledge from various databases of proteins and domains for inferring DDIs. Sharing some common points in ILP framework in bioinformatics with [24], this paper concentrates on discovering knowledge of domain-domain interactions. To this end, we first examine seven most informative genome databases, and extract more than a hundred thousand possible and necessary ground facts on protein domains. We then employ inductive logic programming (ILP) to infer efficiently DDIs. We carry out a comparative evaluation of findings for DDIs and learning methods in terms of sensitivity and specificity. By analyzing various produced rules, we found many interesting relations between DDIs and protein functions, biological pathways, conserved motifs and pattern sites.

The remainder of the paper is organized as follows. In Section 2, we present our proposed methods to predict DDIs using ILP and multiple genome databases. Then the evaluation is given in Section 3. Finally, some concluding remarks are given in Section 4.

2 Method

In this section, we describe our proposed method to predict domain-domain interactions from multiple genome databases. Two main tasks of the method are: (1) Generating background knowledge[1] from multiple genome databases and (2) Learning DDI predictive rules by ILP from generated domain and protein data.

We first describe these tasks in the next section, Section 2.1, then present our proposed framework using ILP to exploit extracted background knowledge for DDI prediction (Section 2.2).

2.1 Generating Background Knowledge from Multiple Genome Databases

Unlike previous work mentioned in Section 1, we chose and extracted data from seven genome databases to generate background knowledge with an abundant number of ground facts and used them to predict DDI. Figure 1 briefly presents these seven databases.

We integrated domain data and protein data from seven genome databases: four domain databases (Pfam database, PROSITE database, PRINTS database, and InterPro database) and three protein databases from UniProt database, MIPS database, Gene Ontology.

Extract domain and protein data from multiple genome databases. The first issue faced is, what kinds of genome databases are suitable for DDI prediction. When choosing data, we are concerned on two points. First is biological role of that data in domain-domain interaction, and second is the availability of that data.

Denote by D the set of all considered protein domains, d_i a domain in D, p_k a protein that consists of some domains d_is, and P the set of such proteins. A domain pair (d_i, d_j) that interacts with each other is denoted by d_{ij}, otherwise by $\neg d_{ij}$. In fact, whether two domains d_i and d_j interact depends on: (i) the domain features of d_i and d_j and, (ii) the protein features of some proteins p_ks consisting of d_i and d_j [25]. Denote by df_t^m a domain feature tth extracted from the domain

---

[1] The term 'background knowledge' is used here in terms of language of inductive logic programming.

1. **Pfam** [6]: Pfam contains a large collection of multiple sequence alignments and profile hidden Markov models (HMM) covering the majority of protein domains.

2. **PRINTS** [7]: A compendium of protein fingerprints database. Its diagnostic power is refined by iterative scanning of a SWISS-PROT/TrEMBL composite.

3. **PROSITE** [18]: Database of protein families and domains. It consists of biologically significant sites, patterns and profiles.

4. **InterPro** [4]: InterPro is a database of protein families, domains and functional sites in which identifiable features found in known proteins can be applied to unknown protein sequences.

5. **Uniprot** [21]: UniProt (Universal Protein Resource) is the world's most comprehensive catalog of information on proteins which, consists of protein sequence and function data created by combining the information in Swiss-Prot, TrEMBL, and PIR.

6. **MIPS** [3]: The MIPS Mammalian Protein-Protein Interaction Database is a collection of manually curated high-quality PPI data collected from the scientific literature by expert curators.

7. **Gene Ontology (GO)** [19]: The three organizing principles of GO are molecular function, biological process and cellular component. This database contains the relations between GO terms.

Fig. 1. Description of genome databases used

database M. With different domains, one feature df_t^m may have different values. For example, the domain site and pattern feature extracted from PROSITE database have some values like *Casein kinase II phosphorylation site* or *Anaphylatoxin domain signature*. Denote by pf_r^l a protein feature rth extracted from protein database L. Also in different domains, one protein feature pf_r^l may have different values. For example, GO term feature extracted from GO database have some values like *go0006470* or *go0006412*. The extracted domain/protein features are mentioned as biologically significant factors in domain-domain interactions [11], [20], [25], etc. The combination of both domain features and protein features constructed the considerable background knowledge associated with DDIs.

Algorithm 1 shows how to extract data (values) of domain features df_t^ms and protein features pf_r^ls for all domains $d_i s \in D$ from multiple data sources. Pfam domain accessions are domain identifiers and ORF (open reading frame) names are protein identifiers. We know that one protein can have many domains and one domain can belong to many proteins. Then, each protein identifier is mapped with the identifiers of its own domains. As the result, protein feature values are assigned to domains.

This paper concentrates on predicting DDIs for *Saccharomyces cerevisiae* – a budding yeast, as the *Saccharomyces cerevisiae* database is available. To map proteins and their own domains, the interacting proteins in DIP database [17], well-known yeast PPI database, are selected. If one protein has no domain, features of that protein are not predictive for domain-domain interactions. If one domain does not belong to any interacting proteins in DIP database, it seems not to have any chance to interact with others. Thus, we excluded all proteins and domains which did not have matching partners (Step 5, Step 6). Having extracted interacting proteins from DIP database, mapping data are more reliable and meaningful. After mapping proteins and their domains, the values of all domain/protein features are extracted (from Step 8 to Step 12).

Algorithm 1. Extracting protein and domain data from multiple sources

---

Input:

       Set of domains $D \supset \{d_i\}$.

       Multiple genomic data used for extracting background knowledge $(S^{Pfam}, S^{InterPro}, S^{PROSITE}, S^{PRINTS}S^{Uniprot}, S^{MIPS}, S^{GO})$.

Output:

       Set of domain feature values $Feature^{domain}$.

       Set of protein features values $Feature^{protein}$.

1: $Feature^{domain} := \emptyset$; $Feature^{protein} := \emptyset$; $P := \emptyset$.
2: Extract all interacting proteins p_ks from DIP database; $P := P \cup \{p_k\}$.
3: **for all** proteins p_ks $\in P$ and domains d_is $\in D$.
4:    Mapping proteins p_ks with their own domains d_is
       by the protein identifiers and the domain identifiers.
5:    **if** a domain d_i does not belong to any protein p_k **then** $D := D \backslash \{d_i\}$.
6:    **if** a protein p_k does not consist of any domain d_i **then** $P := P \backslash \{p_k\}$.
7: **for each** $d_i \in D$
8:    Extract all values $df_t^m.values$ for domain features d_t^ms from domain database M
       $(\forall M \in (S^{Pfam}, S^{InterPro}, S^{PROSITE}, S^{PRINTS}))$.
9:    **if** $df_t^m.value \notin Feature^{domain}$ **then**
       $Feature^{domain} := Feature^{domain} \cup \{d_t^m.value\}$.
10:    Extract all values $pf_r^l.values$ of protein feature pf_r^ls from protein database L
       $(\forall L \in S^{Uniprot}, S^{MIPS}, S^{GO}))$.
11:    **if** $pf_r^l.value \notin Feature^{domain}$ **then**
       $Feature^{protein} := Feature^{protein} \cup \{pf_r^l.value\}$.
12: **return** $Feature^{domain}, Feature^{protein}$.

Generating background knowledge. The data which we extracted from seven databases have different structures: numerical data (for example, the number of motif), text data (for example, protein function category), mixture of numerical and text data (for example, protein keywords, domain sites). The extracted data (the values of all domain/protein features) are represented in form of predicates.

Aleph system [1] is applied to induce rules. Note that Aleph uses *mode declarations* to build the bottom clauses, and a simple mode type is one of : (1) the input variable (+), (2) the output variable (−), and (3) the constant term (#). In this paper, target predicate is interact_domain(domain, domain). The instances of this relation represent the interaction between two domains. For background knowledge, all domain/protein data are shortly denoted in form of different predicates. Table 1 shows the list of predicates used as background knowledge for each genomic data. With the twenty background predicates, we obtained totally 100,421 ground facts associated with DDI prediction.

Extracted domain features (from four databases: Pfam, InterPro, PROSITE and PRINTS) are represented in the form of predicates. These predicates describe domain structures, domain characteristics, domain functions and protein-domain relations. Among them, there are some predicates which are the relations between accession numbers of two databases, for example, *prints(pf00393,pr00076)*. Data from different databases are bound by these predicates. In PRINTS database,

Table 1. Predicates used as background knowledge in various genomic data

Genomic data	Background knowledge predicates	#Ground fact
Pfam	prosite(+Domain,-PROSITE_Domain)	1804
	A domain has a PROSITE annotation number	
	interpro(+Domain,-InterPro_Domain)	2804
	A domain has an InterPro annotation number	
	prints(+Domain,-PRINTS_Domain)	1698
	A domain has a PRINTS annotation number	
	go(+Domain,-GO_Term)	2540
	A domain has a GO term	
InterPro	interpro2go(+InterPro_Domain,-GO_Term)	2378
	Mapping of InterPro entries to GO	
PROSITE	prosite_site(+Domain,#prosite_site)	2804
	A domain contains PROSITE significant sites or motifs	
PRINTS	motif_compound(+Domain,#motif_compound)	3080
	A domain is compounded by number of conserved motifs	
Uniprot	haskw(+Domain,#Keyword)	13164
	A domain has proteins keywords	
	hasft(+Domain,#Feature)	8271
	A domain has protein features	
	ec(+Domain,#EC)	2759
	A domain has coded enzyme of its protein	
	pir(+Domain,-PIR_Domain)	3699
	A domain has a Pir annotation number	
MIPS	subcellular_location(+Protein,#Subcellular_Structure)	10638
	A domain has subcellular structures in which its protein is found.	
	function_category(+Domain,#Function_Category)	11975
	A domain has the protein categorized to a certain function category	
	domain_category(+Domain,#Domain_Category)	5323
	A domain has proteins categorized to a certain protein category	
	phenotype_category(+Domain,#Phenotype_Category)	8066
	A domain has proteins categorized to a certain phenotype category	
	complex_category(+Domain,#Complex_Category)	7432
	A domain has proteins categorized to a certain complex category	
GO	is_a(+GO_Term,-GO_Term)	1009
	is_a relation between two GO terms	
	part_of(+GO_Term,-GO_Term)	1207
	part_of relation between two GO terms	
Others	num_int(+Domain,#num_int)	804
	A domain has a number of domain-domain interactions	
	ig(+Domain,+ Domain, #ig)	8246
	Interaction generality is the number of domains that interact with just two considered domains	
Totals		100,421

motif_compound information gives the number of conserved motifs found in proteins and domains. The number of motifs is important in understanding the conservation of protein/domain structures in the evolutionary process [7]. We generated predicate motif_compound(+Domain,#motif_compound). This predicate is predictive for DDI prediction and gives information about the stability of DDIs (example rules are shown and analyzed in Section 3.2). For example: $motif\_compound(pr00517, compound(8))$, where $pr00517$ is the accession numbers in PRINTS database and *compound(8)* is the number of motifs.

Protein domains are the basic elements of proteins. Protein features have a significant effect on domain-domain interactions. These protein features (extracted from three databases Uniprot, MIPS and GO) are showed in the form of predicates. These predicates describe function categories, subcellular locations, GO terms, etc. They give the relations between DDIs and promising protein features. For example, most interacting proteins are in the same complexes [14].

The domains of these interacting proteins can interact with each other. As a result, if some domains belong to the some proteins categorized in the same complex, they can be predicted to have some domain-domain interactions. The predicate complex_category(+Domain,#Complex_Category) means that a domain has proteins categorized to a certain complex category. For example, $complex\_category(pf00400, transcription\_complexes)$, where $pf00400$ is Pfam accession number and $transcription\_complexes$ is complex category name.

2.2 Learning DDI Predictive Rules by ILP from Generated Domain and Protein Data

There have been many ILP systems that are successfully applied to various problems in bioinformatics, such as protein secondary structure prediction [13], protein fold recognition [12], and protein-protein interaction prediction [24]. The proposed ILP framework for predicting DDIs from multiple genome databases is described in Algorithm 2.

Algorithm 2. Discovering rules for domain-domain interactions

---

Input:

    The domain-domain interactions database $InterDom$ Number of negative examples $(\neg d_{ij})$ N

    Multiple genomic data used for extracting background knowledge $(S^{Pfam}, S^{InterPro}, S^{PROSITE}, S^{PRINTS} S^{Uniprot}, S^{MIPS}, S^{GO})$

Output: Set of rules R for domain-domain interaction prediction.

1: $R := \emptyset$.
2: Extract positive examples set $S_{interact}$ from $InterDom$.
3: Generate negative examples $\neg d_{ij}$s by selecting randomly N domain pairs from D where $\neg d_{ij} \notin S_{interact}$.
4: **for each** domain $d_i \in D$
5:     **call** Algorithm 1 to generate values for features d_t^ms from domain database M $(\forall M \in (S^{Pfam}, S^{InterPro}, S^{PROSITE}, S^{PRINTS}))$ and protein features pf_r^ls from protein database L $(\forall L \in (S^{Uniprot}, S^{MIPS}, S^{GO}))$.
6:     Integrate all domain features df_t^ms and protein features pf_r^ls for generating background knowledge
7: Run Aleph to induce rules r.
8: $R := R \cup \{r\}$.
9: **return** R.

In the framework, the common procedure of ILP method is presented. Step 2 and Step 3 are for generating positive and negative examples (see Section 3). In Steps 4 to 7, we extracted background knowledge including both domain features and protein features (see Section 2.1). Aleph system [1] is applied to induce rules in Step 8. Aleph is an ILP system that uses a top-down ILP covering algorithm, taking as input background information in the form of predicates, a list of modes declaring how these predicates can be chained together, and a designation of one predicate as the head predicate to be learned. Aleph is able to use a variety of search methods to find good clauses, such as the standard methods of breadth-first search, depth-first search, iterative beam search, as well as heuristic methods requiring an evaluation function. We use the default evaluation function *coverage* (the number of positive and negative examples covered by the clause) in our work.

3   Evaluation

3.1   Experiment Design

In this paper, we used 3000 positive examples from InterDom database. Inter-Dom database consists of DDIs of multiple organisms [16]. Positive examples are domain-domain interactions in InterDom database which have score threshold over 100 and no false positives. The set of interacting pairs $S_{interact}$ in Algorithm 2 consists of these domain-domain interactions. Because there is no database for non domain-domain interaction, the negative examples $\neg d_{ij}$s are randomly generated. A domain pair $(d_i, d_j) \in D$ is considered to be a negative example, if the pair does not exist in the interaction set. In this paper, we chose different numbers of negatives (500, 1000, 2000, 3000 negative examples). To validate our proposed method, we conducted a 10-fold cross-validation test, comparing cross-validated sensitivity and specificity with results obtained by using AM [23] and SVM method. The AM method calculates a score d_{kl} for each domain pair (D_k, D_l) as the number of interacting protein pairs containing (D_k, D_l) divided by the number of protein pairs containing (D_k, D_l).

In the approach of predicting protein-protein interactions based on domain-domain interactions, it can be assumed that domain-domain interactions are independent and two proteins interact if at least one domain pairs of these two proteins interact. Therefore, the probability p_{ij} that two proteins P_i and P_j interact can be calculated as

$$p_{ij} = 1 - \prod_{D_k \in P_i, D_l \in P_j} (1 - d_{kl})$$

We implemented the AM and SVM methods in order to compare them with our proposed method. We use the same database applying ILP to input AM and SVM. The probability threshold is set to 0.05 for the simplicity of comparison. For SVM method, we used SVM^{light} [9]. The linear kernel with default values of the parameters was used. For Aleph, we selected *minpos* = 3 and *noise* = 0, i.e. the lower bound on the number of positive examples to be covered by an acceptable clause is 3, and there are no negative examples allowed to be covered by an acceptable clause. These parameters are the smallest that allow us to induce rules with biological meaning. We also used the default evaluation function *coverage* which is defined as $P - N$, where P, N are the number of positive and negative examples covered by the clause.

3.2   Analysis of Experimental Results

Table 2 shows the performance of Aleph compared with AM and SVM methods. Most of our experimental results had higher sensitivity and specificity compared with AM and SVM. The sensitivity of a test is described as the proportion of true positives it detects of all the positives, measuring how accurately it identifies positives. On the other hand, the specificity of a test is the proportion of true negatives it detects of all the negatives, and thus is a measure of how accurately it identifies negatives. It can be seen from Table 2 that the proposed method showed a considerably high sensitivity and specificity given a certain number of negative examples. The number of negative examples should be chosen neither too large nor too small to avoid an imbalanced learning problem.

The performance of method in terms of specificity and sensitivity are also statistically tested in terms of confidence intervals. Confidence intervals give us an estimate of the amount of error involved in our data. To estimate 95% confidence interval for each calculated specificity and sensitivity, we used t distribution. The 95% confidence intervals are shown in Table 2.

Table 2. Performance of Aleph compared with AM and SVM methods. The sensitivity and specificity are obtained for each randomly chosen set of negative examples. The last column demonstrates the number of rules obtained using our proposed method, with the minimum positive cover set to 3.

# Neg	Sensitivity			Specificity			# Rules
	AM	SVM	Aleph	AM	SVM	Aleph	
500	0.49±.027	**0.86±.010**	0.83±.016	0.54±.074	0.24±.004	**0.61±.075**	127
1000	0.57±.018	0.63±.074	**0.78±.042**	0.44±.033	0.49±.009	**0.68±.042**	173
2000	0.50±.015	0.32±.014	**0.69±.027**	0.50±.021	0.73±.015	**0.80±.018**	196
3000	0.49±.021	0.22±.017	**0.62±.027**	0.53±.022	0.81±.013	**0.84±.010**	235
Avg.	0.51±.020	0.51±.029	**0.73±.028**	0.50±.038	0.57±.010	**0.73±.036**	

Besides comparing cross-validated sensitivity and specificity, cross-validated accuracy and precision are considered. And all of your experiment results had higher accuracy and precision compared with AM and SVM. The average accuracy (0.76) and precision (0.82) of Aleph are higher than both AM method (0.50 and 0.61 respectively) and SVM method (0.62 and 0.70 respectively).

The experimental results have shown that ILP approach potentially predicts DDIs with high sensitivity and specificity. Further more, the inductive rules of ILP encouraged us to discover lots of comprehensive relations between DDIs and domain/protein features. Analysing our results in comparison with information in biological literatures and books, we found that ILP induced rules could be applied to the further related studies in biology.

The simplest rule covering many examples of positives is the self-interact rule. Many domains tend to interact with themselves (86 domain-domain interactions among positive examples). This phenomenon is reasonable because indeed lots of proteins interact with themselves, and they consist of many of the same domains. Figure 2 shows some other induced rules.

Rule 1 [Pos cover = 15 Neg cover = 0]
  $interact\_domain(A, B) : -ig(A, B, C), C = 5,$
  $function\_category(B, transcription),$
  $protein\_category(A, transcription\_factors).$

Rule 2 [Pos cover = 20 Neg cover = 0]
  $interact\_domain(A, B) : - \quad num\_int(B, C), gteq(C, 20),$
  $complex\_category(A, scf\_comlexes).$

Rule 3 [Pos cover = 51 Neg cover = 0]
  $interact\_domain(A, B) : - \quad interpro(B, C), interpro(A, C), interpro2go(C, D).$

Rule 4 [Pos cover = 23 Neg cover = 0]
  $interact\_domain(A, B) : - \quad prints(B, C), motif\_compound(C, compound(8)),$
  $function\_category(A, protein\_synthesis).$

Rule 5 [Pos cover = 31 Neg cover = 0]
  $interact\_domain(A, B) : - \quad prints(B, C),$
  $motif\_compound(C, compound(13)), haskw(A, cell\_cycle).$

Rule 6 [Pos cover = 29 Neg cover = 0]
  $interact\_domain(A, B) : - \quad num\_int(A, C), C = 7,$
  $function\_category(B, metabolism), haskw(B, thread\_structure).$

Rule 7 [Pos cover = 32 Neg cover = 0]
  $interact\_domain(A, B) : - \quad ig(A, A, C), C = 3,$
  $function\_category(B, cell\_type\_differentiation),$
  $phenotype\_category(A, nucleic\_acid\_metabolism\_defects).$

Rule 8 [Pos cover = 15 Neg cover = 0]
  $interact\_domain(A, B) : - \quad phenotype\_category(B, conditional\_phenotypes)$
  $hasft(A, domain\_rna\_binding\_rrm).$

Rule 9 [Pos cover = 16 Neg cover = 0]
  $interact\_domain(A, B) : - \quad prosite(B, C),$
  $prosite\_site(C, tubulin\_subunits\_alpha\_beta\_and\_gamma\_signature).$

Rule 10 [Pos cover = 37 Neg cover = 0]
  $interact\_domain(A, B) : - \quad go(B, C), is\_a(C, D), hasft(A, chain\_bud\_site\_$
  $selection\_protein\_bud5).$

Fig. 2. Some induced rules obtained with *minpos* = 3

In the set of induced rules, there are (1) rules of only domain features (*i.e.* Rule 9), rules of only protein features (*i.e.* Rule 8) and especially rules of mixture of both domain features and protein features (*i.e.* Rule 4, Rule 5). In rules, the coverage values presented are the average predictive coverage on the 10 folds.

Related to *motif compound* feature in domain, we found that the more motifs a domain has, the more interactions the domain has with other domains. This means that domains which have many conserved motifs tend to interact with others. And the interactions of these domains play an important role in forming stable domain-domain interactions in particular and protein-protein interactions in general [11]. Rule 4 shows that if we have two domains - one of them with eight motifs, and the other one belonging to proteins categorized in *protein_synthesis* function category, then the two domains interact.

Discovering the rules related to domain sites and domain signatures with predicate `prosite_site(domain,#prosite_site)`, we found some significant sites in domain joining in the domain-domain interactions. Rule 9 shows the relation between the accession numbers in Pfam database and PROSITE database, and then the signature information of domain in PROSITE database. This rule means that if one domain belongs to both Pfam database and PROSITE database and has *tubulin_subunits_alpha_beta_and_gamma_signature*, then it can interact with others. The rules like Rule 9 can be applied to understand protein-protein interaction interfaces and protein structures [20].

Rule 6 is an example which infers the relation between DDIs and biological pathways. From this rule, if we have an interacting domain pair, one of them has seven domain-domain interactions, and the other domain belongs to one protein which has keyword *thread_struture*, we can say that that protein functions in a certain metabolic pathway.

Thanks to inductive rules of ILP, we found a lot of relations between DDIs and different domain and protein features. We expect that the combination of these rules will be very useful for understanding DDIs in particular and protein structures, protein functions and protein-protein interactions in general.

4   Conclusion

We have presented an approach using ILP and multiple genome databases to predict domain-domain interactions. The experimental results demonstrated that our proposed method could produce comprehensible rules, and at the same time, performed well compared with other work on domain-domain interaction prediction. In future work, we would like to investigate further the biological significance of novel domain-domain interactions obtained by our method, and apply the ILP approach to other important tasks, such as determining protein functions, protein-protein interactions, and the sites, and interfaces of these interactions using domain-domain interaction data.

References

1. A.Srinivasan. http://web.comlab.ox.ac.uk/oucl/research/areas/machlearn/Aleph/.
2. X.W. Chen and M. Liu. Prediction of protein-protein interactions using random decision forest framework. *Bioinformatics*, 21(24):4394–4400, 2005.
3. Comprehensive Yeast Genome Database. http://mips.gsf.de/genre/proj/yeast/.
4. InterPro database concerning protein families and domains. http://www.ebi.ac.uk/interpro/.
5. M. Deng, S. Mehta, F. Sun, and T. Chen. Inferring domain-domain interactions from protein-protein interactions. *Genome Res.*, 12(10):1540–1548, 2002.

6. Protein families database of alignments and HMMs. http://www.sanger.ac.uk/Software/Pfam/.

7. Protein figerprint. http://umber.sbs.man.ac.uk/dbbrowser/PRINTS/.

8. D. Han, H.S.Kim, J.Seo, and W.Jang. A domain combination based probabilistic framework for protein protein interaction prediction. In *Genome Inform. Ser. Workshop Genome Inform*, page 250259, 2003.

9. Thorsten Joachims. http://svmlight.joachims.org/.

10. R.M. Kim, J. Park, and J.K. Suh. Large scale statistical prediction of protein - protein interaction by potentially interacting domain (PID) pair. In *Genome Inform. Ser. Workshop Genome Inform*, pages 48–50, 2002.

11. H. S. Moon, J. Bhak, K.H. Lee, and D. Lee. Architecture of basic building blocks in protein and domain structural interaction networks. *Bioinformatics*, 21(8):1479–1486, 2005.

12. M.Turcotte, S.H.Muggleton, and M.J.E.Sternberg. Protein fold recognition. In *Proc. of the 8th International Workshop on Inductive Logic Programming (ILP-98)*, pages 53–64, 1998.

13. S. Muggleton, R.D. King, and M.J.E. Sternberg. Protein secondary structure prediction using logic-based machine learning. *Protein Eng.*, 6(5):549–, 1993.

14. S. K. Ng and S. H. Tan. Discovering protein-protein interactions. *Journal of Bioinformatics and Computational Biology*, 1(4):711–741, 2003.

15. S.K. Ng, Z. Zhang, and S.H Tan. Integrative approach for computationally inferring protein domain interactions. *Bioinformatics*, 19(8):923–929, 2003.

16. S.K Ng, Z Zhang, S.H Tan, and K. Lin. InterDom: a database of putative interacting protein domains for validating predicted protein interactions and complexes. *Nucleic Acids Res*, 31(1):251–254, 2003.

17. Database of Interacting Proteins. http://dip.doe-mbi.ucla.edu/.

18. PROSITE: Database of protein families and domains. http://kr.expasy.org/prosite/.

19. Gene Ontology. http://www.geneontology.org/.

20. D. Reichmann, O. Rahat, S. Albeck, R. Meged, O. Dym, and G. Schreiber. From The Cover: The modular architecture of protein-protein binding interfaces. *PNAS*, 102(1):57–62, 2005.

21. Universal Protein Resource. http://www.pir.uniprot.org/.

22. R. Riley, C. Lee, C. Sabatti, and D. Eisenberg. Inferring protein domain interactions from databases of interacting proteins . *Genome Biology*, 6(10):R89, 2005.

23. E. Sprinzak and H. Margalit. Correlated sequence-signatures as markers of protein-protein interaction. *Journal of Molecular Biology*, 311(4):681–692, 2001.

24. T.N. Tran, K.Satou, and T.B.Ho. Using inductive logic programming for predicting protein-protein interactions from multiple genomic data. In *PKDD*, pages 321–330, 2005.

25. K. Wilson and J.Walker. *Principle and Techniques of Biochemistry and Molecular Biology*. Cambridge University Press, 6 edition, 2005.

26. J. Wojcik and V. Schachter. Protein-protein interaction map inference using interacting domain profile pairs. *Bioinformatics*, 17(suppl-1):S296–305, 2001.

Clustering Pairwise Distances with Missing Data: Maximum Cuts Versus Normalized Cuts*

Jan Poland and Thomas Zeugmann

Division of Computer Science
Hokkaido University, Sapporo 060-0814, Japan
{jan, thomas}@ist.hokudai.ac.jp
http://www-alg.ist.hokudai.ac.jp/~{jan, thomas}

Abstract. Clustering algorithms based on a matrix of pairwise similarities (kernel matrix) for the data are widely known and used, a particularly popular class being spectral clustering algorithms. In contrast, algorithms working with the pairwise distance matrix have been studied rarely for clustering. This is surprising, as in many applications, distances are directly given, and computing similarities involves another step that is error-prone, since the kernel has to be chosen appropriately, albeit computationally cheap. This paper proposes a clustering algorithm based on the SDP relaxation of the max-k-cut of the graph of pairwise distances, based on the work of Frieze and Jerrum. We compare the algorithm with Yu and Shi's algorithm based on spectral relaxation of a norm-k-cut. Moreover, we propose a simple heuristic for dealing with missing data, i.e., the case where some of the pairwise distances or similarities are not known. We evaluate the algorithms on the task of clustering natural language terms with the Google distance, a semantic distance recently introduced by Cilibrasi and Vitányi, using relative frequency counts from WWW queries and based on the theory of Kolmogorov complexity.

1 Introduction

Let a set of n objects or data points, $x_1, \ldots, x_n$, be given. We might not know anything about the objects, but assume that their pairwise distances $d_{ij} = d(x_i, x_j)$ are known. Here, $d \colon M \times M \to \mathbb{R}$ is a distance measure over a set M, i.e., $d(x, y) \geq 0$, $d(x, y) = d(y, x)$ for all $x, y \in M$, and $d(x, y) = 0$ iff $x = y$. Then we can *cluster* the data, i.e., assign the x_i to k distinct groups such that the distances within groups are small and the distances between the groups are large. This is done as follows. Construct a graph from the pairwise distances and choose an algorithm from the large class of recently published methods based on graph-theoretic cut criteria. As the cuts are usually $\mathcal{NP}$-hard to optimize, appropriate relaxations have been subject to intensive research. Two types of relaxations are particularly important:

---

* This work was supported by JSPS 21st century COE program C01. Additional support has been provided by the MEXT Grand-in-Aid for Scientific Research on Priority Areas under Grant No. 18049001.

N. Lavrač, L. Todorovski, and K.P. Jantke (Eds.): DS 2006, LNAI 4265, pp. 197–208, 2006.

(1) *Spectral* methods, where the top eigenvectors of the graph's adjacency matrix are used to project the data into a lower dimensional space. This gives rise to new theoretical investigations of the popular spectral clustering algorithms.

(2) *Semi-definite programming* (*SDP*), where the discrete constraints of the cut criterion are replaced by continuous counterparts. Then convex solvers can be used for the optimization.

Surprisingly, all of the clustering approaches suggested so far work on a graph of *similarities* rather than distances. This means that, given the distances, we need one additional step to obtain similarities from distances, e.g., by applying a Gaussian kernel. This also involves tuning the kernel width, a quantity which the clustering algorithm is quite sensitive to. Hence, it is natural to avoid this step by using a cut criterion that directly works with the distance graph, e.g., max-cut. We follow Frieze and Jerrum [4] and solve the max-cut problem via an SDP relaxation. We compare this method with a representative of spectral clustering algorithms, namely the spectral relaxation of the normalized cut criterion [12].

As a second contribution of this paper, we propose a simple heuristic for dealing with *missing data*, i.e., the case where some of the pairwise distances d_{ij} are unknown. Then, our aim is to substitute the missing d_{ij} by a value which is most likely to leave the values of the cuts intact. This turns out to be the *mean* of the observed d_{ij}.

One motivation for considering missing data is given by the application we shall use to test the algorithms: Clustering of natural language terms using the Google distance. The Google distance [2] is a means of computing the pairwise distance of any searchable terms by just using the relative frequency count resulting from a web search. The Google API provides a convenient way for automating this process, however with a single key (which is obtained by prior registration) the maximum amount of daily queries is currently limited to 1000. Hence, by querying an incomplete sparse distance matrix rather than a full one, one can speed up considerably the overall process, as we shall demonstrate.

The paper is structured as follows. In the next two sections, we introduce the two algorithms based on max-k-cut and norm-k-cut relaxations, respectively, and recall some theory. In Section 4 we address the missing data problem. Section 5 confronts the two algorithms, looking on the exact cut criteria rather than the relaxations, and compares the computational resources required. Section 6 describes the Google distance. In Section 7 we present experimental results with the Google distance. Relation to other work is discussed and conclusions are given in Section 8.

2 Max-k-Cut

Given a fully connected, weighted graph $G = (V, D)$ with vertices $V = \{x_1, \ldots, x_n\}$ and edge weights $D = \{d_{ij} \geq 0 \mid 1 \leq i, j \leq n\}$ which express pairwise distances, a *k-way-cut* is a partition of V into k disjoint subsets $S_1, \ldots, S_k$. Here k is assumed to be given. We define the predicate $A(i, j) = 0$ if x_i and x_j happen to be in the same subset, i.e., if $\exists \ell [1 \leq \ell \leq k, \ 1 \leq i, j \leq n$ and $i, j \in S_\ell]$, and $A(i, j) = 1$, otherwise. The weight of the cut $(S_1, \ldots, S_k)$ is defined as

$$\sum_{i,j=1}^{n} d_{i,j} A(i,j) \ .$$

The *max-k-cut* problem is the task of finding the partition that maximizes the weight of the cut. It can be stated as follows: Let $a_1, \ldots, a_k \in \mathcal{S}^{k-2}$ be the vertices of a regular simplex, where

$$\mathcal{S}^d = \{x \in \mathbb{R}^{d+1} \mid \|x\|_2 = 1\}$$

is the d-dimensional unit sphere. Then the inner product $a_i \cdot a_j = -\frac{1}{k-1}$ whenever $i \neq j$. Hence, finding the max-k-cut is equivalent to solving the following integer program:

$$\text{IP : maximize } \frac{k-1}{k} \sum_{i<j} d_{ij}(1 - y_i \cdot y_j)$$

$$\text{subject to } y_j \in \{a_1, \ldots, a_k\} \text{ for all } 1 \leq j \leq n.$$

Frieze and Jerrum [4] propose the following semidefinite program (SDP) in order to relax the integer program:

$$\text{SDP : maximize } \frac{k-1}{k} \sum_{i<j} d_{ij}(1 - v_i \cdot v_j)$$

$$\text{subject to } v_j \in \mathcal{S}^{n-1} \text{ for all } 1 \leq j \leq n \text{ and}$$

$$v_i \cdot v_j \geq -\frac{1}{k-1} \text{ for all } i \neq j \text{ (necessary if } k \geq 3).$$

The constraints $v_i \cdot v_j \geq -\frac{1}{k-1}$ are necessary for $k \geq 3$ because otherwise the SDP would prefer solutions where $v_i \cdot v_j = -1$, resulting in a larger value of the objective. We shall see in the experimental part that this indeed would result in invalid approximations. The SDP finally can be reformulated as a convex program:

$$\text{CP : minimize } \sum_{i<j} d_{ij} Y_{ij} \tag{1a}$$

$$\text{subject to } Y_{jj} = 1 \text{ for all } 1 \leq j \leq n \text{ and} \tag{1b}$$

$$Y_{ij} \geq -\frac{1}{k-1} \text{ for all } i \neq j \text{ (necessary if } k \geq 3) \text{ and} \tag{1c}$$

$$Y = (Y_{ij})_{1 \leq i,j \leq n} \text{ satisfies } Y \succeq 0. \tag{1d}$$

Here, for the matrix $Y \in \mathbb{R}^{n \times n}$ the last condition $Y \succeq 0$ means that Y is positive semidefinite. Efficient solvers are available for this kind of optimization problems, such as CSDP [1] or SeDuMi [10]. In order to implement the constraints $Y_{ij} \geq -\frac{1}{k-1}$ with these solvers, positive slack variables Z_{ij} have to be introduced together with the equality constraints $Y_{ij} - Z_{ij} = -\frac{1}{k-1}$.

Finally, for obtaining the partitioning from the vectors v_j or the matrix Y, Frieze and Jerrum [4] propose to sample k points $z_1, \ldots, z_k$ randomly on $\mathcal{S}^{n-1}$, representing the groups, and assign each v_j to the closest group, i.e., the closest z_j. They show approximation guarantees generalizing those of Goemans and

Williamson [5]. In practice however, the approximation guarantee does not necessarily imply a good clustering, and applying the k-means algorithm for clustering the v_j gives better results here. We use the kernel k-means (probably introduced for the first time by [9]) which directly works on the scalar products $Y_{ij} = v_i \cdot v_j$, without need of recovering the v_j. We recapitulate the complete algorithm:

Algorithm. Clustering as an SDP relaxation of max-k-cut
Input: Distance matrix $D = (d_{ij})$.
1. Solve the SDP via the CP (1a) through (1d).
2. Cluster the resulting matrix Y using kernel k-means.

3   Normalized k-Cut

The normalized cut criterion has emerged as one of the most widely accepted cut criteria for clustering. It is defined on a graph $G = (V, W)$ of pairwise similarities rather than distances: $W = \{w_{ij} \mid w_{ij} \in [0,1], 1 \le i, j \le n\}$. Here, we identify the edges of G with their weights given by the similarities. For a k-way-cut, i.e., a partition of V into k disjoint subsets $S_1, \ldots, S_k$, the norm-k-cut criterion is defined as (cf. Yu and Shi [12])

$$\frac{1}{k}\sum_{\ell=1}^{k}\frac{\sum_{i\in S_\ell, j \notin S_\ell} w_{ij}}{\sum_{i\in S_\ell, j\in V} w_{ij}} = 1 - \frac{1}{k}\sum_{\ell=1}^{k}\frac{\sum_{i\in S_\ell, j\in S_\ell} w_{ij}}{\sum_{i\in S_\ell, j\in V} w_{ij}} =: 1 - \mathrm{knassoc}(S_1, \ldots, S_k), \quad (2)$$

where $\mathrm{knassoc}(S_1, \ldots, S_k)$ is called the *k-way normalized associations criterion*. Therefore, minimizing the norm-k-cut value is equivalent to maximizing the norm-knassoc value. Following [12], this is the task we consider. For a vector $v = (v_1, \ldots, v_n)$ we write $\mathrm{Diag}(v)$ to denote the matrix $M = (m_{ij})_{1\le i,j\le n}$ with $m_{ii} = v_i$ and $m_{ij} = 0$ for all $i \ne j$. Furthermore, for a matrix $M = (m_{ij})_{1\le i,j\le n}$ we write $\mathrm{diag}(M)$ to denote the vector $(m_{11}, \ldots, m_{nn})$.

Optimizing (2) can be restated as solving the following integer program

$$\mathrm{IP} : \text{maximize } \tfrac{1}{k}\mathrm{tr}(Z^T W Z)$$
$$\text{subject to } Z = X(X^T \Sigma X)^{-\frac{1}{2}}, \tag{3}$$
$$\Sigma = \mathrm{Diag}\Big(\big(\textstyle\sum_{i=1}^{n} w_{ij}\big)_{1\le j\le n}\Big), \text{ and}$$
$$X \in \{0,1\}^{n\times k} \text{ such that } \sum_{\ell=1}^{k} X(j, \ell) = 1 \text{ for all } 1 \le j \le n.$$

Relaxing the constraints on Z and passing to continuous domain, we need to solve the following continuous program

$$\mathrm{ContP} : \text{maximize } \tfrac{1}{k}\mathrm{tr}(Z^T W Z) \tag{4}$$
$$\text{subject to } Z^T \Sigma Z = I_k,$$

where I_k is the k-dimensional identity matrix and Σ is defined as above. The space of all optima of (4) can be described with the help of a spectral decomposition of the Laplacian $L = \Sigma^{-\frac{1}{2}} W \Sigma^{-\frac{1}{2}}$

$$L = \Sigma^{-\frac{1}{2}} W \Sigma^{-\frac{1}{2}} = U \Lambda U^T, \tag{5}$$

where U is orthogonal and Λ is diagonal. Let $\Lambda^* \in \mathbb{R}^{n \times k}$ denote the part of Λ containing the largest k eigenvalues (such that $\Lambda^*(i,j) = 0$ unless $i = j$), and $U^* \in \mathbb{R}^{n \times k}$ be the corresponding eigenvectors, then

$$\mathcal{Z} = \{\Sigma^{-\frac{1}{2}} U^* R \mid R^T R = I_k\} \tag{6}$$

describes the space of solutions of (4).

For some relaxed solution $Z \in \mathcal{Z}$, the corresponding $X \in \mathbb{R}^{n \times k}$ can be obtained by inverting (3):

$$\tilde{X} = \mathrm{Diag}(\mathrm{diag}(ZZ^t))Z.$$

Then one can reconstruct an integer solution X by applying an EM procedure and alternatingly optimizing the rotation matrix R and the discretization X:

1. For given R and resulting Z and $\tilde{X}$, let $X(i, \ell) = 1$, if $\tilde{X}(i, \ell) \geq \tilde{X}(i, m)$ for all $1 \leq m \leq k$, and let $X(i, \ell) = 0$, otherwise.
2. For given X and $\tilde{X}$, compute a singular value decomposition $X^T \tilde{X} = \tilde{U} \tilde{\Lambda} \tilde{U}^T$ and let $R = \tilde{U} \tilde{U}^t$.

We recapitulate the algorithm:
Algorithm. Clustering as a spectral relaxation of norm-k-cut [12]
Input: Similarity matrix $W = (w_{ij})$.
1. Solve the eigenvalue problem (5).
2. Use the described EM procedure to optimize the discretization.

Note that the EM procedure used in this algorithm is different from, but nevertheless closely related to the k-means algorithm used in the algorithm based on max-k-cut and many other spectral clustering algorithms.

4   Missing Data

Assume that either the distance matrix D or the similarity matrix W is not fully specified, but a portion of the off-diagonal entries is missing. One motivation for considering this case could be the desire to save resources by computing only part of the entries (e.g., for the Google distance discussed below, normal user registration permits only a limited amount of queries a day). Suppose that $M \in \{0,1\}^{n \times n}$ is a matrix such that $\mathrm{diag}(M) = 0$ and $M(i,j) = 1$ if and only if $D(i,j)$ (or $W(i,j)$, respectively) is not missing. Assume that the diagonal of D is zero and that of W is one, and denote the ith column of a matrix X by $X[i]$. Define the mean of the observed values,

$$\bar{D} = \frac{1}{\sum_{i,j} M(i,j)} \sum_{i=1}^{n} D[i]^T M[i] \quad \text{or} \quad \bar{W} = \frac{1}{\sum_{i,j} M(i,j)} \sum_{i=1}^{n} W[i]^T M[i].$$

Then, replacing the missing entries in D with the value $\bar{D}$, the resulting distance matrix D is an unbiased estimate for the original full matrix, if the positions of the missing values are sampled from a uniform distribution. Hence, the resulting max-k-cut criterion for each partition is an unbiased estimate for the criterion respective to the original matrix, and this is the best we can do to achieve our goal that the optimal k-way-cuts of the original and the completed matrix are the same.

Also in the case of a similarity matrix W, the missing values should be replaced by the mean of the observed values. Asymptotically for $n \to \infty$, this also yields an unbiased estimate for the norm-k-cut criterion. However, the reasoning is more difficult here, since the norm-k-cut criterion is a sum of quotients, and for two random variables X and Y, we have $E[X/Y] \neq E[X]/E[Y]$. Still, the actual values of numerator and denominator are close to their expectations, as one can verify using concentration inequalities, e.g., Hoeffding's inequality. Then, for large n, with high probability the quotient is close to the corresponding quantity for the original (full) similarity matrix.

5   Max-k-Cut Versus Norm-k-Cut

In this section, we compare the max-cut and norm-cut criteria on distance and similarity matrices that are small enough to allow for a brute-force computation of the exact criteria. We start from 10×10 matrices D_0 and W_0 consisting of two blocks of each size 5,

$$
D_0 = \begin{pmatrix} 0 \cdots 0 & 1 \cdots 1 \\ \vdots \ddots \vdots & \vdots \ddots \vdots \\ 0 \cdots 0 & 1 \cdots 1 \\ 1 \cdots 1 & 0 \cdots 0 \\ \vdots \ddots \vdots & \vdots \ddots \vdots \\ 1 \cdots 1 & 0 \cdots 0 \end{pmatrix} \quad \text{and} \quad W_0 = \begin{pmatrix} 1 \cdots 1 & 0 \cdots 0 \\ \vdots \ddots \vdots & \vdots \ddots \vdots \\ 1 \cdots 1 & 0 \cdots 0 \\ 0 \cdots 0 & 1 \cdots 1 \\ \vdots \ddots \vdots & \vdots \ddots \vdots \\ 0 \cdots 0 & 1 \cdots 1 \end{pmatrix}.
$$

From these matrices, distance matrices D and similarity matrices W are obtained by (1) perturbing the value by Gaussian noise of varying amplitude, (2) making the matrices symmetric and rescaling them to the interval $[0, 1]$, (3) removing a fraction of the off-diagonal values and replacing them by the mean of the remaining values. Another matrix we use for the norm-cut criterion is a kernel matrix obtained from the distance matrix using a Gaussian kernel, $W^D = \exp(-\frac{1}{2\sigma^2}D^2)$ (all operations are meant in the pointwise sense here). Since the values of the distance matrix are normalized to $[0, 1]$, we use a fixed $\sigma = \frac{1}{3}$. The missing values of W^D are replaced by the mean of the observed values in $\bar{W}^D$.

All values displayed in Figures 1 through 3 below are means of 1500 independent samples. Figure 1 shows that, when using the max-cut criterion, the relative number of experiments that result in a different clustering than the originally intended one, grows if either the noise amplitude or the fraction of missing values increases. Of course this was expected. The max-cut criterion even yields always the correct clustering if both noise amplitude and missing data fraction are sufficiently low.

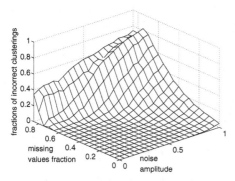

Fig. 1. Average fraction of incorrect clusterings by max-k-cut on a noisy distance matrix with missing data

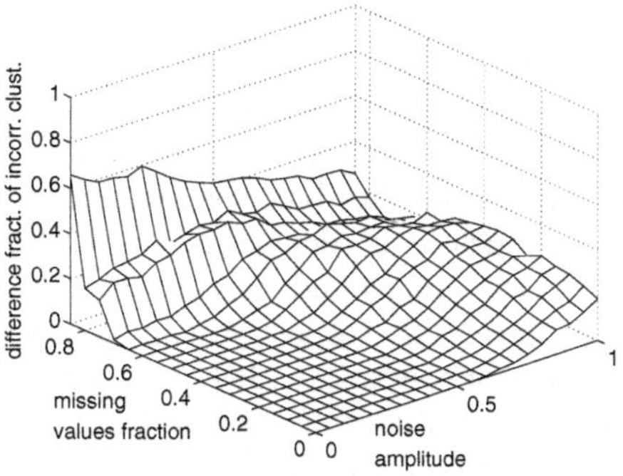

Fig. 2. Difference of the average fraction of incorrect clusterings by norm-k-cut relative to max-k-cut, where the similarity matrix W^D was obtained from the distance matrix D as $W^D = \exp(-\frac{1}{2\sigma^2}D^2)$

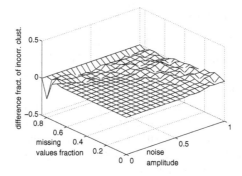

Fig. 3. Average fraction of incorrect clusterings by norm-k-cut: Difference of a $W^D = \exp(-\frac{1}{2\sigma^2}D^2)$ matrix to a directly generated similarity matrix W

The same holds in principle for the norm-cut criterion, both for the directly generated similarity matrices W and for those matrices W^D derived from the distance matrix by means of the Gaussian kernel. However, in Figure 2, where the average difference of the error rates of the norm-cut clustering of W^D to the max-cut clustering of D is displayed, we can see: The norm-cut clustering always produces a higher error rate. The error rate is even more significantly higher for large fractions of missing values.

Did we introduce this increased error artificially by the additional transformation with the Gaussian kernel? Figure 3 indicates that this is not the case, as it shows nowhere a significantly positive value. Precisely, Figure 3 displays the difference of the error rates of the norm-cut clusterings of W^D relative to the directly generated matrices W.

Next, we turn to the computational resources required by the algorithms. Both max-cut and norm-cut are $\mathcal{NP}$-hard to optimize, so let us look at the relaxations. The spectral decomposition of a $n \times n$ matrix can be done in time $O(n^3)$, and if only the top k eigenvectors are desired, the effort can be even reduced to $O(kn^2)$ by an appropriate Lanczos method. Therefore the norm-cut/spectral algorithm has quadratic or (depending on the implementation) at most cubic complexity.

On the other hand, solving the SDP in order to approximate max-cut is more expensive. The respective complexity is $O(n^3 + m^3)$ (see [1]), where m is the number of constraints. If $k = 2$, then $m = n$ and the overall complexity is cubic. However, for $k \geq 3$, we need $m = O(n^2)$ constraints, resulting in an overall computational complexity of $O(n^6)$.

Finally, we remark that the analysis of the eigenvalues of the similarity matrix can yield a quite useful criterion to automatically determine the number k of clusters, in case that k is not known. We do not know of a corresponding method based on the distance matrix. We shall not further discuss this issue here and assume in the following that k is known.

6   The Google Distance

Our sample application for the subsequent simulations will be clustering of natural language terms using the Google distance. This distance has been suggested by Cilibrasi and Vitányi [2] as a semantical distance function on pairs of words or terms. For instance, for most of today's people, the terms "Claude Debussy" and "Béla Bartók" are much tighter related than "Béla Bartók" and "Michael Schumacher". The World Wide Web represents parts of the world we live in as a huge collection of documents, mostly written in natural language. We briefly describe the derivation of the Google distance, starting with concepts from Kolmogorov complexity (Algorithmic Information Theory).

Let us fix a universal Turing machine (which one we fix is not relevant, since each universal machine can interpret each other by using a "compiler" program of constant length). For concreteness, we assume that its program tape is binary, such that all subsequent logarithms referring to program lengths are w.r.t. base 2 (which is also not relevant for our algorithms). The output alphabet is ASCII. Then, the (prefix) Kolmogorov complexity of a character string x is defined as

$$K(x) = \text{length of the shortest self-delimiting program generating } x,$$

where by the requirement "self-delimiting" we make sure that the programs form a prefix-free set and therefore the Kraft inequality holds:

$$\sum_x 2^{-K(x)} \leq 1 \text{ , where } x \text{ ranges over all ASCII strings}$$

The Kolmogorov complexity is a well-defined quantity regardless of the choice of the universal Turing machine, up to an additive constant.

If x and y are ASCII strings and x^* and y^* are their shortest (binary) programs, respectively, we can define $K(y|x^*)$, which is the length of the shortest self-delimiting program generating y where x^*, the program for x, is given.

$K(x|y^*)$ is computed analogously. Thus, we may follow [8] and define the *universal similarity metric* as

$$d(x,y) = \frac{\max\left\{K(y|x^*), K(x|y^*)\right\}}{\max\left\{K(x), K(y)\right\}} \tag{7}$$

This can be interpreted as (approximately) the ratio by which the complexity of the more complex string decreases, if we already know how to generate the less complex string. The similarity metric is almost a metric according to the usual definition, as it satisfies the metric (in)equalities up to order $1/\max\left\{K(x), K(y)\right\}$.

Given a collection of documents like the World Wide Web, we can define the probability of a term or a tuple of terms just by counting relative frequencies. That is, for a tuple of terms $X = (x_1, x_2, \ldots, x_n)$, where each term x_i is an ASCII string, we set

$$p^{\text{www}}(X) = p^{\text{www}}(x_1, x_2, \ldots, x_n) = \frac{\text{\# web pages cont. all } x_1, x_2, \ldots, x_n}{\text{\# relevant web pages}}. \tag{8}$$

Conditional probabilities can be defined likewise as

$$p^{\text{www}}(Y|X) = p^{\text{www}}(Y \diamond X)/p^{\text{www}}(X) \, ,$$

where X and Y are tuples of terms and $\diamond$ denotes the concatenation. Although the probabilities defined in this way do not satisfy the Kraft inequality, we may still define complexities

$$K^{\text{www}}(X) = -\log\left(p^{\text{www}}(X)\right) \text{ and } K^{\text{www}}(Y|X) = K^{\text{www}}(Y \diamond X) - K^{\text{www}}(X). \tag{9}$$

Then we use (7) in order to define the *web distance* of two ASCII strings x and y, following Cilibrasi and Vitányi [2], as

$$d^{\text{www}}(x,y) = \frac{K^{\text{www}}(x \diamond y) - \min\left\{K^{\text{www}}(x), K^{\text{www}}(y)\right\}}{\max\left\{K^{\text{www}}(x), K^{\text{www}}(y)\right\}} \tag{10}$$

We query the page counts of the pages by using the Google API, so we call d^{www} the Google distance. Since the Kraft inequality does not hold, the Google distance is quite far from being a metric, unlike the universal similarity metric above.

The "number of relevant web pages" arising in (8) will be estimated by hand for all of the subsequent simulations. Actually, the clustering is not very sensitive to this value. On the other hand, Google no longer publishes its database size, and the full database size would not be the most appropriate value anyway for many applications. E.g., if we cluster words in a not so common language, then the index size relative to this language might be more appropriate.

7   Experimental Results with the Google Distance

We evaluate both clustering algorithms from Sections 2 and 3 on a set of natural language terms clustering tasks. We used the following datasets, which are all available at http://www-alg.ist.hokudai.ac.jp/datasets.html .

Table 1. Empirical comparison of the algorithms on the basic data sets (without removing additional data)

name	data set information size	#clusters	missing data	clustering errors and comp. time max-cut/SDP	norm-cut/spectral
people2	50	2	0%	0 (4 sec)	0 (0.07 sec)
people3	75	3	0%	0 ($\sim$ 90 sec)	2 (0.13 sec)
people4	100	4	0%	1 ($\sim$ 876 sec)	5 (0.2 sec)
people5	125	5	0%	4 ($\sim$ 2544 sec)	8 (0.35 sec)
alt-ds	64	2	0%	1 (3 sec)	1 (0.1 sec)
math-med-fin	60	3	0%	1 ($\sim$ 36 sec)	1 (0.1 sec)
finance-cs-j	30	2	0%	4 (1.8 sec)	1 (0.05 sec)
phil-avi-d	198	2	50%	5 (12 sec)	6 (2 sec)
math-cuisine	600	2	70%	23 (137 sec)	22 (16.6 sec)

The dataset people2 contains the names of 25 famous classical composers and 25 artists (i.e., two intended clusters), people3 contains all names from people2 plus 25 bestseller authors, people4 is extended by 25 mathematicians, and people5 additionally contains 25 classical composers. The dataset alt-ds contains not terms in natural language, but rather titles and authors' last names from (almost all of) the papers from the ALT 2004 and DS 2004 conferences. Furthermore we use the datasets math-med-fin containing 20 terms each from the mathematical, medical, and financial terminology, finance-cs-j contains 20 financial and 10 computer science terms in Japanese, phil-avi-d has 98 terms from philately and 100 terms from aviation in German, and math-cuisine has 254 mathematical and 346 cuisine-related terms (in English). The distance matrices of the last two data sets are not fully given: in phil-avi only 50% of the entries are known, in math-cuisine it is only 30%.

For the norm-cut based algorithm, we need to convert the distance matrix to a similarity matrix. We do this by using a Gaussian kernel $W^D = \exp(-\frac{1}{2\sigma^2}D^2)$ and set the width parameter $\sigma = \bar{D}/\sqrt{2}$, which gives good results in practice. Another almost equally good choice is $\sigma = \frac{1}{3}$, which can be justified by the fact that the Google distance is scale invariant and mostly in $[0, 1]$.

Table 1 shows the number of clustering errors, i.e., the number of data points that are sorted to a different group than the intended one, respectively, on the data sets just described. One can see that both algorithms perform well in principle, in fact many of the "errors" displayed are in reality ambiguities of the data, e.g., the only misclustering in the math-med-fin data set concerns the term "average" which was intended to belong to the mathematical terms but ended up in the financial group.

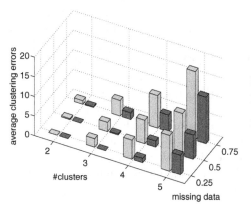

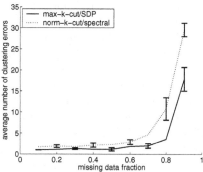

Fig. 4. Comparison of the max-cut/SDP (dark bars) and the norm-cut/spectral algorithm (light bars) with variable fraction of missing data and variable number of clusters and data set size, on the data sets `people2-people5`

Fig. 5. Comparison with variable fraction of missing data on the `math-med-fin` set

We remark that the constraints (1c) and the resulting huge SDP size were really necessary in order to get reasonable results: Without these constraints, e.g., clustering the `people5` data set with the SDP algorithm, the resulting average number of errors is 36.

Looking on the computation times in Table 1 (measured on a 3 Ghz Pentium IV), the spectral method is clearly much faster than the SDP, in particular for $k = 3$ or more clusters. Here the quadratic number of constraints in the SDP and the resulting 6th order computation time are really expensive. Actually, the available SDP software (CSDP, SeDuMi) do not even work at all with much larger problems if $k \geq 3$.

Next we consider a situation with varying fraction of missing data, shown in Figure 4 for the data sets `people2-people5`. Here the max-cut/SDP algorithm consistently outperforms the norm-cut/spectral algorithm, in particular if the number of clusters or the fraction of missing data grows. The same can be observed on the `math-med-fin` as shown in Figure 5. Note that both algorithms work quite well until about 70% missing data, after that the error increases sharply. Both figures are based on 20 independent samples of missing data each, where the missing data locations were sampled in a balanced way such that each row and column of the distance matrix has the same fraction of missing values. Figure 5 also displays 95%-confidence bars based on the standard normal assumption.

8   Relations to Other Work and Conclusions

There are many papers on clustering based on similarity matrices, in particular spectral clustering. It seems that norm-(k-)cut is quite established as an ideal criterion here, but there are different, such as min-max cut [3]. But also SDP has been used in connection with spectral clustering and kernels: [11] propose a SDP relaxation for norm-k-cut clustering based on a similarity matrix, while [6] and [7] use SDP for completion and learning of kernel matrices, respectively.

To our knowledge, the present work is the first one to use a distance matrix and a max-(k-)cut criterion for similar clustering tasks, which is natural in many applications where distances are given instead of similarities. We have seen that a SDP relaxation works quite well and yields results which tend to be superior to the spectral clustering results, in particular if the fraction of missing values grows. However, the SDP relaxation is expensive for $k = 3$ or more clusters. Thus we conclude with the open question of how to obtain a more efficient relaxation of max-k-cut, for instance a spectral one.

References

[1] B. Borchers and J. G. Young. Implementation of a primal-dual method for sdp on a shared memory parallel architecture. March 27, 2006.

[2] R. Cilibrasi and P. M. B. Vitányi. Automatic meaning discovery using Google. Manuscript, CWI, Amsterdam, 2006.

[3] C. H. Q. Ding, X. He, H. Zha, M. Gu, and H. D. Simon. A min-max cut algorithm for graph partitioning and data clustering. In *ICDM '01: Proceedings of the 2001 IEEE International Conference on Data Mining*, pages 107–114. IEEE Computer Society, 2001.

[4] A. Frieze and M. Jerrum. Improved algorithms for max k-cut and max bisection. *Algorithmica*, 18:67–81, 1997.

[5] M. X. Goemans and D. P. Williamson. .879-approximation algorithms for MAX CUT and MAX 2SAT. In *STOC '94: Proceedings of the twenty-sixth annual ACM symposium on Theory of computing*, pages 422–431. ACM Press, 1994.

[6] T. Graepel. Kernel matrix completion by semidefinite programming. In *ICANN '02: Proceedings of the International Conference on Artificial Neural Networks*, pages 694–699. Springer-Verlag, 2002.

[7] G. R. G. Lanckriet, N. Cristianini, P. Bartlett, L. E. Ghaoui, and M. I. Jordan. Learning the kernel matrix with semidefinite programming. *JMLR*, 5:27–72, 2004.

[8] M. Li, X. Chen, X. Li, B. Ma, and P. M. B. Vitányi. The similarity metric. *IEEE Transactions on Information Theory*, 50(12):3250–3264, 2004.

[9] B. Schölkopf, A. Smola, and K.-R. Müller. Nonlinear component analysis as a kernel eigenvalue problem. *Neural Computation*, 10(5):1299–1319, 1998.

[10] J. Sturm. Using SeDuMi 1.02, a MATLAB toolbox for optimization over symmetric cones. *Optimization Methods and Software*, 11(12):625–653, 1999.

[11] E. P. Xing and M. I. Jordan. On semidefinite relaxation for normalized k-cut and connections to spectral clustering. Technical Report UCB/CSD-03-1265, EECS Department, University of California, Berkeley, 2003.

[12] S. X. Yu and J. Shi. Multiclass spectral clustering. In *ICCV '03: Proceedings of the Ninth IEEE International Conference on Computer Vision*, pages 313–319. IEEE Computer Society, 2003.

Analysis of Linux Evolution Using Aligned Source Code Segments

Antti Rasinen, Jaakko Hollmén, and Heikki Mannila

Helsinki Institute of Information Technology, Basic Research Unit
Helsinki University of Technology, Laboratory of Computer and
Information Science, P.O. Box 5400, FI-02015 TKK, Finland
`Antti.Rasinen@hut.fi, Jaakko.Hollmen@hut.fi, Heikki.Mannila@hut.fi`

Abstract. The Linux operating system embodies a development history of 15 years and community effort of hundreds of voluntary developers. We examine the structure and evolution of the Linux kernel by considering the source code of the kernel as ordinary text without any regard to its semantics. After selecting three functionally central modules to study, we identified code segments using local alignments of source code from a reduced set of file comparisons. The further stages of the analyses take advantage of these identified alignments. We build module-specific visualizations, or descendant graphs, to visualize the overall code migration between versions and files. More detailed view can be achieved with chain graphs which show the time evolution of alignments between selected files. The methods used here may also prove useful in studying large collections of legacy code, whose original maintainers are not available.

1 Introduction

In data analysis, or data mining [5], one is usually interested in analyzing problems with little or no *a priori* information about the problem. Taking software as a collection of data — the approach taken in this paper — we aim to analyze the source code with data mining methods and explore the evolution of the software system itself.

We study the source code of the Linux operating system. Its open development process allows us to sample versions throughout the fifteen years of development. The freely available source code and the large number of literature (e.g. [11], [8], [1]) about Linux provides us a way to validate our results. After determining the suitable versions and modules for further analysis, we map the similar segments in the source code. This approach is inspired by the synteny maps [3] used in genetic research.

A study [10] in 1999 examined two other free software systems: the Apache web server and the bash shell. They used common source code quality metrics such as cyclomatic complexity, number of lines of code, number of individual function calls and others. Their approach differs from ours as we do not use predefined metrics. Another study [4] in 2000 showed that the growth of Linux has been super-linear during the the period from 1994 to 2000. Their analysis was based on the sizes of the source code files.

N. Lavrač, L. Todorovski, and K.P. Jantke (Eds.): DS 2006, LNAI 4265, pp. 209–218, 2006.

The rest of the paper is organized as follows. In Section 2, we describe the Linux operating system and its evolution. The alignment algorithms used in the paper are described in Section 3. The alignments are further analyzed to form graphs, described in Section 4. A closer look at the file-specific evolution is given by chains, described in Section 5. The results are discussed in Section 6. The paper is summarized in Section 7.

2   Linux Operating System and Its Evolution

Linux is a popular open-source operating system developed by hundreds of core developers around the world. It runs on several platforms and supports a wide range of hardware devices. Linux is *free*; it is available for no cost and there are very few limitations concerning its use and distribution.

Linux began as a hobby project for a Finnish student Linus Torvalds, but it matured quickly into a usable operating system for Intel-based PCs. Version 1.0 was released in March 1991 and version 2.0 in June 1996. Versions with an even minor version number, e.g. 2.2.*x*, are called *stable branches*. The latest stable version is 2.6.16.9 at the time of writing, April 2006. For the purposes of our study, we chose six versions of Linux: 1.0, 1.2.0, 2.0.1, 2.2.0, 2.4.0, and 2.6.0. These are the first versions of the stable branches of the kernel. They are separated from each other by at least a year. The size of the Linux source code has almost doubled at the release of each new stable branch. The size of the 2.6-branch is over 200 megabytes. A vast majority of the Linux code base consists of device drivers. In the latest 2.6-branch, the size of the subdirectory `drivers` is approximately half of the total size.

However, the evolution of device drivers is very limited and isolated. Once a driver has been written, the only changes will be bug fixes. There is also little interaction between driver source code files — the main exception is when a driver is based on another driver for older hardware. The Linux source code distribution is divided into subdirectories. Most of these subdirectories contain a part of the source code for the kernel, such as the aforementioned `drivers`. We call such subdirectories *modules*. Note that parts of the kernel that are loaded dynamically at run-time are also called modules.

In this work we will examine the changes in Linux by studying how source code has changed within and between files. Thus, we will use the source code itself as our data.

To narrow our data, we chose such *modules* that they contain most of the essential functionality of an operating system kernel: memory management, process management, scheduling, virtual memory *et cetera*. We were left with three modules: `ipc`, `kernel`, and `mm`. Another reason for limiting ourselves to a subset of the entire source code distribution is the computational complexity of the algorithm we intend to use. This is discussed in Section 3.1.

Each release of Linux has roughly doubled the size of the source code distribution as shown by Figure 1. Two of the three modules chosen for this work, `ipc` and `mm`, follow the same general pattern. The module `kernel` behaves differently

and shrinks between releases. This is related to the large number of changes between kernel series 1.x and 2.x. Several elements in the module are transferred
to other subdirectories or under architecture-specific directories.

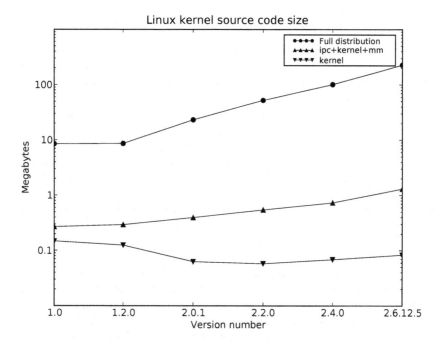

Fig. 1. Comparison of growth rates for the full distribution, modules studied in this
work (`ipc`, `mm` and `kernel`) and the kernel module alone. The full distribution includes
the source code, header files, documentation etc.; the modules have been cleaned up
and they contain only the C-language source code. The combined size of `ipc`, `kernel`
and `mm` modules follows the general exponential, while `kernel` module, when examined
alone, remains almost constant. Note that sizes are plotted on a logarithmic scale.

3 Measuring Similarity Between Source Code Files

To compare files from different versions with each other, we need a measure
of similarity. To build such a measure, we first considered the source code files
as strings (text representation). There are several methods that assign a *global*
similarity score for two strings, such as the Needleman-Wunsch [7] algorithm or
the Jaro distance [6].

Our strings also have an internal structure. The basic building blocks of a computer program, functions, variable definitions, loops *et cetera* are very mobile.
Functions can be moved from file to file with ease or commonly used structures can be isolated into a new file. These are the kind of changes we are
interested in.

Fitting these two requirements together is easier than it seems. Figuring out the internal structure can help us in deciding the overall similarity. There is a class of string matching algorithms, known as *local alignment* algorithms, that are suited for our approach.

3.1 String Alignment Algorithms

Local alignment algorithms isolate similar segments from the two input strings. In this context, similar does not mean identical; several algorithms allow gaps or mismatches between letters. This fit our needs well, since it allowed us to ignore differences that were "too small", such as an added space or newline.

There are several local alignment algorithms available. We chose to use one of the simplest, the Smith-Waterman algorithm [9]. It is based on the earlier Needleman-Wunsch algorithm [7]. The algorithm uses dynamic programming [2] to compute a score table, from which the best-matching alignment can be extracted. More precisely, we can extract the relevant substrings from the two input strings.

Ordinary Smith-Waterman algorithm only returns the highest scoring alignment between the two strings. We modified the algorithm to compute more than one alignment. After we have found the longest alignment, the rows and columns representing it are marked as visited. This partitions the score table into several subtables, which represent different combinations of the remaining substrings.

As an example, consider strings "ACDC-COMMON-DCDC" and "Beatles-COMMON-Rolling Stones". After we remove the longest common substring, "-COMMON-", we are left with pairs ("ACDC", "Beatles"), ("ACDC", "Rolling Stones"), ("DCDC", "Beatles") and ("DCDC", "Rolling Stones"). We need to to recompute the scores in the subtables corresponding to these pairs with Smith-Waterman algorithm. Note that when an alignment is found in a subtable, we must partition all such subtables that contain the rows and columns associated with that alignment.

We set a minimum limit for the length of the alignments in order to keep them meaningful. Without a limit the modified algorithm would continue until both strings are exhausted. This would result in a large number of length-1 alignments. When string s_1 in file F_1 is aligned with string s_2 in file F_2, we say that F_1 and F_2 *share an alignment*.

Given two strings with lengths M and N, both the time and memory complexity are both $O(MN)$. A quick analysis of the computational requirements shows that it is too expensive to go through each possible pair of files. Our initial starting point was for each of the three modules to compare each file with every other file, except with those in the same version. There were 10724 such pairs. To reduce that number, we chose ten pairs that clearly were related and compared their alignments with ten randomly chosen pairs. The distributions of the alignment lengths were clearly different for the two sets of files. The longest alignments in the randomly chosen set had a score less than 150, which corresponds to approximately 80 characters. In comparison, each of the pairs in the hand-picked set had alignments above the score of 150 and the longest alignments had a score over 20000. Computing the score of the longest alignment is faster than computing all the alignments. Thus as a preliminary step, we pruned out pairs whose longest mutual alignment had a score less than 150. This resulted in 769 pairs.

3.2   Similarity from Alignments

All the alignments produced by the Smith-Waterman algorithm have a score associated with them. These can be used to build several different similarity measures for the files themselves by transforming them into vectors. Consider the files F_1 and F_2 who share alignments with scores $\{a_1, a_2, \ldots, a_n\}$. The Euclidian norm can be transformed into a similarity function $S_2(F_1, F_2) = \sum_{k=1}^{n} a_k$. For our work we considered the 1-norm and the ∞-norm, which correspond to taking a sum and the maximum of the set of scores, respectively.

Alignments with lower score are often less "meaningful" than the ones with a higher score, which might distort the results. For this reason we occasionally considered only alignments above a certain cutoff score, such as 150. Note that this does not affect the ∞-norm. It is worth pointing out, that these similarity measures are not complete. There may be files that do not share any alignments; for such files the similarity is undefined.

4   Descendant Graphs of Modules

We used the similarity information to construct a *descendant graph*. As a similarity measure, we used a 1-norm and discarded alignments under score 150. With the descendant graph we examined how code is reorganized between consecutive versions. We plotted all the files in a given module (`ipc`, `kernel`, or `mm`) and connected each file F to its "descendants", that is, files in the next version that share an alignment with F. The similarity between the files is denoted by the weight of the link—the heaviest links belong to the top quartile, while the lightest links belong to the bottom quartile. Examples of a descendant graph are in Figures 2 and 3.

The descendant graph is a very useful tool for an initial exploration of the source code evolution. For example, a file that has been split into two or more parts, is easily recognizable. There are several such examples in Figure 2. Files that have been rewritten between versions lack a link between them, which makes them also very noticeable. More details can be found rapidly with relatively small effort. A good way to use the descendant graph is to find starting points for a more in-depth analysis, such as studying the source code by hand.

On the module level the descendant graph allows for a very quick way to identify those modules that have been subject to higher amounts of reorganization. Figures 2 and 3 differ significantly, with code "flowing" differently in the two modules.

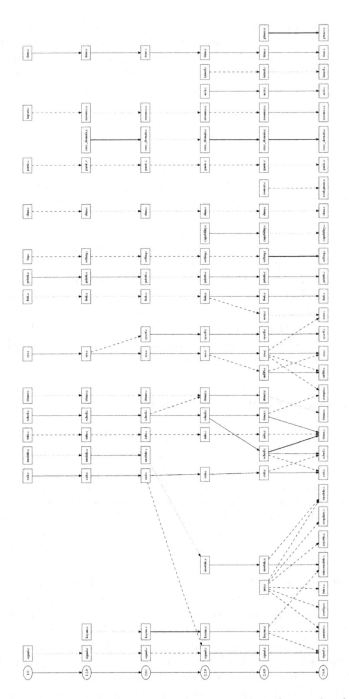

Fig. 2. The descendant graph for the module `kernel`. Note the simple relationships between files in different versions.

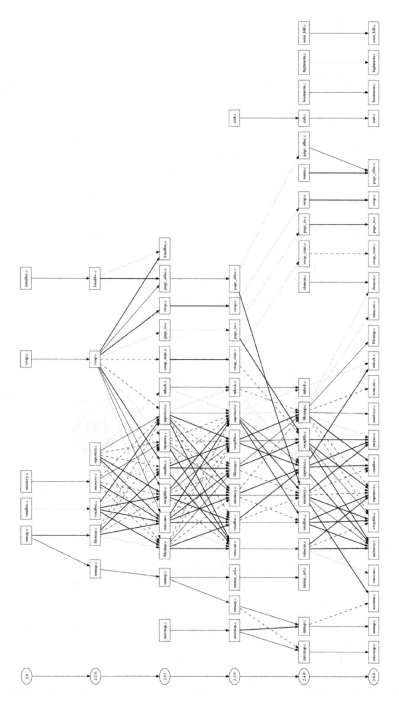

Fig. 3. The descendant graph for the module mm. The interactions are more diverse in this example. Note especially the strongly connected central core.

5   Detailed Analysis with a Chain Graph

The descendant graph is only intended for a global view at the module level.
It is a high-level tool specifying the files and versions where significant change
occurs, but it lacks information about the changes themselves. This is analogous
to the difference between global and local alignment algorithms.

We could improve the descendant graph by explicitly revealing the internal
structure of the file; that is, show the locations of the alignments within the
files and how they are connected to the corresponding alignment in an other file.
This resembles the HistoryFlow visualization presented in [12]. Unfortunately
the new edges in the graph severely degrade the clarity of the visualization. On
a large or complex descendant graph this negates the benefits given by the more
detailed view.

Our solution was to restrict the visualization to a subgraph of the original
graph. At the moment, our software only allows linear chains—a file has at most
one child and one parent—but it would not be difficult to extend it to more
general subgraphs. In this work chains were found to be sufficiently clear. An
example of two chains is shown in Figure 4.

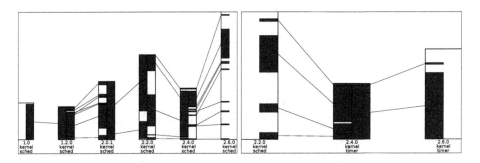

Fig. 4. The file `sched.c` is split into two files (`sched.c` and `timer.c`) between versions
2.2.0 and 2.4.0

The chain graph shows the relative size of the files as well as the location
of the alignments. Each file is split into left and right halves, which contain
the alignments with the preceding and the succeeding versions, respectively.
Alignments are linked together with a line.

As discussed above, the chain graph complements the descendant graph.
Studying individual alignments allows for us to recognize different patterns of
change. As an example, Figure 4 shows how parts of `sched.c` in version 2.2.0
being moved into a new file, `timer.c`. In addition to the aforementioned *split*,
other commonly occurring changes were *gradual change*, *rewrite*, and *fusion*. In
gradual change the file stays mostly unchanged. A rewrite is almost the complete
opposite, as most of the file is being programmed from scratch. It is not straightforward to detect rewrites, as the two files do not share many alignments. This
can be avoided with an automatic system that checks similarly named files in

different versions and reports when there are suspiciously few alignments. Fusion means the merger of two files. It is rarer than the other common change patterns. In all the cases we encountered, a small file is inserted into a much larger file.

6   Discussion

We believe that the methodology introduced in this paper can be used in exploratory analysis of large software systems, where little knowledge is available about the system evolution. As an example, consider a system built over several years by a contractor. It is possible that when the client wishes to have an update release of the software, the original vendor may be bankrupt, its key people may have left the company, or another vendor may have won the bid instead. In such cases the new vendor must quickly assimilate the architecture of the current system.

Good project management and rigorous documentation will ease the task of the new vendor. In addition, there are already several tools and methods for analysing the current version of a software system, such as call graphs, similarity analysers, software metric suites and instrumentation tools. Unfortunately project hand-offs do not always proceed smoothly, in which case the knowledge transfer process may be incomplete.

The current architecture of a software system is a product of evolution. The key decisions made several releases ago may have left their mark in "vestigial" routines, code files or even modules, especially when the project environment does not support refactoring. In this case architects, designers and developers would in our view benefit from our methods. A large-scale map would help explain the most difficult features of the existing code base. It is worth pointing out that our aim is not to replace the existing tools, but to augment them.

The above scenario is loosely based on a real-life project in which one of the authors participated. Informal personal discussions with software developers indicated that most of them had encountered similar projects.

The method here is intended as a proof of concept. As a result, there are several areas where our work could be improved. The computation of alignments is quite simplistic and as a result it is very computationally intensive.

7   Summary and Conclusions

We have examined the structure and evolution of the Linux kernel. We considered the source code as ordinary text and analyzed it with data analysis methods. First we identified local alignments of source code from a reduced set of files and versions. Based on the alignments, we built descendant graphs to visualize the evolution of the modules. A more closer look on the code migration between different versions can be achieved by the introduced chain graphs. Both visualizations are built using the alignments of the source code.

We have knowingly used the most primitive representation of the source code, namely the textual representation without any use of the semantics of the programming language or without any use of software metrics. Arguably, a tokenized version of the source code could reveal more functional patterns in the source modules. We intend to look at this issue in our further research.

Also, other systems than Linux could be explored to gain more insight to the generality of the software evolution. Trial runs with real-life projects would also give valuable information about the benefits and drawbacks of our methods. This requires improvements to the user interface, as the current approach requires large amounts of expertise.

References

1. M. Beck, H. Böhme, M. Dziadzka, U. Kunitz, R. Magnus, C. Schröter, and D. Verworner. *Linux Kernel Programming, 3rd ed.* Pearson Education Ltd., 2002.
2. Richard Bellman. On the theory of dynamic programming. In *Proceedings of the National Academy of Sciences of the United States of America*, volume 38, pages 716–719, Aug 1952.
3. Ethan A. Carver and Lisa Stubbs. Zooming in on the human-mouse comparative map: Genome conservation re-examined on a high-resolution scale. *Genome Research*, 7(12):1123–1137, December 1997.
4. Michael W. Godfrey and Qiang Tu. Evolution in open source software: A case study. *16th IEEE International Conference on Software Maintenance (ICSM'00)*, 0:131–142, 2000.
5. David J. Hand, Heikki Mannila, and Padhraic Smyth. *Principles of Data Mining.* MIT Press, 2001.
6. M. A. Jaro. Advances in record linking methodology as applied to the 1985 census of tampa florida. *Journal of the American Statistical Society*, 64:1183–1210, 1989.
7. Saul B. Needleman and Christian D. Wunsch. A general method applicable to the search for similarities in the amino acid sequence of two proteins. *Journal of Molecular Biology*, 48:443–453, March 1970.
8. David A. Rusling. The Linux kernel. Published in Web at `http://www.tldp.org/LDP/tlk/tlk.html`. Last visited on March 25th 2006.
9. T. F. Smith and M. S. Waterman. Identification of common molecular subsequences. *Journal of Molecular Biology*, 147:195–197, 1981.
10. Ladan Tahvildari, Richard Gregory, and Kostas Kontogianni. An approach for measuring software evolution using source code features. *Sixth Asia-Pacific Software Engineering Conference (APSEC'99)*, 00:10–17, 1999.
11. Linus Torvalds. Linux: a portable operating system. Master's thesis, University of Helsinki, 1997.
12. Fernanda B. Viégas, Martin Wattenberg, and Kushal Dave. Studying cooperation and conflict between authors with history flow visualizations. In *Conference on Human Factors in Computing Systems*, 2004.

Rule-Based Prediction of Rare Extreme Values

Rita Ribeiro[1] and Luís Torgo[2]

[1] LIACC - University of Porto, R. Ceuta, 118, 6o, 4050-190 Porto, Portugal
`rita@liacc.up.pt`
[2] FEP/LIACC - University of Porto, R. Ceuta, 118, 6o, 4050-190 Porto, Portugal
`ltorgo@liacc.up.pt`

Abstract. This paper describes a rule learning method that obtains models biased towards a particular class of regression tasks. These tasks have as main distinguishing feature the fact that the main goal is to be accurate at predicting rare extreme values of the continuous target variable. Many real-world applications from scientific areas like ecology, meteorology, finance,etc., share this objective. Most existing approaches to regression problems search for the model parameters that optimize a given average error estimator (e.g. mean squared error). This means that they are biased towards achieving a good performance on the most common cases. The motivation for our work is the claim that being accurate at a small set of rare cases requires different error metrics. Moreover, given the nature and relevance of this type of applications an interpretable model is usually of key importance to domain experts, as predicting these rare events is normally associated with costly decisions. Our proposed system (R-PREV) obtains a set of interpretable regression rules derived from a set of bagged regression trees using evaluation metrics that bias the resulting models to predict accurately rare extreme values. We provide an experimental evaluation of our method confirming the advantages of our proposal in terms of accuracy in predicting rare extreme values.

1 Introduction

In data mining there are several prediction problems for which the rare instances of the concept to be learned are the most important ones. Forecasting large changes on stock prices, ecological or meteorological catastrophes, are a few examples of applications where we are faced with this kind of problems. In all these applications, domain experts are specially interested in having accurate and interpretable predictions of such rare events as these are usually associated with costly actions/decisions. The work presented in this paper addresses this kind of applications in a regression context. These problems present difficult challenges to learning methods as we are trying to model a concept that is rare and less represented than other common concepts in the used data sets.

Predictive data mining tasks fall in two categories: classification, where the target variable is discrete; and regression, where the target variable is continuous. Within classification, this kind of problems is a well-known subject of research

N. Lavrač, L. Todorovski, and K.P. Jantke (Eds.): DS 2006, LNAI 4265, pp. 219–230, 2006.
© Springer-Verlag Berlin Heidelberg 2006

and is related to the problem of unbalanced class distributions [14]. According to some studies (e.g. [5,12,14]), the existence of a minority class brings an additional difficulty to the traditional classification methods which are biased to the prediction of the most common values by evaluation criteria such as accuracy.

Most existing work on prediction of rare events within data mining is related to classification tasks. Still, the same type of problems appear in the context of regression. As in classification, one should change the evaluation criteria used by the traditional regression methods that are biased to the prediction of the most common values. In a previous work [9], we have handled this type of problems by proposing a new splitting criterion for CART-like regression trees [1]. Based on this previous work, we now present a rule-based regression system for the prediction of rare and extreme values of a continuous target variable. Compared to the former, this new system improves on both accuracy and interpretability due to the modular characteristics of rule-based systems.

2   Background

Predicting rare events has been receiving an increasing attention from the data mining community. This interest stems from both the important associated applications, and from the fact that learning a concept based on cases that are rare is a non-trivial task for traditional learning methods. Standard machine learning methods are biased to the prediction of the most common values and usually assume that all the prediction errors have the same "cost".

Within classification tasks, one of the proposed approaches to handle rare cases is to use misclassification costs (e.g. [10]). This allows the errors committed at some subset of cases belonging to a rare class to be more penalized and thus models will be biased towards avoiding these errors. Moreover, on these problems with an unbalanced class distribution, some authors [5,13,14] have shown that evaluating models by classification accuracy is not adequate. In this context, they proposed different performance metrics based on ROC curves or in measures like precision and recall. The Two-Phase Rule Induction method proposed by Joshi et al. [5] is an example of a rule induction system biased towards the minority class which induces the rule set in two steps considering recall and then precision. Given the clear tradeoff between precision and recall, some works propose the use of *F-measure* [7], one of the measures which combines those two, as shown in Equation 1.

$$F = \frac{(\beta^2 + 1) \cdot precision \cdot recall}{\beta^2 \cdot precision + recall} \tag{1}$$

where $0 \leq \beta \leq 1$, controls the relative importance of recall to precision.

Nevertheless, all these classification approaches are not directly applicable to the type of problems we wish to address here because our target variables are continuous. Although there are several works that handle regression problems through a classification approach (e.g. [4,8,15]), these approaches do not fully meet our target applications requirements. Our goal is not only to capture the rare and extreme values, but also to be able to predict them as accurately as possible in a numeric perspective. This means that the degree of extremeness of the target variable is also relevant for domain experts, as different actions can be taken according to that degree. One could argue that by having several classes associated to these different degrees of extremeness would overcome this difficulty and allow classification methods to be applied. Still, we argue that this would split an already low populated class associated to rare events into even less frequent classes, thus making the problem even harder. Moreover, this would always be a coarse approximation of an inherently continuous prediction problem.

3 Our System: **R-PREV**

Given the requirements of our target applications, our goal was to develop a system capable of accurately predict rare extreme values of a continuous target variable in an interpretable way.

In regression methods, as in most learning tasks, the model parameters estimation process is guided by some preference criterion. The most common choices are estimators of the true average prediction error (e.g. mean squared error) of the models. These performance metrics are calculated over all the range of values of the target variable and thus will tend to bias the models to maximize performance over the most common values, as these will have a stronger impact on the overall mean error. This type of preference criteria is not suitable when the interest resides on the performance on a special subset of values that are not very frequent. This is the case of our target applications: the obtained model should perform specially well over the rare and extreme values of the continuous target variable.

The Rule-based Prediction of Rare Extreme Values system (R-PREV) we propose in this paper is based on the trees obtained by a system we have described in a previous work [9]. As such, we will now provide a detailed description of main features of this later system that we will refer as "Base Tree". This consists of a regression tree induction system based on CART [1] but with a different splitting criterion that enables the induction of trees biased towards the prediction of rare and extreme values. The main idea is to use the *F-measure* presented in Equation 1 as splitting criterion for the tree growth. The first step to allow the use of this metric is to provide a formal definition of what is a rare extreme value. In cases where no domain knowledge is available to define this notion, we have used the statistical notion of outlier, given by the box-plot [2], to establish the two thresholds that define the rare extreme high and low values of the target variable. The default thresholds are the so-called adjacent values of the box-plot of a continuous variable. The upper-adjacent value, thr_H, is defined as the largest observation that is less or equal to the 3rd quartile plus $1.5r$, where r is the interquartile range, i.e. the difference between the 3rd and 1st quartiles of the target variable. In an equivalent way, the lower adjacent value, thr_L, is the smallest observation that is greater or equal to the 1st quartile minus $1.5r$.

Once we have these two thresholds, obtained either by existing domain knowledge or by using the information of box-plots, we can define our rare extreme values as,

$$RE = \{y \in Y \mid y > thr_H \vee y < thr_L\}$$
$$RE_H = \{y \in Y \mid y > thr_H\} \qquad (2)$$
$$RE_L = \{y \in Y \mid y < thr_L\}$$

Depending on the application, we may have either RE_H or RE_L empty. In Figure 1, there is an example of a box-plot obtained for a continuous variable representing the median values of houses in Boston residential areas, using the well-known Boston Housing data set [11]. The circles are the rare extreme values determined by the upper adjacent and lower adjacent values represented by the two horizontal lines outside the box. In this particular case, we only have high rare extreme values, that is, residential areas with extremely expensive houses which distinguish themselves from the rest. According to these thresholds, in this dataset there are no low extremes, i.e. extremely cheap houses.

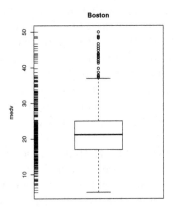

Fig. 1. Example of a box-plot for the 'medv' attribute of Boston dataset

Once we have the concept of rare extreme value defined, it is necessary to specify how to calculate recall and precision in a regression context[1], as they are required for obtaining the *F-measure* (c.f. Equation 1).

Let $\hat{y}_i$ be the prediction obtained for the case $\langle x_i \rangle$ whose target true value is y_i. We can define the following two sets:

- $\widehat{Y}_{RE} = \{\hat{y}_i \in \widehat{Y} \mid y_i < thr_L \vee y_i > thr_H\}$, i.e., $\widehat{Y}_{RE}$ is the set of $\hat{y}$ predictions of the model for the rare extreme value cases;
- $\widehat{Y}_{\widehat{RE}} = \{\hat{y}_i \in \widehat{Y} \mid \hat{y}_i < thr_L \vee \hat{y}_i > thr_H\}$, i.e., $\widehat{Y}_{\widehat{RE}}$ is the set of $\hat{y}$ predictions of the model that are rare extreme values.

---

[1] These measures are originally defined in a classification context.

Given these sets, we define recall as the proportion of rare extreme values in our data that are predicted as such (i.e. covered) by our model, by the following equation,

$$recall = \frac{|\{\hat{y}_i \in \widehat{Y}_{RE} \mid \hat{y}_i < thr_L \vee \hat{y}_i > thr_H\}|}{|\widehat{Y}_{RE}|} \tag{3}$$

Regarding precision, if we use its standard definition it would be defined as the proportion of predicted rare extreme values that are really extremes (c.f. Equation 4).

$$precision_{stand} = \frac{|\{\hat{y}_i \in \widehat{Y}_{\widehat{RE}} \mid y_i < thr_L \vee y_i > thr_H\}|}{|\widehat{Y}_{\widehat{RE}}|} \tag{4}$$

However, because we are in a regression context, we adapted the concept of precision so that the amplitude of the differences between predictions and true values is taken into account (c.f. [9]). In this context, we have proposed and used the following definition of precision,

$$precision = 1 - NMSE_{\widehat{RE}} \tag{5}$$

where $NMSE_{\widehat{RE}}$ is the normalized squared error of the model for the cases predicted as rare extreme values,

$$NMSE_{\widehat{RE}} = \frac{\sum\limits_{\hat{y}_i \in \widehat{Y}_{\widehat{RE}}} (\hat{y}_i - y_i)^2}{\sum\limits_{\hat{y}_i \in \widehat{Y}_{\widehat{RE}}} (\overline{Y} - y_i)^2} \tag{6}$$

Suppose that in some application $thr_H = 10$. If we have a test case with a true value of 12, the proposed definition allows us to signal a prediction of 11 as much better (more precise) than a prediction of 30 for the same test case. Thus, with this proposed definition of precision, we are able to bias the models to be accurate in the degree of extremeness. The use of a normalized metric like $NMSE$ ensures that precision varies between 0 and 1 like recall. For rare situations where $NMSE$ goes above 1, which means that the model is performing worse than the naive average model, we consider that the precision of the model is 0.

As mentioned before the tree growth procedure is guided by a split criterion based on precision and recall, namely the *F-measure* (c.f. Equation 1). The best split will be the one that maximizes the *F-measure* in one of the partitions generated by the split. The tree continues its growing process until the *F-measure* value of a split goes above some pre-specified threshold f^2, or until there are no

---

[2] Experiments carried out in a previous work [6], have shown that the "best" setting for the threshold f is domain dependent and thus for achieving top performance for a particular problem, some tunning process is recommended. In the context of the experiments of this paper we have used a the default value of 0.7, which these experiments have shown as a generally reasonable setup in many domains.

more rare extreme values in the current node partition. Full details on the growth of these trees can be obtained in [9].

Interpretability is of key importance to our target applications as the predicted rare events are usually associated with costly decisions. In this context, we have decided to select a rule-based formalism to represent our models. Rules are usually considered to have greater explanatory power then trees, mainly due to their modular characteristics.

We have obtained a set of rules based on trees generated by the "Base Tree" system we have just described. A set of rules can be easily obtained from a regression tree. Each path from the root of the tree to a leaf is transformed into a rule of the form:

$$\textbf{if } cond_1 \wedge cond_2 \wedge ... \wedge cond_n \textbf{ then } v_i$$

where each $cond_k$ is a test over some predictor variable in the considered path i, and v_i is the value of the leaf at the end of that path.

Trees generate a mutually exclusive partition of the input space of a problem. This means that given any test case, only one rule will cover it. We have decided to obtain our set of rules from a set of trees obtained through a bagging process in order to both decrease the variance component of the error of the resulting model and also to eliminate this mutually exclusivity property, thus ensuring a higher modularity of each rule in the final model. We start by obtaining a pre-specified number of stratified bootstrap samples, so that for each sample we can have a similar distribution function for the target variable. For each sample we run the regression tree method referred above and then transform it into a set of rules. This process is repeated for all trees obtained from the bootstrap samples.

Once we get this large set of rules, R, originated from different trees we try to simplify it in two forms: individually, using some simple logical simplifications of the conditions on each rule; and globally by eliminating some rules from this set using the information regarding their specificity and *F-measure*. As a result of this process we obtain an ordered rule set, also known as a decision list.

We measure the specificity of a rule by the number of cases that are uniquely covered by that rule,

$$spec(r) = | \ \{\langle \mathsf{x}_i \rangle \ | \ cover(R, \langle \mathsf{x}_i \rangle) = \{r\}\} \ | \tag{7}$$

where R is the entire rule set and $cover(R, \langle \mathsf{x}_i \rangle)$ is the set of rules that cover the case $\langle \mathsf{x}_i \rangle$.

We want to retain rules with high specificity because they represent knowledge that is not captured by any other rule. Regarding the remaining rules (whose specificity is zero), we order them by their evaluation criterion (*F-measure* score) and then select the top k rules according to a user-specified margin parameter, m.

This means that the final theory, T, is given by the rules belonging to the initial set, R, ordered by their *F-measure*, such that,

$$T = \{ \ r \in R \ | \ spec(r) > 0 \vee F(r) > (1 - m) \cdot F(r_{top}) \ \} \tag{8}$$

where $F(r)$ is the *F-measure* of the rule r, r_{top} is the rule with the best *F-measure* and m is the margin parameter.

Notice that if we want a theory formed only by rules that have some specificity then we can set the margin parameter (m) to 0. This is also the setting that leads to a smaller theory, as larger values will increase the number of rules.

This rule selection process removes the complete coverage property that results from the mutual exclusivity of a tree, which means that there may exist a test case that is not covered by any of the rules in the final theory, T. For these situations we have added a default rule at the end of our ordered set of rules, T. This default rule basically predicts the mean value of the target variable (c.f. Equation 9).

$$T_f = T \cup \{ \text{ if } null \text{ then } \overline{Y} \} \tag{9}$$

Obtaining a forecast for a test case $\langle x_i \rangle$ involves averaging all the predictions of the rules satisfied by $\langle x_i \rangle$, weighted by their respective F-measure,

$$\hat{y}_i = \frac{\sum_{k=1}^{|C|} F(C[k]) \cdot predict(C[k], \langle x_i \rangle)}{\sum_{k=1}^{|C|} F(C[k])} \tag{10}$$

where $C = cover(T_f, \langle x_i \rangle)$.

The algorithm of R-PREV can be summarized by the steps given in Figure 2.

---

1. establish thr_L and thr_H for the target variable Y;
2. specify β and f parameters;
3. specify a margin parameter m;
4. generate n bootstrap stratified samples;
5. $R = null$;
6. for each i from 1 to n
   (a) $t_i = \text{BaseTree}(sample_i, thr_L, thr_H, \beta, f)$;
   (b) obtain the rule set R_i from the tree t_i;
   (c) perform logic simplifications over R_i;
   (d) $R = R \cup R_i$;
7. sort the rules in set R by their F-measure values;
8. obtain the final theory:
   $T_f = \{ r \in R \mid spec(r) > 0 \vee F(r) > (1 - m) \cdot F(r_{top}) \} \cup \{ \text{ if } null \text{ then } \overline{Y} \}$

Fig. 2. R-PREV main algorithm

4   An Analysis of System R-PREV

In this section we analyze the performance of our proposal with respect to the two main features that distinguish it from our previous work [9]: the improved accuracy at forecasting rare extreme values; and the interpretability advantages of its rule-based formalism.

4.1   Predictive Accuracy

We have carried out a set of experiments with the goal of estimating the performance of our proposed system in the task of predicting rare extreme values, when compared to related regression methods. Namely, we have compared several variants of our system (R-PREV) with different settings in terms of the margin parameter ($m = 0\%, 25\%, 50\%, 75\%, 100\%$), with: the base system used to obtain the trees ("BaseTree" [9]); a CART-like regression tree; and a bagged CART-like regression tree (BaggCART). The selection of competitors was carried out with the goal of having a better understanding of the gains caused by each of the added features of our system when compared to a simple CART-like tree.

The experiments were carried out on a set of real-world problems, some of which are commercial applications. The methods were tested over 24 data sets using 10 repetitions of a 10-fold cross-validation procedure. Regarding the methods that use bagging we have used 50 bootstrap stratified samples.

Taking into account recent results reported in [3] regarding the comparison of multiple models over several data sets, we have used the Friedman test and the post-hoc Nemenyi test for asserting the statistical significance of the observed differences in performance.

We have estimated the performance of the different methods by means of the *F-measure* values as this statistic is better at characterizing the performance in rare extreme values. We have used a β value of 0.5 for the *F-measure* calculation. This choice is justified by the fact that, given that we are addressing a numeric prediction task, precision is always the most important factor. With $\beta = 0.5$, we are giving it doubled importance relatively to recall. For each dataset, we calculated the mean value of F obtained by each method over all repetitions. In order to check whether the systems can be considered equivalent, we applied the Friedman test. This test ranks internally the obtained results for each dataset over all the compared methods and obtains rank data like the one shown in Table 1.

In these experiments we have used the default parameters of all systems as our goal was not to optimize their performance on each individual problem. Therefore, some individual results may not be as good as possible. This is particularly noticeable in our R-PREV system, as previous experiments [6] have revealed a certain sensitivity to the setting of parameter f.

Regardless of this, the Friedman test applied over the rank data, reported a significant difference between the 8 compared methods at a significance level of 5%. Given this result, we proceed by applying the post-hoc Nemenyi test to the rank data in order to compare all methods to each other. The results of this test are better visualized by the CD (critical difference) Diagram proposed by [3] and presented in Figure 3. This diagram represents the information on the statistical significance regarding every pairwise comparison between methods, which means that each method is represented by 7 symbols as we have 8 methods being compared. The methods are plotted at their respective average ranking value in terms of the X axis (notice that CART and BaggCART have the same average ranking). The vertical axis position has no meaning. A dotted line connecting two symbols has the meaning that according to the Nemenyi test, the methods are significantly different at a 5% level. Bold lines indicate

Table 1. Ranking of the different regression methods over the set of datasets

datasets	R-PREV $m = 0$	R-PREV $m = 25$	R-PREV $m = 50$	R-PREV $m = 75$	R-PREV $m = 100$	Base Tree	CART	BaggCART
servo	8.0	7.0	4.0	3.0	5.0	6.0	2.0	1.0
triazines	7.5	4.0	3.0	2.0	1.0	5.0	6.0	7.5
algae1	7.5	6.0	3.0	2.0	1.0	4.0	5.0	7.5
algae2	6.5	6.5	3.0	2.0	1.0	4.0	6.5	6.5
algae3	7.5	7.5	3.0	2.0	1.0	5.0	4.0	6.0
algae4	7.5	7.5	4.0	2.0	1.0	6.0	3.0	5.0
algae5	7.0	7.0	3.0	2.0	1.0	4.0	5.0	7.0
algae6	7.5	7.5	5.0	2.0	1.0	6.0	3.0	4.0
algae7	8.0	7.0	3.0	2.0	1.0	6.0	4.0	5.0
machine-cpu	8.0	6.0	7.0	1.0	2.0	4.0	5.0	3.0
china	7.0	7.0	3.0	2.0	1.0	5.0	4.0	7.0
sard0	8.0	7.0	6.0	2.0	1.0	5.0	4.0	3.0
sard2	8.0	4.0	3.0	2.0	1.0	6.0	7.0	5.0
sard3	8.0	4.0	3.0	1.0	2.0	6.0	5.0	7.0
sard4	7.5	3.0	4.0	2.0	1.0	7.5	6.0	5.0
sard5	7.0	3.0	2.0	5.0	1.0	7.0	7.0	4.0
sard0-new	8.0	7.0	5.0	2.0	1.0	6.0	4.0	3.0
sard1-new	8.0	4.0	3.0	2.0	1.0	6.0	7.0	5.0
Boston	8.0	6.0	7.0	4.0	2.0	5.0	3.0	1.0
onekm	7.0	5.0	1.0	2.0	3.0	4.0	7.0	7.0
cw-drag	8.0	7.0	5.0	4.0	3.0	6.0	1.0	2.0
co2-emission	8.0	5.0	4.0	2.0	1.0	3.0	6.0	7.0
acceleration	8.0	6.0	5.0	2.0	1.0	3.0	7.0	4.0
available-power	8.0	7.0	6.0	4.0	5.0	3.0	2.0	1.0
avg.ranks	7.65	5.88	3.96	2.33	1.62	5.1	4.73	4.73

that the difference in average ranking is not statistically significant at the same confidence level. An ideal performance would be a method whose symbols are at the right most position of the graph (lowest average ranking), and are connected only by dotted lines to every other method (all pairwise comparisons are statistically significant). Thus, we can observe that R-PREV with $m = 100$ and with $m = 75$ are clearly the two best methods with a very high statistical significance in almost all pairwise comparisons. In particular, they are significantly better than CART, BaggCART and "BaseTree". Still, we should also remark that R-PREV with $m = 0$ is significantly worse than all standard CART related methods. This clearly indicates the importance of having more rules contributing to the predictions, but unfortunately also means that our model needs theories with more rules (thus less interpretable) to achieve top performance.

In summary, these results provide clear evidence that R-PREV can achieve very competitive predictive performance in terms of accuracy at predicting rare extreme values measured by the F statistic.

4.2   Interpretability

As we have seen in the previous section, in order to achieve top performance we need a larger number of rules. Still, for each test case only a reduced number of rules is used to obtain the prediction and domain experts can analyze these before taking any action associated with rare events. This was a property that we were seeking when developing our system: provide domain experts with comprehensible explanations of the system predictions.

Critical Difference Diagram

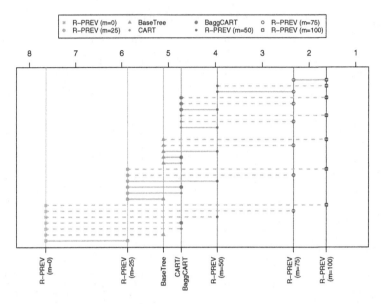

Fig. 3. The Critical Difference Diagram obtained from our experiments

In order to illustrate this comprehensibility issue we have selected the Boston Housing data set. This selection was guided by the fact that this domain concerns a topic (housing prices as a function of socio-economical factors) that is easily understandable by non-expert readers. Other data sets would require domain knowledge in order to comment the interpretability and/or reasonability of the rules obtained by R-PREV.

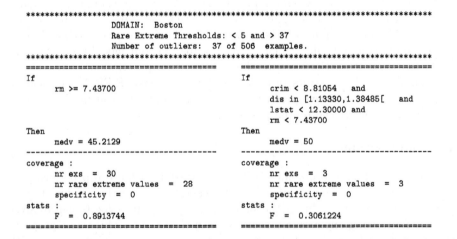

Fig. 4. Example of two of top best rules obtained by R-PREV for Boston domain

Figure 4 shows two of the top most valuable rules obtained by R-PREV for the Boston Housing data set. These two rules give some interesting insights regarding the more expensive areas in Boston. One of the rules tells us that an area where houses have more than 7 rooms tends to have a very high median price of houses. The second rule says that areas with a low crime rate, near the city center, with a small percentage of lower status population and with small houses, are also quite expensive. Both seem to capture quite common sense knowledge and would probably be regarded as correct by an expert of this domain, which would then mean that this expert would easily "accept" the model predictions. This kind of knowledge cannot be captured by standard methods like CART, because the models they obtain are focused on being accurate at the more frequent cases and not the rare extreme values like ours.

5   Conclusions

In this paper we have described a rule-based regression system, called R-PREV, conceived to address a particular class of problems that occur in several real-world applications. The applications we envisage have as main objective to produce accurate and interpretable predictions of rare extreme values of a continuous target variable. We claim that this particularity makes these problems hard to solve by the standard regression methods as they are biased to achieve a good performance on the most common values of a target variable. A different evaluation criterion is needed to overcome this limitation.

In this paper we present an extension of our previous approach to this class of problems. The extension was developed with the goal of improving both the accuracy and the interpretability of the models. The experimental evaluation we have carried out provides clear evidence that R-PREV outperforms a set of other systems with a high degree of statistical confidence. Regarding interpretability, which is crucial in most applications we are addressing, the use of a rule-based formalism leads to highly interpretable models, as we have shown by some examples.

Acknowledgements

This work was partially supported by FCT project MODAL (POSI/4049/2001) co-financed by POSI and by the European fund FEDER, and by a PhD scholarship given by FCT (SFRH/BD/1711/2004) to Rita Ribeiro.

References

1. L. Breiman, J. Friedman, R. Olshen, and C. Stone. *Classification and Regression Trees.* Statistics/Probability Series. Wadsworth & Brooks/Cole Advanced Books & Software, 1984.
2. W. Cleveland. *Visualizing data.* Hobart Press, 1993.
3. Janez Demšar. Statistical comparisons of classifiers over multiple data sets. *Journal of Machine Learning Research*, 7:1–30, January 2006.

4. Nitin Indurkhya and Sholom M. Weiss. Solving regression problems with rule-based ensemble classifiers. In *KDD '01: Proceedings of the seventh ACM SIGKDD international conference on Knowledge discovery and data mining*, pages 287–292, New York, NY, USA, 2001. ACM Press.

5. Mahesh V. Joshi, Ramesh C. Agarwal, and Vipin Kumar. Predicting rare classes: Comparing two-phase rule induction to cost-sensitive boosting. In *Proceedings of the Sixth European Conference, PKDD 2002*, pages 237–249, 2002.

6. R. Ribeiro. Prediction models for rare phenomena. Master's thesis, Faculty of Economics, University of Porto, Portugal, February 2004.

7. C. Van Rijsbergen. *Information Retrieval*. Dept. of Computer Science, University of Glasgow, 2nd edition, 1979.

8. L. Torgo and J. Gama. Regression using classification algorithms. *Intelligent Data Analysis*, 1(4), 1997.

9. L Torgo and R. Ribeiro. Predicting outliers. In N. et al. Lavrac, editor, *Proceedings of Principles of Data Mining and Knowledge Discovery (PKDD-03)*, volume 2838 of *LNAI*, pages 447–458. Springer-Verlag, 2003.

10. P. Turney. Types of cost in inductive learning. In *Proceedings of the Workshop on cost-sensitive learning at the 17th ICML*, pages 15–21, 2000.

11. UCI Machine Learning Repository - http://www.ics.uci.edu/ mlearn/MLSummary.html.

12. Gary Weiss and Haym Hirsh. Learning to predict rare events in categorical time-series data. In *AAAI Workshop on Predicting the Future: AI Approaches to Time-Series Problems*, volume WS-98-07, pages 83–90. AAAI Press, 1998.

13. Gary Weiss and Haym Hirsh. Learning to predict extremely rare events. In *AAAI Workshop on Learning from Imbalanced Data Sets*, volume WS-00-05, pages 64–68. AAAI Press, 2000.

14. Gary Weiss and Foster Provost. The effect of class distribution on classifier learning: an empirical study. Technical Report Technical Report ML-TR-44, Department of Computer Science, Rutgers University, 2001.

15. Sholom M. Weiss and Nitin Indurkhya. Rule-based machine learning methods for functional prediction. *Journal of Artificial Intelligence Research*, 3:383–403, 1995.

A Pragmatic Logic of Scientific Discovery

Jean Sallantin[1], Christopher Dartnell[2], and Mohammad Afshar[3]

[1] LIRMM, UMR 5506
161 rue Ada, 34392 Montpellier Cedex 5 - France
js@lirmm.fr
[2] EURIWARE
44 Rue des Vindits, 50130 Cherbourg-Octeville - France
christopher.dartnell@euriware.fr
[3] Ariana Pharmaceuticals
Pasteur Biotop, 28 rue Dr Roux Paris 75724 - France
m.afshar@arianapharma.com

Abstract. To the best of our knowledge, this paper is the first attempt to formalise a pragmatic logic of scientific discovery in a manner such that it can be realised by scientists assisted by machines. Using Institution Agents, we define a dialectic process to manage contradiction. This allows autoepistemic Institution Agents to learn from a supervised teaching process. We present an industrial application in the field of Drug Discovery, applying our system in the prediction of pharmaco-kinetic properties (ADME-T) and adverse side effects of therapeutic drug molecules.

1 Introduction

Scientific discovery is a collective process made possible by the tracability of judgment, positive and negative results, theories and conjecture through their publication and evaluation within a community. Without this tracability, scientific results will not last long enough to influence others, and there would be no science. This tracability is the key to localise points of debate between members of a community, to open new research fields, to put forward problems and paradoxes that need further investigation and the establishment of a consensual frame of reference. This collective process leads to a social organisation in which some members specialise in publishing, refuting, or proving results, and have gained credit which defines them as a reference in the community.

The logic of scientific discovery presented by Popper [1] or Lakatos [2], and discussed by the Vienna Circle puts forward the elaboration of norms and the break-points taking place during the formation of scientific theories. However, to formalise scientific discovery, one has to define logically notions such as paradox, postulate, result and conjecture, which was not possible whithout using a logical system allowing to reason non trivially in presence of contradictions. Moreover, the process of scientific discovery is a collective process that can only be formalized by taking interaction into account in a constructive way, as in multi-agent theories. Finally, scientific discovery is an interactive adaptive process, and it is only very recently that Angluin's works on machine learning theory gave a formal basis to the convergence analysis of such a process. To

N. Lavrač, L. Todorovski, and K.P. Jantke (Eds.): DS 2006, LNAI 4265, pp. 231–242, 2006.

the best of our knowledge, this paper is the first attempt to merge these three domains in order to formulate a pragmatic logic of scientific discovery.

In section 2, we propose a cubic model to express judgment about statements in the context of scientific discovery. We then show that the set of judgments is closed when the underlying logic is a paraconsistent logic C1. In section 3, we assume that this logic is applied independently by different institutions, and we present their properties fixed by their interaction protocol: the respect of a hierarchy, the pair evaluation, and finally the auto evaluation. This enables to tune these institutions in order to match a specific context on knowledge construction and representation. In section 4, we assess the learnability of scientific theories by scientists assisted by learning machines during an interaction following this protocol, and we present an industrial application in the field of Drug Discovery, applying our system in the prediction of pharmaco-kinetic properties (ADME-T) and adverse side effects of therapeutic drug molecules.

2   Logical Expectations: Cube of Judgments

We assume that the form of reasoning used in science is the same for every institution and every scientist. This form is given by a modality attributed to a statement beyond the following: paradox, proof, refutation, result, conjecture, postulate, contingent, and possible. This set of modalities is assumed complete and closed by negation. In this section, we define with these modalities the cube of judgments and we have to work with paraconsistent logic.

2.1   Square of Modalities

The figure 1 expresses Aristotle's *square of modalities*. Aristotle's logic is said to be *ontic* since every modality is expressed from a single modality $\Box$ and negation $\neg$ and the square of oppositions is closed by doubling this negation: $\Box = \neg\neg\Box$. The top modalities (Necessary, Impossible) are used to express universal statements whereas the lower modalities (Possible, Contingent) are used to express particular statements.

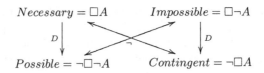

$$Necessary = \Box A \qquad\qquad Impossible = \Box\neg A$$

$$Possible = \neg\Box\neg A \qquad\qquad Contingent = \neg\Box A$$

Fig. 1. Aristotle's square of modalities

We can make a parallel with Scientific Discovery and the theory of proof and refutation as follow:

– "'A is necessary'" = A is proven: $\Box A$
– "'A is impossible'" = "'A is refuted'": $\Box\neg A$
– "'A is possible'" = "'A has not been refuted'": $\neg\Box\neg A$
– "'A is contingent'" = "'A has not been proven'": $\neg\Box A$

To link these modalities, epistemic logic uses axioms: the axiom D describes the vertical relations between necessary and possible, and between impossible and contingent". By following two different paths on the square of oppositions, we can reach the same point, and we define consistency constraints by considering that these two paths lead to the same result:

- What is necessary is possible and therefore is not impossible.
- What is impossible is contingent and therefore is not necessary.

In intuitionistic logic, the negation of a concept A is not a concept but an application from this concept into a *contradiction*, which is a statement both true and false ($A \land \neg A$). In the same way, a paradox is a statement which is both proven and refuted. For instance, a bike without $wheel \lor frame \lor handlebar$ would be contradictory. Classical logic becomes trivial in the presence of a single contradiction, following the principle of contradiction: *given two contradictory propositions, one being the negation of the other, only one of them is false*. On the opposite, paraconsistent logic allows reasoning in a non trivial way in the presence of contradictions [3] [4] [5].

2.2   Paraconsistent Logic

Paradoxes have often been at the source of scientific discoveries, and have often lead to new approaches and revisions of the frame of reference. This only happened when the whole theory used to explain the concerned domain did not completely collapse under the weight of its contradictions, and that is why we need to use paraconsistent logic to formalise a logic of scientific discovery. Paraconsistent logic uses different negations associated with different levels of contradiction to allow reasoning in the presence of contradictions as in classical logic with no contradictory statement. Given a theory T, we call *'formal antinomy'* any meta-theoretical result showing that T is trivial. A *'formal paradox'* is the derivation of two contradictory results of T. Paraconsistent logic can be paradoxical without being antinomic: an *informal paradox* is an acceptable argument for which premises are acceptable (they seem true), argument is acceptable (valid), and the conclusion unacceptable (seems false).

To achieve our goal of producing a complete judgment system, taking into account contradictions, we need to complete the set of modalities with those of paradox and conjecture, hypothesis and result. The square of oppositions then becomes a cube of judgments for which the square is a diagonal plane as shown in figures 2 and 3.

Definition 1. *The **cube of judgments** $Cube = (\Box, \neg)$ is the set of ontic modalities derivable from a modality $\Box$ and a negation $\neg$.*

Property 1. *In a paraconsistent logic C1, this cube of judgments is complete and closed by negation.*

This property, highlighted by the diagonal planes of the cube on figure 2 is given by the following two principles of abstraction that caracterise C1 logic [6], from which a paraconsistent interpretation of the Morgan's laws can be verified:

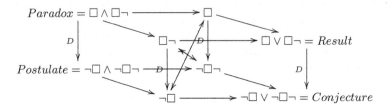

Fig. 2. Square of deontic judgments as a diagonal plane of the $Cube$

The weak principle of abstraction: If two propositions are not contradictory, then none of the logical relations between them is contradictory:

– What is not a non contradictory postulate is a *result*:

$$\neg : \neg\Box \wedge \neg\Box\neg \longrightarrow \Box \vee \Box\neg.$$

– What is not a non contradictory paradox is a *conjecture*:

$$\neg : \Box \wedge \Box\neg \longrightarrow \neg\Box \vee \neg\Box\neg$$

The strong principle of abstraction: Out of two propositions, if one is not contradictory, then none of the logical relations between them is contradictory:

– What is not a non contradictory conjecture is a *paradox*:

$$\neg : \neg\Box \vee \neg\Box\neg \longrightarrow \Box \wedge \Box\neg.$$

– What is not a non contradictory result is a *postulate*:

$$\neg : \Box \vee \Box\neg \longrightarrow \neg\Box \wedge \neg\Box\neg.$$

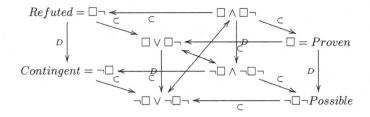

Fig. 3. Square of oppositions resulting from C1 logic, as a diagonal plane of the $Cube$

This cube of judgments expresses a set of modalities closed by negation that can judge any statement, object, or situation, formulated in the language upon which this logic is applied.

3   The Institution Agent Social Game

Section 2 presented the properties of a closed system producing judgments and taking into account contradiction. Such a system can be used to model the decision process of an agent holding incomplete knowledge, and we call such an agent an *"Institution Agent" (IA)*.

Definition 2 (IA). *An IA is an agent using the Cube to judge statements.*

We assume that the logic used during this decision process is the same for every IA, and we focus on the adaptation and the interaction of IAs sharing a vocabulary and trying to build a common language or frame of reference with this vocabulary.

Three logical properties are needed to qualify this interaction protocol and to add a logical control to the adaptation process:

- deontic: an IA must be able to attribute credits to another IA, to interact, and to teach another IA,
- defeasible: Lower IAs must be able to adapt their behavior to the norms imposed by the higher ones,
- autoepistemic: an IA can be seen as composed by at least two interacting IAs and can therefore learn its own hierarchy of norms and auto-adapt.

In this section, we suppose that each IA can be represented by a particular normative system resulting from its own experience and adaptation during an interaction with other IAs.

Definition 3. *We call a **Normative System (NS)** the couple $(L, Cube)$ formed by:*

- *L: a language formed by a hierarchy of concepts and the relations between them*
- *Cube: a cube of judgments*

3.1   Deontic Logic

Often used in multi-agent systems to constrain an agent's behaviour, annotable deontic logic uses modalities expressing obligation, interdiction, advice, and warning. According to Frege's definition, these statements express a judgment, ie. the recognition of the type of truth of the statement [7]. Imputations (gains or losses, risk estimation) are used to estimate the risk incurred in a given situation to decide what action to take or what behaviour to adopt. A modality and an imputation have to be used to express statements of the following form: "The obligation to respect the speed limit is attached to a imputation of x". A credit value can also be associated to IAs, ordering them hierarchically, to define which one is the most qualified to rule in a given context, for example by defining a social organisation as a government with a parliament, a senate, ...

Scientific discovery is a collective process, and needs interaction between researchers to exchange their points of view and judgments. That's how IAs interact: by exchanging judgments about statements. More exactly, by asking another IA if it agrees with a particular judgment: "this statement is a conjecture, is it not?", to which the answer is "yes" or "no, it is a result". Exchanging judgments creates the negation in the common

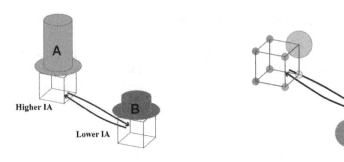

Fig. 4. IA's credit   **Fig. 5.** Exchanging judgments

frame of reference (language), and the revision of the normative system associated with one IA or the other. Two judgments are especially important: *judging one's conjecture as being a paradox*, and *judging one's postulate as already being a result*. KEM^{TM}, presented in section 4.2, illustrates this control by a scientist over the IA assisting him.

3.2 Defeasible Logic

It is possible to link two NS by respecting a defeasible logic to take into account a hierarchy of Institution Agents. The resulting hierarchy of IAs has to be brought together with the transitivity axiom, that stands as follows: "What is necessary in a normative system of proof and refutation is also necessary in a lower normative system". In other words, no one should be unaware of the law, no one should go against a superior law. [8] gives a concrete usage of defeasible logic, that allows us to order rules and to supervise an IA, for example with another higher IA, as illustrated on figure 6.

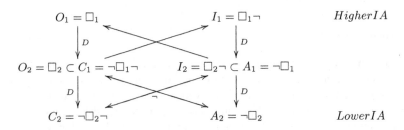

Fig. 6. Normative system hierarchy

- Every Obligation of a lower IA belongs to the superior IA's advice.
- Every Interdiction of a lower IA belongs to the superior IA's warnings.

The middle line shows the conditions according to which an IA can be supervised by another one. The violation of this constraint ($O_2 = \Box_2 \subset I_1 = \Box_1\neg$ or $I_2 = \Box_2\neg \subset O_1 = \Box_1$) can put forward contradictions between the two IA's normative system. We present in section 4 how IAs can learn and adapt their normative systems. Finding a contradiction, and trying to eliminate it, leads to the initiation of a transaction between the two IAs, during which they adapt their normative system. When no contradiction remains, a new IA can be created, formed by the association of the two precedent IAs, and this process ensures the tracability of all the events leading to an IA's creation.

3.3 Autoepistemic Logic

Aristotle distinguishes endophasy as an inner dialog. This is a constructive manner to build an intelligent agent as the result of an auto-adaptation. The inner IAs can be interpreted as managing believes, desires or intentions (BDI), for example. By applying the dialectic and deontic interaction presented in section 3.1, an IA is able to acquire its own NS, which prepares an efficient learning, and even enable self learning from examples.

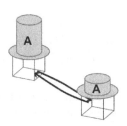

Fig. 7. Autoepistemic dialog   **Fig. 8.** IA formation

In this section, we presented how an interaction process and a hierarchical control can be used to build an agent able to adapt its defeasible deontic and autoepistemic Normative System.

4   Learnability

To estimate the complexity of an IA's creation, we embrace machine learning theories, and we discuss the learnability of a normative system by an Institution Agent. We illustrate various learning methods as decision trees or version spaces, then we show that this system is related to Angluin's theories on learning monotonous functions by querying, and learning from different teachers [9] [10] [11]. Finally, we present an industrial application dedicated to Drug Discovery.

4.1   Learning a Scientific Theory

Definition 4. *A **scientific theory** is an application : $F : L \longrightarrow \Omega$ such that $(L, F(L))$ is a normative system, associating to every statement $x \in L$ a scientific judgment $F(x) \in \Omega$.*

Definition 5. *We call T^{cube} the lattice obtained by "forgetting" the negation links coming from the weak and strong principles of abstraction (section 2.2). T^0 (figure 9 is the truth lattice underlying a classical logic. T^1 and T^2 (respectively in bold and italic characters on figure 10)are the lower and upper sub-lattices of T^{cube}.*

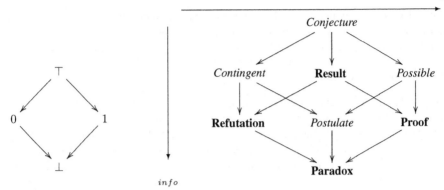

Fig. 9. T^0: the lattice underly-   **Fig. 10.** T^{cube}: the lattice underlying the cube's logic
ing classic logic

Definition 6. *A **scientific theory learned by an IA** is an application :* $F_{IA} : L \longrightarrow$ T^{Cube} *such that* $(L, F_{IA}(L))$ *is a normative system, associating to every statement* $x \in L$ *a scientific judgment* $F_{IA}(x) \in T^{cube}$.

Remark 1. If we assume that a *paradox* (bottom of the lattice) is more informative than a *conjecture* (top of the lattice) and that a *proof* (right) is more true than an *refutation* (left), this lattice can be oriented following a vertical axis representing the information level and an horizontal axis representing the truth level.

Remark 2. T^1 represents the modalities used by the teacher during Angluin's protocol [11]. The interactive process used in the following cases is an interaction using only membership queries. The furtherance of science can never be the result of an isolated scientist who cannot verify the interest of his theory. Reference theories of machine learning use as well Equivalence Queries EQs, which should compare two scientific theories $F_{I.A._1}$ and $F_{I.A._2}$. The protocol defined in section 3, depends on the use of EQs in which case putting forward a contradiction answered an EQ: an interaction between two hierarchicaly ordered IAs allows the confrontation of two non comparable theories through the confrontation of their hypotheses and conjectures on the one hand, with paradoxes and results on the other.

Property 2. *Since* T^{cube} *is a modular lattice, a scientific theory* F_{IA} *is learnable in a polynomial time using membership queries.*

The following cases show the generality of this approach.

Case 1. *Given a set* L *of boolean and real variables, a **scientific theory learned by a decision tree** is an application* $F_{DT} : L \longrightarrow T^0$ *such that* $(L, F_{DT}(L))$ *is a normative system.*

Case 2. *Given a set* L *of boolean variables, a **scientific theory learned by a version space** is an application* $F_{VS} : L \longrightarrow T^1$ *such that* $(L, F_{VS}(L))$ *is a normative system.*

Case 3. *Given a set* L *of boolean variables, a **scientific theory learned by a galois lattice** is an application :* $F_{GL} : L \longrightarrow T^{Cube}$ *such that* $(L, F_{GL}(L))$ *is a normative system.*

Case 4. *Given a set L of boolean and real variables, given a set of results coming from a decision tree method,* **a scientific theory learned by** *DT/GL is an application : $F_{DT/GL} : L \longrightarrow T^{0,2}$ such that $(L, F_{DT/GL}(L))$ is a normative system.*

All these cases of scientific theories are monotonous functions and are therefore learnable. However, only cases 3 and 4, which take into account dialectical aspects required to manage the norms and the ruptures in scientific discovery, are learnable by an IA. The following section develops the case 4.

4.2   Dialectic Protocol and Application in Drug Design

A real application of learning in scientific discovery, is from collaboration with Ariana Pharmaceuticals in Drug design [12].

KEM^{TM} can suggest specific molecular modifications to achieve multiple objectives, after analysing a multi-parametric database. Data mining is performed with an Institution Agent using DT/GL to learn. KEM^{TM} is an Institution Agent resulting from the interaction of an $IA_{DT/GL}$ and a expert scientist, who has in mind his own normative system. To teach KEM^{TM} how to learn his normative system, the expert scientist describes each example by way of a set of non paradoxical results. KEM^{TM} learns from these examples a scientific theory, and the scientist uses $(x, F_{DT/GL}(x))$ as a rational mirror of his own normative system. In a dialectic way, KEM^{TM} evolves and adapts to create a new IA from the learning process.

To assist the learning process, KEM^{TM} selects an hypothesis that is not a paradox and, more specifically, KEM^{TM} selects a conjecture within this hypothesis that is not a result. Then the scientist admits new examples to eliminate the conjecture as a result or modifies his own normative system to eliminate the hypothesis as a paradox. Such a method has been tried and succesfully tested in a legal context where the "learners" are humans, to build efficient normative systems [13] [14].

Designing novel therapeutic molecules is a challenging task since one needs not only to select an active molecule, the molecule needs also to be absorbed, needs to be stable within the body (i.e. not metabolized too rapidly) and finally it needs to have low toxicity and side effects. This is called improving the ADME-T profile (Absorption, Distribution, Metabolism, Excretion and Toxicity).

In this example we focus on the prediction of Absorption, a key issue in drug design since this is one of the important and early causes of failure in the drug discovery process. Indeed molecules need to be absorbed before they can perform any desired activity. Absorption is a complex process involving both passive (diffusion) and active (through transporter proteins) accross cellular membranes. For passive transport, molecules need to be soluble (hydrophilic) in water and at the same time they need to be greacy (hydrophobic) to penetrate cellular membranes that are formed of lipids. This contradicting requirement is modulated by active transport, where molecules need to be recognized (i.e. complementarity of shape and charge) by a another molecule (transporter) that helps them through membranes.

Although no one can for sure predict the absorption of a new molecule, a number of empirical rules are known. This is an interesting context for applying our IA since our key requirement is to capture knowledge from the experimental data and then evolve and improve this model in a consistent manner.

To illustrate our approach we focus on a set of 169 molecules for which the absorption in man has been experimentally evaluated (4 classes. 0 not absorbed, 3 highly absorbed). These molecules are described using a set of physico chemical properties such as molecular radius, different calculated measures of their total polar surface accessible to water (*TPSA* and *VSA POL*), their hydrophobicity (*SLOGP*), presence of halogens etc.

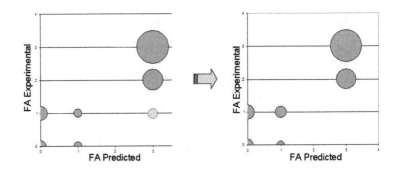

Fig. 11. Predictions *A* and *B*

Initially, the system learns from the dataset a set of rules linking the structure of the molecule to the absorption. The quality of the prediction is tested in a subsequent stage on a novel set of molecules. The results are shown on prediction *A* in figure 11. Ideally the predictions should be on the diagonal. An error of one class is tolerated. However, it is clear that for one molecule, the error is larger (ie experimental : class 1 vs predicted class 3).

Figure 12 shows that the molecule (Ranitidine) has been predicted with *fraction absorbed in man 3* i.e. highly absorbed. However, if the user forces *fraction absorbed in man 3* to be false, the system shows that this contradicts a learned rule *VSA pol 2* → *fraction absorbed in man 3*. At this stage the user realises that indeed this rule was true for the learning set, however this is not generally true and it can be eliminated. Once this rule has been eliminated, the user goes back to predicting once more the test set and results are shown in Figure 11, prediction *B*. As expected, the results have been improved. The important point is that the improvement has been done in a controlled way under the user's supervision.

In scientific discovery, there are in general no Oracles who can say a priori whether a prediction is correct or not. Experimentalists design a hypothesis that is consistent with existing empirical data and then set about to test it. We beleive that the key for a computational system is to adhere to the same process i.e. build up an explanation / reasons for suggesting for predicting an outcome. If the system is able to provide enough arguments, the user will "trust" it and try the experience.

KEM^{TM} is an Institution Agent resulting from a process combining both human and machine learning. It is very interesting to log the various adaptations of the learned normative system coming from the addition of examples or normative theories and to

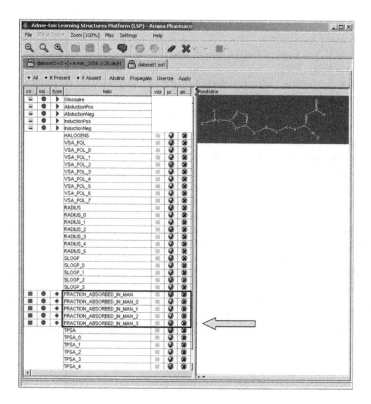

Fig. 12. KEM^{TM}

analyse process of the formation of such an IA. This method also give a compliance record of the various processes chosen or rejected in the formation of the resulting IA.

5 Conclusion

We propose a pragmatic logic to manage scientific judgment. This set of judgments is closed by negation when using paraconsistent logic C1. Using Institution Agents, we define a dialectic process to manage contradiction. This allows autoepistemic Institution Agents to learn from a supervised teaching process. This methodology is now tried and tested in various domains: in drug design, in Law[14], and even in mathematical games [15].

References

1. Popper, K.R.: Conjectures and Refutations: The Growth of Scientific Knowledge. Harper and Row (1963)
2. Lakatos, I.: Proofs and Refutations. Cambridge University Press (1976)
3. da Costa, N.C.A., Beziau: La logique paraconsistante. Le concept de preuve à la lumière de l'intelligence artificielle (1999) 107–115

4. da Costa, N.C.A.: Paraconsistent mathematics. In: Frontiers of paraconsistent logics, RSP research study press (2000)
5. da Costa, N.C.A.: Logiques classiques et non classiques : essai sur les fondements de la logique. Masson (1997)
6. Beziau, J.Y.: La logique paraconsistante. Logiques classiques et non classiques, essai sur les fondements de la logique (1997)
7. Pozza, C.D., Garola, C.: a pragmatic interpretation of intuitionistic propositional logic. Erkenntnis **43** (1995) 81–109
8. Nakamatsu, K., Kato, T., Suzuki, A.: Basic ideas of defeasible deontic traffic signal control based on a paraconsistent logic program evalpsn. Advances in Intelligent Systems and Robotics (2003)
9. Angluin, D.: Queries and concept learning. Machine Learning **2** (1988) 319–342
10. Angluin, D.: Queries revisited. Theoretical Computer Science **313** (2004) 175–194
11. Angluin, D., Krikis: Learning from different teachers. Machine Learning **51** (2003) 137–163
12. Afshar, M., Lanoue, A., Sallantin, J.: New directions: multidimensionnal optimization in drug discovery. Comprehensive Medicinal Chemistry 2 **4** (2006)
13. Nobrega, G.M.D., Cerri, Sallantin: A contradiction driven approach to theory information: Conceptual issues pragmatics in human learning, potentialities. Journal of the Brazilian Computer Society **9** (2003) 37–55
14. Sallantin, J.: La découverte scientifique assistée par des agents rationnels. Revue des sciences et technologie de l'information (2003) 15–30
15. Dartnell, C., Sallantin, J.: Assisting scientific discovery with an adaptive problem solver. In: Discovery Science. (2005) 99–112

Change Detection with Kalman Filter and CUSUM

Milton Severo[1,2] and João Gama[1,3]

[1] Fac. of Economy, University of Porto, Portugal
[2] Department of Hygiene and Epidemiology, Fac. of Medicine,
University of Porto, Portugal
[3] LIACC, University of Porto
milton@med.up.pt, jgama@liacc.up.pt

Abstract. In most challenging applications learning algorithms acts in dynamic environments where the data is collected over time. A desirable property of these algorithms is the ability of incremental incorporating new data in the actual decision model. Several incremental learning algorithms have been proposed. However most of them make the assumption that the examples are drawn from a stationary distribution [13]. The aim of this study is to present a detection system (DSKC) for regression problems. The system is modular and works as a post-processor of a regressor. It is composed by a regression predictor, a Kalman filter and a Cumulative Sum of Recursive Residual (CUSUM) change detector. The system continuously monitors the error of the regression model. A significant increase of the error is interpreted as a change in the distribution that generates the examples over time. When a change is detected, the actual regression model is deleted and a new one is constructed. In this paper we tested DSKC with a set of three artificial experiments, and two real-world datasets: a Physiological dataset and a clinic dataset of Sleep Apnoea. Sleep Apnoea is a common disorder characterized by periods of breathing cessation (apnoea) and periods of reduced breathing (hypopnea) [7]. This is a real-application where the goal is to detect changes in the signals that monitor breathing. The experimental results showed that the system detected changes fast and with high probability. The results also showed that the system is robust to false alarms and can be applied with efficiency to problems where the information is available over time.

1 Introduction

In most challenging applications learning algorithms acts in dynamic environments where the data is collected over time. A desirable property of these algorithms is the ability of incremental incorporating new data in the actual decision model. Several incremental learning algorithms have been proposed to deal with this ability (e.g., [5, 12, 6]). However most learning algorithms, including the incremental ones, assume that the examples are drawn from a stationary distribution [13]. In this paper we study learning problems where the process

N. Lavrač, L. Todorovski, and K.P. Jantke (Eds.): DS 2006, LNAI 4265, pp. 243–254, 2006.

generating data is not strictly stationary. In most of real world applications, the target concept could gradually change over time. The ability to incorporate this concept drift is a natural extension for incremental learning systems.

In many practical problems arising in quality control, signal processing, monitoring in industrial plants or biomedical, the target concept may change rapidly [2]. For this reason, it is essential to construct algorithms with the purpose of detecting changes in the target concept. If we can identify abrupt changes of target concept, we can re-learn the concept using only the relevant information. There are two types of approaches to this problem: methods where the learning algorithm includes the detection mechanism, and approaches where the detection mechanism is outside (working as a wrapper) of the learning algorithm. The second approach has the advantage of being independent of the learning algorithm used. There are also several methods for solving change detection problems: time windows, weighting examples according their utility or age, etc [9]. In the machine learning community few works address this problem. In [15] a method for structural break detection is presented. The method is an intensive-computing algorithm not applicable for our proposes of processing large datasets.

The work presented here follows a time-window approach. Our focus is determining the appropriate size of the time window. We use a Kalman filter [14, 18] that smooths regression model residuals associated with a change detection CUSUM method [2, 4, 10]. The Kalman filter is widely used in aeronautics and engineering for two main purposes: for combining measurements of the same variables but from different sensors, and for combining an inexact forecast of system's state with an inexact measurement of the state [17]. When dealing with a time series of data points $x_1, x_2, ..., x_n$ a filter computes the best guess for the point x_{n+1} taking into account all previous points and provides a correction using an inexact measurement of x_{n+1}.

The next section explains the method structure of the proposed system. The experimental evaluation is presented in section 3. In this section we apply our system to estimate the airflow of a person with Sleep Apnoea. We use the online change detection algorithm to detect changes in the airflow. Last section presents the conclusions and lessons learned.

2   Detection System in Regression Models with Kalman Filter and CUSUM

In this paper we propose a modular detection system (DSKC) for regression problems. The general framework is shown in figure 2. The system is composed by three components: a regression learning algorithm, a Kalman filter [14] and a CUSUM [2, 4]. At each iteration, the system first component, the learning algorithm, receives one unlabeled example, x_i, and then the actual model predicts, $\hat{y}_i$. After the model forecast, it receives an input from the environment, y_i and calculates the residual $r_i = |y_i - \hat{y}_i|$. The system uses r_i and the Kalman filter error estimate of the actual model, $\hat{r}_{i-1}$, to compute a residual for the dispersion, $rd_i = |r_i - \hat{r}_{i-1}|$. The Kalman filter, the system second component, receives both

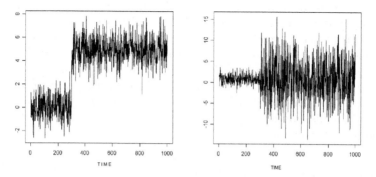

Fig. 1. Two types of concept drift: changes in the mean and changes in dispersion

residuals and updates the learning algorithm state estimate. The state estimate is form by mean error, $\hat{r}_i$, and the dispersion error, $\hat{r}d_i$. Normally, a learning algorithm will improve the predictions with the arrival of new examples, mainly in the initial learning stage. For that reason, it is very important to provide a run-time estimation of the residuals. In general, run-time estimation is provided by simple mechanism, such as auto regressive or auto regressive moving average or Kalman filter. The advantages of the last filter are: allows to adaptively tune the filter memory to faster track variations in the estimation and allows to improve the accuracy of the estimation by exploiting the state update laws and variance of the estimation [14].

The proposed system detects changes in mean error of the actual model and changes in the respective dispersion. Figure 1 illustrates the two types of changes we are interested in. The pair $(r_i, \hat{r}_i)$ is transmitted to the mean CUSUM and the pair $(rd_i, \hat{r}d_i)$ is transmitted to the dispersion CUSUM. Both CUSUM's compare the values they receive. A change occurs if significant differences between both values received or significant differences between consecutive residuals are found. If the change is an increase in the error mean or an increase in the dispersion, the system gives an order to erase the actual learning model and start to

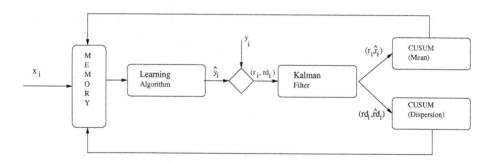

Fig. 2. Detection System with Kalman Filter and CUSUM Framework

construct a new model using the new examples. If the change is a decrease in the error mean or a decrease in the dispersion error, then the system gives an order to the Kalman filter to weight the new residuals heavier, thus the filter can follow the mean error and dispersion error faster. If no significant differences are found, the new example is incorporated into the learning model.

The proposed architecture is general. It can work with any regression learning algorithm and any loss function. The main assumption is that a change is reflected in the distribution of the examples, leading to an increase of the error of the actual regression model.

3   Experimental Evaluation

In this section we describe the evaluation of the proposed system. We used three artificial datasets and two real-world datasets. A real-world physiological [1] dataset was used to evaluate if our DSKC provides a reliable estimate of the learning model error. We used three artificial datasets and one real-world dataset of Sleep Apnoea to evaluate the efficiency of the DSKC. An artificial data allows us to perform controlled experiments. The experiments were designed in a way that we know when the change occurs. To test the generality of the proposed methodology, we used two distinct learning algorithms: a regression tree and a linear regression model [1]. Four performance measures were used to evaluate the efficiency of the DSKC: number of false alarms (FA) and true alarms (TA), mean number of examples for detection of the changes (MNE) and normalized mean absolute error (NMAE). The median test [3] was used to compare the normalized mean absolute error with and without change detection for each dataset and learning algorithm.

3.1   Physiological Dataset

In this section we study the impact in boosting the discriminative power of a signal given by our system components: the Kalman filter and the CUSUM. We use the physiological dataset [1] that was collected using BodyMedia wearable body monitors. These continuous data are measures of 9 sensors and an indication of the physical activities of the user. The dataset comprises several months of data from 18 subjects. We divided the dataset by subject and age and used only the sets related to changes between sleep and awake. We measured the discrimination between sleep and awake phases using the original sensors, using only the Kalman filter estimate of a sensor, and using the estimate of the sensor from Kalman filter with CUSUM. We used the sensors 7 and 9 because they were, respectively, the sensors with the larger and the smaller discriminative power. The

---

[1] In the set of experiments reported here, we use batch versions of both algorithms that train a new model at every iteration using all examples from the last change detected or since the beginning until that moment. The focus of the paper is change detection in regression problems. It is expected that the main conclusions apply to incremental versions of the algorithms.

Table 1. Area under the ROC curve for the sensor, the Kalman filter estimate (KF) and Kalman filter with CUSUM estimate (KFC)

Area Under the ROC Curve						
	sensor 7			sensor 9		
ID	KFC	KF	sensor	KFC	KF	sensor
1	0.88	0.93	0.92	0.78	0.81	0.73
2	0.95	0.95	0.87	0.81	0.76	0.66
3	0.96	0.96	0.89	0.83	0.77	0.67
4	0.95	0.97	0.92	0.65	0.60	0.58
5	0.96	0.96	0.91	0.72	0.66	0.59
6	0.92	0.94	0.91	0.85	0.82	0.75
7	0.92	0.92	0.89	0.85	0.81	0.73
8	0.91	0.89	0.87	0.60	0.57	0.50
9	0.88	0.89	0.86	0.68	0.65	0.56
10	0.81	0.93	0.85	0.59	0.58	0.55
11	0.94	0.99	0.99	0.62	0.63	0.57
12	0.95	0.97	0.97	0.68	0.64	0.56
13	0.94	0.97	0.93	0.85	0.84	0.83
14	0.93	0.97	0.97	0.68	0.58	0.53
15	0.92	0.92	0.85	0.62	0.59	0.53
16	0.95	0.97	0.94	0.53	0.50	0.547
Median	**0.94**	**0.95**	**0.91**	**0.68**	**0.65**	**0.57**

discrimination power was measured using the area under the ROC curve. The results show (table 1) that the discrimination power increases when we applied the Kalman filter or the Kalman filter with CUSUM to the original sensors. This fact is more evident when the sensor exhibit less discriminative power. The less discriminative sensor is sensor 9. In that case, the improvement verified with the Kalman filter plus the CUSUM is 5.9% with a p-value $p < 0.002$. The improvement decreases for sensor 7, where the Kalman filter alone has better results $(2.2\%, p < 0,018)$. These results suggest that the use of the Kalman filter with CUSUM provides a reliable estimate of the learning model error.

3.2   Artificial Datasets

The three artificial datasets used were composed by 3000 random examples. Two random changes were generated between the 30th and the 2700th examples.

1. The first dataset has five normally distributed attributes with mean 0 and standard deviation 50. The dependent variable is a linear combination of the five attributes with an white noisy with standard deviation 10. New coefficients of linear combination where built at every change. The five coefficients of the linear combination where generated by a uniform distribution over $[0, 10]$.

2. The second dataset has two uniformly distributed attributes over $[0, 2]$. The dependent variable is a linear combination of the first attribute sine, and the

cosine of the second attribute. We add, to each attribute, white noise with standard deviation 1. As in the previous artificial dataset, new coefficients were built at every change.

3. The third dataset is a modified version of the dataset that appear in the MARS paper [8]. This dataset has 10 independent predictor variables $x_1, ...,$ x_{10} each of which is uniformly distributed over $[0,1]$. The response is given by

$$y = 10 \sin (\pi x_1 x_2) + 20 (x_3 - 0,5)^2 + 10x_4 + 5x_5 + \epsilon \cdot \qquad (1)$$

A permutation of the predictor variables was made at each change.

For each type of dataset we randomly generated 10 datasets.

3.3 Results on Artificial Datasets

Tables 2, 3 and 4 show the results for respectively, dataset one, two and three. We can observe that DSKC is effective with all learning algorithms. Overall we detected 73% ($CI95\% = [64\% - 81\%]$) of all true changes with no false alarms ($CI95\% = [0\% - 6\%]$). The results show that the proportion of true changes varies between 50% for the third dataset with linear regression and 90% for the same dataset but with regression trees; the mean number examples needed for detection varies from 8.3 to 42.13.

Table 2. Results for the first artificial dataset

	Regression Trees					Linear Models				
	No Detection	Detection				No Detection	Detection			
	NMAE	NMAE	TA	FA	MNE	NMAE	NMAE	TA	FA	MNE
1	0.75	0.71	1	0	45.0	0.30	0.10	2	0	3.0
2	0.73	0.62	2	0	38.0	0.25	0.11	1	0	9.0
3	0.85	0.66	2	0	27.5	0.51	0.11	2	0	5.0
4	0.68	0.67	1	0	16.0	0.29	0.12	1	0	4.0
5	0.66	0.63	2	0	45.0	0.40	0.13	2	0	19.5
6	0.68	0.64	2	0	40.5	0.31	0.10	2	0	2.0
7	0.79	0.57	2	0	9.5	0.30	0.21	1	0	6.0
8	0.73	0.59	1	0	51.0	0.28	0.08	2	0	26.5
9	0.73	0.69	1	0	43.0	0.22	0.08	2	0	4.5
10	0.84	0.76	1	0	10.0	0.38	0.09	2	0	3.5
η	0.73	0.65	2	0	39.2	0.30	0.11	2	0	4.8
$\bar{x}$	0.74	0.65	1.5	0	32.6	0.32	0.11	1.7	0	8.3

We found significant differences ($p < 0.05$) between the use and not use of our detection system for the normalized mean absolute error, except for the 2nd dataset with linear regression model. The mean normalized error decreased for all datasets with the use of our DSKC. We observed that when the second change occurs relatively closed to the first change or when the first change occurs relatively closed to the beginning of the experience, the change was not detected.

Table 3. Results for the second artificial dataset

	Regression trees				Linear models				
	No Detection	Detection			No Detection	Detection			
	NMAE	NMAE	TA	FA	MNE	NMAE	NMAE	TA	FA	MNE
1	0.57	0.56	1	0	8.0	0.85	0.82	1	0	23.0
2	0.45	0.33	1	0	5.0	0.88	0.88	2	0	48.0
3	0.45	0.42	1	0	78.0	0.89	0.89	1	0	35.0
4	0.53	0.47	2	0	10.5	0.81	0.82	2	0	95.0
5	0.39	0.37	1	0	13.0	0.90	0.90	2	0	84.5
6	0.61	0.36	2	0	9.0	0.83	0.83	0	0	–
7	0.48	0.39	1	0	39.0	1.00	1.00	1	0	5.0
8	0.54	0.45	2	0	8.0	1.00	1.00	2	0	89.5
9	0.45	0.41	2	0	10.0	0.84	0.85	1	0	27.0
10	0.45	0.41	2	1	33.5	0.87	0.87	1	0	83.0
η	0.47	0.41	1.5	0	10.2	0.87	0.87	1.0	0	48.0
$\bar{x}$	0.49	0.42	1.5	0	21.4	0.89	0.88	1.3	0	54.4

Table 4. Results for the third artificial dataset

	Regression trees				Linear models				
	No Detection	Detection			No Detection	Detection			
	NMAE	NMAE	VA	FA	MNE	NMAE	NMAE	VA	FA	MNE
1	0.80	0.67	2	0	21.5	0.71	0.59	1	0	38.0
2	0.73	0.69	1	0	33.0	0.73	0.73	0	0	–
3	0.84	0.66	2	0	23.0	0.68	0.57	1	0	65.0
4	0.86	0.67	2	0	28.0	0.71	0.58	2	0	55.0
5	0.82	0.66	2	0	33.0	0.68	0.57	1	0	19.0
6	0.71	0.68	1	0	14.0	0.75	0.59	2	0	54.5
7	0.86	0.68	2	0	39.0	0.69	0.59	1	0	25.0
8	0.80	0.66	2	0	50.5	0.63	0.60	1	0	41.0
9	0.87	0.68	2	0	20.5	0.88	0.88	0	0	–
10	0.82	0.68	2	0	17.5	0.67	0.59	1	0	39.0
η	0.82	0.67	2	0	25.5	0.70	0.59	1.0	0	40.0
$\bar{x}$	0.81	0.67	1.8	0	28.0	0.63	0.62	1.0	0	42.1

As we can see in table 5, the proportion of changes detected was 25%, when the number of examples between changes is less than 332, against 89%, when there are more than 532 examples. The association between the number of examples required by the learning algorithm and detection or not detection of the change is significant ($p < 0.001$).

3.4 Sleep Apnoea Dataset

After measuring the performance of our detection system in the artificial datasets, we evaluate the performance in a real problem where change points

Table 5. Association between the number of examples read and the ability to detect versus not detect changes

Number Examples	Not Detected	Detected	p
[1 − 332[	9 (75.0%)	3 (25.0%)	< 0.001
[332 − 532[	11 (45.8%)	13 (54.2%)	
[532 − ∞[	8 (11.1%)	64 (88.9%)	

Table 6. Results for dataset of Sleep Apnoea

	Regression Trees			Linear Model		
	No Detection	Detection		No Detection	Detection	
	NMAE	NMAE	TA	NMAE	NMAE	TA
1	0.940	0.923	18	0.981	0.974	17

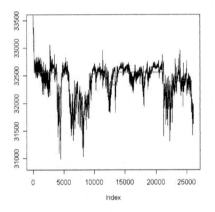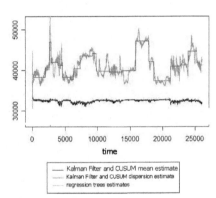

Fig. 3. The regression tree segmentation of the mean airflow (left) and airflow dispersion (right) filter by Kalman Filter and CUSUM

and change rates are not known. For such we applied our system to a dataset from patients with Sleep Apnoea. Sleep Apnoea is a common disorder characterized by periods of breathing cessation (apnoea) and periods of reduced breathing (hyponea). The standard approach to diagnoses apnoea consists of monitoring a wide range of signals (airflow, snoring, oxygen saturation, heart rate...) during patient sleep. There are several methods for quantifying the severity of the disorder, such as measuring the number of Apnoeas and Hypopnoea per hour of sleep or measuring the number of breathing events per hour. There is a heterogeneity of methods for defining abnormal breathing events, such as reduction in airflow or oxygen saturation or snoring [7]. It can be seen as pathological, when the number of Apnoeas and Hypopnoea/hour is larger then 20 events per hour [11]. Our goal in this experiment was to evaluate if our detection system could detect abnormal breathing events.

Table 7. Alarms detected by DSKC with regression trees and linear models in sleep Apnoea dataset and the leafs cut point of regression tree applied off-line

	Regression trees	Linear model	Cut-points (Reg. tree)	type of change
1	2017	1965	1962.5	disp,+
2	2600	2682	2631.5	disp,+
3	—	—	2851.5	disp,-
4			3864.5	mean,-
5	4172	4135	—	—
6	—	—	4551.5	mean,+
7	—	—	4595.5	disp,-
8	5835	—	—	—
9	—	—	5875.5	mean,-
10	—	6165	6324.5	disp,+
11	—	7415	7322.5	disp,+
12	—	9287	9202.5	mean,+
13	—	—	9764.5	disp,-
14	10207	10211	—	—
15	11106	11112	—	—
16	11531	—	—	—
17	—	—	11793.5	mean,-
18	12318	12452	—	—
19	13404	13396	13632.5	mean,+
20	14686	14927	—	—
21	15848	15802	15808.5	disp,+
22	—	17762	—	—
23	—	—	17833.5	disp,-
24	18046	—	—	—
25	—	—	18609.5	disp,-
26	20456	20463	—	—
27	—	—	21207.5	mean,-
28	21216	21280	21222.5	disp,+
29	22505	22253	—	—
30	—	—	22743.5	mean,+
31	23139	—	—	—
32	24018	—	23961.5	disp,+
33	24581	24400	—	—
34	—	—	25290.5	disp,-
35	—	—	25733.5	mean,-

The real-world dataset was a set of sleep signals from a patient with Sleep Apnoea. Three of the 7 signals (airflow, abdominal movement signals and snoring) had 16Hz frequency and the other 4 signals (heart rate, light, oxygen saturation and body) had 1Hz frequency. All signal with 16Hz were transform in 1Hz using the mean for each second. The dataset contained 26102 records from 7 signals, which is approximated 7.5 hours of sleep.

3.5   Results on Sleep Apnoea Dataset

Taking into consideration the problem, we built a model to predict the airflow using all other signals as predictor variables. In this dataset, the regression model is evaluated using the normalized mean absolute error statistic. We did not have any indication where the abnormal breathing event would occur. For this reason, we could not evaluate the number of false alarms and true alarms and the mean number of examples for detection of the changes.

We used two regression models: a generalized linear regression and a regression tree[2]. Both learning algorithms employed exhibited slight better results using the detection mechanism than without (table 6). The total number of alarms (TA) were 18 and 17, respectively, for the regression tree and linear regression models. In order to evaluate the agreement between both learning algorithms to detected abnormal breathing events, we compared the alarms proximity between them. We detected 13 pairs of alarms, specifically, 13 alarms detected by the regression tree model occurred in the proximity of 13 alarms detected by the linear regression model (table 7). To validate how well the CUSUM detect changes, we carried out a second set of experiments. We design two datasets. In both datasets, we consider only a single attribute: the time-Id. The target variable, in problem 1, is the mean of the airflow predict by the Kalman Filter. In problem 2, the target variable is the dispersion of the airflow predict by the Kalman Filter. We run a regression tree in both problems and collect the cut-points from both trees. The set of cut-points of the regressions trees and the alarms detected by our DSKC were compared (figure 3). As shown in table 7, there were 7 increases detected in time in the airflow dispersion and all of them were detected at least by one of the learning models applied to Sleep Apnoea Dataset. There were 4 increases detected in time in the airflow, and 2 of them by one of the learning models applied. Despite both learning algorithms investigated exhibited slight better results using the detection mechanism than without, the alarms detected by both models seems to show agreement, which may imply that we have detected true changes in the distribution of examples in the Sleep Apnoea Dataset.

4   Conclusions

In this paper we discussed the problem of maintaining accurate regression models in dynamic, non-stationary environments. The system continuously monitors the residual of the regression algorithm, looking for changes in the mean and changes in the variance. The proposed method maintains a regression model where residuals are filtered by a Kalman filter. A CUSUM algorithm continuously monitors significant changes in the output of the Kalman filter. The CUSUM works as a wrapper over the learning algorithm (the regression model plus the Kalman filter), monitoring the residuals of the actual regression model. If CUSUM detects an increase of the error, a new regression model is learned using only the

---

[2] The GLM and CART versions implemented in [16].

most recent examples. As shown in the experimental section the Kalman filter application to the residuals gives a good on-line estimation of the learning algorithm state. The results of the method for change detection in regression problems show that it can be applied with efficiency when the information is available sequentially over time. An advantage of the proposed method is that it is independent of the learning algorithm. The results of the change detection algorithm mainly depend on the efficiency of the learning algorithm. They also show, that the Kalman filter has a good performance in detecting real changes from noisy data.

Acknowledgments

This work was developed in the context of project ALES II (POSI/EIA/55340/2004) and Project RETINAE(PRIME/IDEIA/70/00078). Thanks to Andre Carvalho and anonymous reviewers for useful comments.

References

1. D. Andre and P. Stone. Physiological data modeling contest. Technical report, University of Texas at Austin, 2004.
2. Michele Basseville and Igor Nikiforov. *Detection of Abrupt Changes: Theory and Applications.* Prentice-Hall, 1993.
3. G. Bhattacharyya and R. Johnson. *Statistical Concepts and Methods.* New York, John Willey & Sons, 1977.
4. Giuseppe Bianchi and Ilenia Tinnirello. Kalman filter estimation of the number of competing terminals in ieee. In *The 22nd Annual Joint Conference of IEEE Computer and Communications*, 2003.
5. Cauwenberghs, Gert, Poggio, and Tomaso. Incremental and decremental support vector machine learning. *In Advances in Neural Information Processing Systems*, 13, 2001.
6. P. Domingos and G. Hulten. Mining high-speed data streams. In *Knowledge Discovery and Data Mining*, pages 71–80, 2000.
7. W. Ward Flemons, Michael R. Littner, James A. Rowley, W. McDowell Anderson Peter Gay, David W. Hudgel, R. Douglas McEvoy, and Daniel I. Loube. Home diagnosis of sleep apnoeas: A systematic review of the literature. *Chest*, 1543-1579:211–237, 2003.
8. J. Friedman. Multivariate adaptive regression splines. *Annals of Statistics*, 19(1):1–141, 1991.
9. J. Gama, P. Medas, and G. Castillo. Learning with drift detection. In *Brazilian AI Conference*, pages 286–295. Springer Verlag, 2004.
10. Eugene Grant and Richard Leavenworth. *Statistical Quality Control.* McGraw-Hill, 1996.
11. G. Guimarães, J.H. Peter, T. Penzel, and A Ultsch. A method for automated temporal knowledge acquisition applied to sleep-related breathing disorders. *Artificial Intelligence in Medicine*, 23:211–237, 2001.
12. C. M. Higgins and R. M. Goodman. Incremental learning using rule-based neural networks. In *International Joint Conference on Neural Networks*, pages 875–880. Seattle, WA, 1991.

13. G. Hulten, L. Spencer, and P. Domingos. Mining time-changing data streams. In *Proceedings of Knowledge Discovery and Data Mining*. ACM Press, 2001.
14. R. E. Kalman. A new approach to linear filtering and prediction problems. *Transaction of ASME - Journal of Basic Engineering*, pages 35–45, 1960.
15. Kwok Pan Pang and Kai Ming Ting. Improving the centered CUSUMs statistic for structural break detection in time series. In *Proc. 17th Australian Join Conference on Artificial Intelligence*. LNCS, Springer Verlag, 2004.
16. R Development Core Team. *R: A language and environment for statistical computing*. R Foundation for Statistical Computing, Vienna, Austria, 2005. ISBN 3-900051-07-0.
17. Raul Rojas. The Kalman filter. Technical report, Freie University of Berlin, 2003.
18. Greg Welch and Gary Bishop. An introduction to the Kalman filter. Technical report, 95-041, Department of Computer Science, Department of Computer Science, University of North Caroline at Chapel Hill, Apr 2004.

Automatic Recognition of Landforms on Mars Using Terrain Segmentation and Classification

Tomasz F. Stepinski[1] Soumya Ghosh[2] and Ricardo Vilalta[2]

[1] Lunar and Planetary Institute, Houston, TX 77058, USA
tom@lpi.usra.edu
[2] Department of Computer Science, University of Houston, Houston, TX 77204, USA
sghosh@uh.edu, vilalta@cs.uh.edu

Abstract. Mars probes send back to Earth enormous amount of data. Automating the analysis of this data and its interpretation represents a challenging test of significant benefit to the domain of planetary science. In this study, we propose combining terrain segmentation and classification to interpret Martian topography data and to identify constituent landforms of the Martian landscape. Our approach uses unsupervised segmentation to divide a landscape into a number of spatially extended but topographically homogeneous objects. Each object is assigned a 12 dimensional feature vector consisting of terrain attributes and neighborhood properties. The objects are classified, based on their feature vectors, into predetermined landform classes. We have applied our technique to the Tisia Valles test site on Mars. Support Vector Machines produced the most accurate results (84.6% mean accuracy) in the classification of topographic objects. An immediate application of our algorithm lies in the automatic detection and characterization of craters on Mars.

1 Introduction

Landforms in Mars are characterized using imagery and altimetry data. Impact craters are among the most studied landforms on Mars. Their importance stems from the amount of information that a detailed analysis of their number and morphology can produce. Visual inspection of imagery data by domain experts has produced a number of catalogs [3,22] that list crater locations and diameters. Automated algorithms for crater detection from imagery data exist [14,16,8,27,28], but none has been deemed adequately accurate to be employed in a scientific study because of poor accuracy (mainly because of false identifications).

In this paper we develop a methodology for automatic identification of landforms on Mars using altimetry (topographic) data. Our goal is to identify craters' floors and their corresponding walls from other landforms. The distinction between floors and walls is important for subsequent calculation of crater geometry [20]. An accurate knowledge of the geometry for a large database of Martian craters would enable studies with a number of outstanding issues and potential discoveries, such as the nature of degradation processes [24], regional variations in geologic material [10], and distribution of subsurface volatiles [11], leading to

N. Lavrač, L. Todorovski, and K.P. Jantke (Eds.): DS 2006, LNAI 4265, pp. 255–266, 2006.

a much better understanding of the composition of the planet's surface and its past climate.

Our strategy departs significantly from that employed in [25,7]. First, we move away from a pixel-based analysis and adopt an object-based analysis [2], which is more appropriate for spatially extended data. Object-based analysis originated in applications of image analysis of remotely sensed data, where an image is segmented into a number of image objects –areas of spectral and/or textural homogeneity. Our approach is based on extending the notion of objects from images to topographic data, and our first task is to segment a given site into topography objects –areas of topographic homogeneity. Second, we move away from clustering (unsupervised learning) and adopt classification (supervised learning) by assigning landform class label to each object. This change assures that the resultant classification corresponds to recognizable landforms. Classification is feasible because the classified units are now objects instead of pixels. Objects have clear topographic meaning and are more suitable than individual pixels for manual labeling. The number of units to be handled is drastically reduced making the classification process viable. Finally, objects are the source of additional information, such as their size, aggregative statistics, and neighborhood information which are used as extra features for classification.

The novelty of our study can be summarized as follows: 1) the automatic segmentation of a terrain into constituent objects is a new concept in the context of terrain analysis; it has applications in both planetary and terrestrial geomorphology. The methodology creates a spatial database otherwise inaccessible to terrain analysis; 2) using classification algorithms to aggregate segmentation objects into larger, physically relevant structures, is a new concept readily available for data analysis techniques (particularly spatial data mining).

2   Background Information

There are currently four space probes on orbits around Mars remotely collecting data about its surface. They generate a deluge of data, but only a small fraction of this data can be interpreted because analysis is performed manually at high "cost" by domain experts. Automation is the only practical solution to the challenge behind processing a significant portion of the –ever increasing– volume of Martian data.

Martian craters, despite having deceptively simple circular appearance, present a formidable challenge for a pattern recognition algorithm. Some craters are degraded by erosion and are barely distinguishable from their background. In a heavily cratered terrain, where an automated detection is most desirable, there is a significant degree of crater overlapping. Finally, crater sizes differ by orders of magnitude. Image-based detection techniques face additional difficulties as the "visibility" of an impact crater depends on the image quality. Other landforms on Mars, as for example valleys, also present challenges for automatic recognition [19].

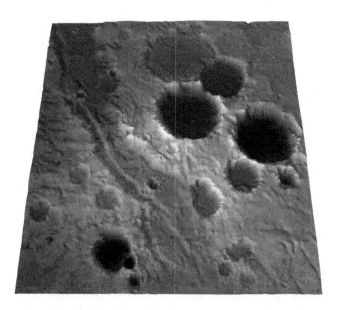

Fig. 1. A perspective view of the Tisia Valles landscape rendered using the DEM. The vertical dimension is exaggerated 10 times, and the north is to the left.

Recently, landform detection algorithms have begun to use topography data as an alternative to images. Mars is the only planet besides the Earth for which global elevation model (DEM) data is available [23]. A DEM is a raster dataset where each pixel is assigned an elevation value. In [25,7] Martian sites are divided into mutually exclusive and exhausting landform categories on the grounds of similarity between pixel-based vectors of terrain attributes. In both studies, landform categories are the result of unsupervised clustering of these vectors, but whereas [25] employs a probabilistic algorithm working under a Bayesian framework, [7] employs a self-organizing map [17]. In principle, these methods avoid the issue of crater identification by automatically categorizing all landforms (including craters) in a given site. The clustering-based division of a site into constituent landforms has the advantage of being an unsupervised, low "cost" process, but it also suffers from the lack of direct correspondence between clusters and generally recognizable landforms. A significant manual post-classification processing is necessary to interpret all results. In addition, some landforms of interest (like, for example, craters' walls) are poorly represented by any single cluster, or even group of clusters.

3 Study Site and Terrain Attributes

Fig. 1. shows a DEM-derived shaded relief of our Tisia Valles test site centered at 46.13°E, 11.83°S. The DEM has a resolution of 500 meters and its dimensions are $N = 385$ rows and $M = 424$ columns. This is a challenging site for landform identification. In a relatively small area many different crater morphologies are present: fresh deep craters with intact walls; old, degraded craters with various degrees of wall erosion; conjoined craters with various degrees of overlap. All these different types of craters are present in a range of sizes. In addition, the site is crossed from south-west to north-east by a broad valley with escarpments on both sites. The valley floor is further sculptured. Smaller scale valleys are present at inter-crater highlands. The same site was previously used in [25].

The DEM carries information about terrain elevation, $z(x, y)$ at the location of each pixel. Terrain attributes are additional raster datasets calculated from the DEM. In this study we use three terrain attributes: slope, curvature, and flood. The slope, $s(x, y)$, is the rate of maximum change of z on the Moore neighborhood[1] of a pixel located at (x, y). The (profile) curvature, $\kappa(x, y)$, measures the change of slope angle and is also calculated using the values of z on the Moore neighborhood of a pixel ($\kappa > 0$ correspond to convex topography, whereas $\kappa < 0$ correspond to concave topography). The flood, $f(x, y)$, is a binary variable; pixels located inside topographic basins have $f(x, y) = 1$, and all other pixels have $f(x, y) = 0$. A vector, $\mathbf{V}(x, y) = \{z, s, \kappa, f\}(x, y)$ describes the topography of the landscape at the level of an individual pixel. We refer to the $(N - 2) \times (M - 2)$ array of vectors $\mathbf{V}$ as the landscape. The landscape is smaller than a DEM because the DEM's edge is eliminated (by removing the two rows and two columns lying on the array boundaries) due to calculations of derivatives using the Moore neighborhood.

4   Segmentation

In the context of image analysis the term segmentation refers to a process of dividing an image into smaller regions having homogeneous color, texture, or both. We have observed that the notion of segmentation can be applied not only to multi-band images, but to all spatially extended datasets including multi-attribute landscapes. A variety of techniques [1,4,13,21] have been proposed to implement image segmentation, and all of them can be, in principle, easily extended to landscape segmentation. In [15] a computationally simple homogeneity index H was proposed, and in [9] this index was combined with the watershed transform [26] for fast, unsupervised segmentation of multi-band images. In this paper we utilize this method for segmentation of multi-attribute landscapes.

The homogeneity measure H is calculated using a square window of width $2K + 1$ (where K is user-defined). Consider a focal pixel (x_c, y_c) having an attribute (for example, an elevation z) $z(x_c, y_c)$. For every pixel in a window we calculate a "separation" vector $\mathbf{d}_i = (x_i - x_c, y_i - y_c)$. From the separation vector we construct a "gradient" vector,

$$\mathbf{g}_i = (z(x_i, y_i) - z(x_c, y_c)) \frac{\mathbf{d}_i}{\|\mathbf{d}_i\|} \tag{1}$$

---

[1] The Moore neighborhood around a focus pixel (x_0, y_0) is a square-shaped area defined by $\{(x, y) : |x - x_0| \leq 1, |y - y_0| \leq 1\}$.

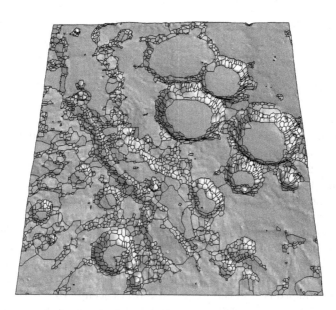

Fig. 2. Segmentation of Tisia Valles landscape into 2631 topographically homogeneous objects

and we use gradient vectors calculated for all pixels in a window to calculate the homogeneity measure H,

$$H = \left\| \sum_{i=1}^{(2K+1)^2} \mathbf{g}_i \right\| \tag{2}$$

A pixel located in the region that is homogeneous with respect to z has a small value of H. On the other hand, a pixel located near a boundary between two regions characterized by different values of z has a large value of H.

A raster constructed by calculating the values of H for all pixels in the landscape can be interpreted as a gray scale image and is refered to as the H-image. We denote the H-image by $\mathbf{H}$. The white areas on $\mathbf{H}$ represent boundaries of homogeneous regions, whereas the dark areas represent the actual regions. The extension of the H-image concept to multiple attributes is straightforward. Let's say that we want to calculate the $\mathbf{H}$ of a landscape on the basis of slope, curvature, and flood attributes. For each pixel we calculate the three individual H values separately and combine them to obtain the overall value of H at that pixel:

$$H = \sqrt{w_s \, H_s{}^2 + w_\kappa \, H_\kappa{}^2 + w_f \, H_f{}^2} \tag{3}$$

where w_s, w_κ, and w_f are weights introduced to offset different numerical ranges of the attributes. All attributes are scaled to have the same range $(0, 1)$, and the weights in (3) correspond to the corresponding scales.

We have segmented the Tisia Valles landscape using three attributes, s, κ and f. Note that we have opted not to use z as a segmentation attribute because no landforms are characterized by their elevations. We use the H-image technique with $K = 2$ to obtain H_s, H_κ, and H_f. Individual H-images are subject to thresholding,

$$\mathbf{H}_k(x, y) = \begin{cases} \mathbf{H}_k(x, y) = \mathbf{H}_k(x, y) & \text{if } \mathbf{H}_k(x, y) > T_k \\ \mathbf{H}_k(x, y) = 0 & \text{otherwise} \end{cases} \tag{4}$$

where $k = s, \kappa, f$, and T_k is an appropriate threshold value introduced to prevent oversegmentation caused by high sensitivity to noise. The thresholded H-images are combined using formula (3) and the combined H-image is segmented using the watershed transform. This procedure (with $T_s = 0.15$, $T_\kappa = 0.92$, and $T_f = 3.9$) segments the site into 2631 topographic objects as shown in Fig. 2. Each object is topographically homogeneous. Note that the number of objects is two orders of magnitude smaller than 161,626 pixels constituting the DEM. The objects are small where topography changes on a small spatial scale, like, for example, on the walls of the craters, or at the escarpments. On the other hand, objects are large when changes occur on only a large spatial scale, like, for example in the inter-crater plain, or on the floors of large craters. The largest object has 15,106 pixels, and the 13 largest objects occupy 75% of the site's area. For each object, $i = 1, \ldots, 2631$, we calculate the mean values, $\bar{s}_i$, $\bar{\kappa}_i$, $\bar{f}_i$, and standard deviation values, σ_i^s, σ_i^κ, σ_i^f, from its constituent pixels. We refer to $\bar{s}_i$, $\bar{\kappa}_i$, and $\bar{f}_i$ simply as slope, curvature, and flood of an object.

Objects making up craters' walls and objects constituting escarpments not associated with craters have similar topographic attributes but are located in different spatial contexts. The object's neighborhood properties provide some information about that context. Ideally, we would like to know classes of object's neighbors, but such information is not available prior to classification. However, a preliminary categorization of objects is possible on the basis of their values of $\bar{s}_i$, $\bar{\kappa}_i$, $\bar{f}_i$. We divide all objects into three categories (low, medium, and high) on the basis of their slope values. Such categorization is used to calculate a neighborhood property of an object i, $\{a_1^s, a_2^s, a_3^s\}_i$, where a_j^s, $j = 1, 2, 3$, is the percentage of the object boundary adjacent to neighbors belonging to slope category j. Similar neighborhood properties, $\{a_1^\kappa, a_2^\kappa, a_3^\kappa\}_i$, $\{a_1^f, a_2^f, a_3^f\}_i$, are calculated on the basis of curvature and flood values, yielding a total of nine attributes corresponding to the spatial context of objects.

5   Classification

We classify topographic objects into six landform classes with clear physical meaning. Class 1 consists of inter-crater plains, a flat terrain that in most cases is homogeneous on relatively large spatial scale. Class 2 consists of craters' floors, a flat terrain inside craters. Class 3 consists of convex craters' walls, whereas class 4 consists of concave craters' walls. Classes 5 and 6 consists of objects that are located on convex and concave non-crater escarpments, respectively. We use the following 12-dimensional feature vector to characterize each segment,

$$\mathbf{u} = \left\{ \bar{s}, \ \bar{\kappa}, \ \bar{f}, \ a_1^s, \ a_2^s, \ a_3^s, \ a_1^\kappa, \ a_2^\kappa, \ a_3^\kappa, \ a_1^f, \ a_2^f, \ a_3^f \right\} \tag{5}$$

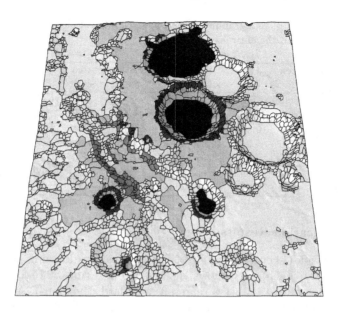

Fig. 3. Training set of 517 labeled topographic objects, 10% gray indicates class 1, black indicates class 2, 20% gray indicates class 3, 80% gray indicates class 4, and 40% gray indicates classes 5 and 6

We note that the object's size, its elevation, and its σ_i^s, σ_i^κ, σ_i^f values are not presently used as part of the feature vector. In general, objects' sizes and elevations have proved to be poor indicators of their class as defined in this study. Inclusion of σ_i^s, σ_i^κ, σ_i^f into the feature vector produces no improvement in classification.

5.1   Training Set

We have manually labeled 517 topographic objects to make up a training set. This set constitutes 20% of all objects and 29% of the site's area. The location of training set objects is shown on Fig. 3. Due to the limitation of illustrating a site using gray scales, classes 5 and 6 are shown employing the same gray shade. The objects were selected for labeling on the basis of geographical coherence (see Fig. 3) but we have also made sure that all landform classes are represented in the training set.

To provide additional information regarding our training set, Table 1 (left side) shows for each class (each row), the number of objects in that class, the fraction of the training set's total area covered by the objects in the class, and the values of the three physical features, s, κ, and f, averaged over objects in each class. The right side of Table 1 provides the same information but for all objects constituting the Tisia Valles site as divided into classes by using a classifier obtained using a Support Vector Machine.

Table 1. Properties of topographic objects in six landform classes averaged over the training set (left) and the entire set (right).

class	object count	area site %	s deg.	κ $\times 10^3$	f	object count	area site %	s deg.	κ $\times 10^3$	f
	517 objects in the training set					All 2631 objects				
1	89	54.0	0.99	-0.1	0.04	957	64.5	0.96	0.12	0.03
2	5	24.0	0.83	0.41	1.0	26	11.5	0.83	0.58	0.91
3	163	9.0	5.24	3.12	0.78	536	8.0	5.2	3.54	0.83
4	145	8.0	4.08	-3.78	0.08	674	10.0	4.11	-4.62	0.13
5	61	2.6	2.49	2.61	0.09	208	2.0	2.02	2.27	0.08
6	54	2.4	2.20	-2.01	0.0	230	4.0	1.78	-1.79	0.0

5.2   Classification Results

Table 2 (left side) shows the result of invoking various learning algorithms on our training set. Each entry shows the average of 5 runs of 10-fold cross validation (numbers in parentheses represent standard deviations). All algorithms follow the implementation of the software package WEKA [29] using default parameters[2]. An asterisk at the top right of a number implies the difference is significantly worse than the first algorithm (Support Vector Machine) at the $p = 0.05$ level assuming a two-tailed t-student distribution.

Table 2 was produced to observe the inherent difficulty associated with differentiating between various Martian landforms. The advantage of relatively complex models over simpler ones points to the need for flexible decision boundaries. The apparent advantage of bagging over the decision tree points to some degree of variance [6]. The right side of Table 2 shows a confusion matrix obtained using the Support Vector Machine model.

Overall, the exclusion errors (i.e., false negatives) are acceptable with the exception of class 6. A significant number of class 6 (concave escarpment) objects are classified as either class 4 (concave crater walls), as we could expect from the local similarity between the two landforms, or as class 1 (inter-crater plains). The largest number of exclusions from class 4 are, as expected, picked up by class 6, but some are picked up by classes 1 and 3. Surprisingly, there is not much confusion between classes 5 and 3. This is probably due to the large difference between the values of the flood attribute in the two classes (see Table 1). The erroneous inclusion errors (i.e., false positives) show similar patterns. Interestingly, class 1 (a big inter-crater plain) suffers a small degree of erroneous inclusions from most other classes that, in general, are characterized by very different values of physical features.

---

[2] SVM uses a cubic kernel with a complexity parameter of $C = 1$. Decision tree uses reduced error pruning with a confidence factor of 0.25 and a stopping criterion for splitting of size 2. Bagging uses bootstrap samples of size equal to the training set and average values over 10 trees. Neural network uses a learning rate of 0.3 and momentum of 0.2 with 500 epochs. k-nearest neighbor uses $k = 5$ and the KKDtree algorithm [5]. Bayesian Network builds a network using the K2 algorithm [12].

Table 2. (Left) Accuracy by averaging over 5 x 10-fold cross-validation for various learning algorithms. An asterisk at the top right of a number indicates a statistically significant degradation with respect to the first algorithm (Support Vector Machine). Numbers enclosed in parentheses represent standard deviations. (Right) Confusion matrix for the Support Vector Machine algorithm, exclusions (false negatives) are in the rows, erroneous inclusions (false positives) are in the columns.

Learning Algorithm	Accuracy Estimation	Confusion Matrix					
		1	2	3	4	5	6
Support Vector Machine	84.61 (4.89)	85	0	1	0	3	0
Neural Network	81.71 (5.09)	0	4	1	0	0	0
Bagging	83.02 (5.30)	2	1	146	10	4	0
Decision Tree	81.01* (5.09)	7	1	8	119	0	10
Bayesian Network	79.42* (5.75)	3	0	4	0	54	0
Nearest Neighbor	78.77* (5.15)	13	0	0	11	1	29

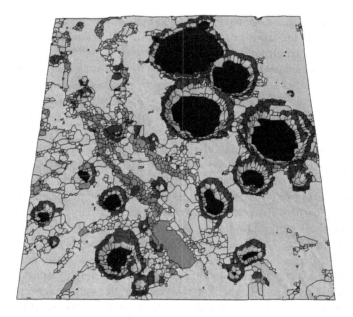

Fig. 4. Classification of all 2631 topographic objects, 10% gray indicates class 1, black indicates class 2, 20% gray indicates class 3, 80% gray indicates class 4, and 40% gray indicates classes 5 and 6

Using a decision function calculated on the basis of the Support Vector Machines model, we have classified all 2631 objects in the Tisia Valles site. The spatial illustration of this classification is shown on Fig. 4 and the numerical results of the classification are given in Table 1 (right side). The values of the three physical features averaged over objects belonging to the same class are very similar to corresponding values in the training set. This reassures us that the training set constitutes a representative sample of the entire site.

Visual inspection of Fig. 4 indicates that classification successfully divided the landscape into its landforms. Large craters are perfectly delineated from the inter-crater plain, and most of the escarpment is also correctly identified. A more in depth examination reveals some inevitable shortcomings. The classifier does not identify the centers of small craters as floors (class 2). Technically, this is correct, because such craters are too small to have flat "floors," but such omissions complicate the process of crater recognition. The smallest craters have centers represented by class 3 objects, which are surrounded by class 4 objects, too large to produce a correct representation of their walls. This is due to insufficient resolution of our segmentation. Finally, there are some limited problems in the correct delineation of small relief escarpments. Overall these shortcomings are minor; most of the complex landscape was correctly interpreted.

6   Conclusions and Future Work

The success of our method depends on two factors, the quality of segmentation and the choice of an appropriate feature vector. Martian terrain requires that the range of topography objects sizes span few orders of magnitude. Although our segmentation achieves this requirement, a larger range is necessary to resolve the smallest craters. Future work will have to address this issue. One possible solution is to avoid a single watershed threshold, and rather use an adaptive threshold with a value that is coupled to the spatial scale of the change in topography.

Our segmentation process can also be viewed as the process of creating a spatial database of topography objects. In spatial data mining, objects are characterized by different types of information: non-spatial attributes, spatial attributes, spatial predicates, and spatial functions [18]. In this paper we have considered only physical features (corresponding to non-spatial attributes) and some neighborhood relations (corresponding to spatial attributes). Using just the physical features produced a 77.4% classification accuracy; this accuracy increases to 84.6% when features based on neighborhood properties are also considered (using a Support Vector Machine). Future studies will employ spatial predicates and spatial functions to further increase our current predictive accuracy.

Finally, we stress that the present classification allows for an automatic identification and characterization of large and medium craters. A study is underway to use our technique to calculate geometries of such craters on Sinai Planum and Hesperia Planum, two locations on Mars that, although similar, are expected to have systematic differences in crater morphologies [20]. Once a good predictive model is found we plan to use it to identify craters along the whole surface of Mars, without any need for creating additional training sets.

Acknowledgements

This work was supported by the National Science Foundation under Grants IIS-0431130, IIS-448542, and IIS-0430208. A portion of this research was conducted at the Lunar and Planetary Institute, which is operated by the USRA under contract CAN-NCC5-679 with NASA. This is LPI Contribution No. 1301.

References

1. Adams, R. Bischof, L.: Seeded Region Growing. IEEE Trans. Pattern Analysis and Machine Intelligence. **16(6)** (1994) 641–647
2. Baatz M., Schäpe, A.: Multiresolution Segmentation - An Optimization Approach for High Quality Multi-Scale Image Segmentation. In: Strobl, J. et al. (eds.): Angewandte Geographische Infor-mationsverarbeitung XII. Wichmann, Heidelberg. (2000) 12-23
3. Barlow, N.G.: Crater Size-Distributions and a Revised Martian Relative Chronology. Icarus. **75(2)** (1988) 285-305
4. Belongie, S., Carson, C., Greenspan, H. Malik, J.: Color- and Texture-based Image Segmentation Using EM and its Application to Content-based Image Retrieval. Proc. of Sixth IEEE Int. Conf. Comp. Vision. (1998) 675-682
5. Bentley J. L.: Multidimensional Binary Search Trees Used for Associative Searching. Communications of ACM, **18**, (9), (1975), 509–517.
6. Breiman, L.: Bagging Predictors. Machine Learning, **24**, (1996) 123–140.
7. Bue, B.D., Stepinski, T.F.: Automated classification of Landforms on Mars. Computers & Geoscience. **32(5)** (2006) 604-614
8. Burl, M.C., Stough, T., Colwell, W., Bierhaus, E.B., Merline, W.J.,Chapman, C.: Automated Detection of Craters and Other Geological Features. Proc. Int. Symposium on Artificial Intelligence, Robotics and Automation for Space, Montreal, Canada. (2001)
9. Chen, Q., Zhou, C., Luo, J., Ming, D.: Fast Segmentation of High-Resolution Satellite Images Using Watershed Transform Combined with an Efficient Region Merging Approach. Proc. 10th Int. Workshop Combinatorial Image Analysis. Lecture Notes in Computer Science, Vol 3322. Springer-Verlag, Berlin Heidelberg New York (2004) 621–630
10. Cintala, M.J., Head, J.W., Mutch, T.A.: Martian Crater Depth/Diameter Relationship: Comparison with the Moon and Mercury. Proc. Lunar Sci. Conf. **7** (1976) 3375-3587
11. Cintala, M.J., Mouginis-Mark, P.J.: Martian Fresh Crater Depth: More Evidence for Subsurface Volatiles. Geophys. Res. Lett. **7** (1980) 329-332
12. Cooper, G.F., Herskovits, E.: A Bayesian Method for the Induction of Probabilistic Networks from Data. Machine Learning **9(4)** (1992) 309–347.
13. Deng, Y., Manjunath, B.S.: Unsupervised Segmentation of Color-Texture Regions in Images and Video. IEEE Trans. Pattern Analysis and Machine Intelligence. **23(8)** (2001) 800–810
14. Honda, R., Iijima, Y., Konishi, O.: Mining of Topographic Feature from Heterogeneous Imagery and Its Application to Lunar Craters. Progress in Discovery Science: Final Report of the Japanese Discovery Science Project, Springer-Verlag, Berlin Heidelberg New York (2002) p. 395
15. Jing, F., Li, M.J., Zhang, H.J., Zhang, B.: Unsupervised Image Segmentation Using Local Homogeneity Analysis. Proc. IEEE International Symposium on Circuits and Systems. (2003) II-456–II-459
16. Kim, J.R., Muller, J-P., van Gasselt, S., Morley, J.G., Neukum, G., and the HRSC Col Team: Automated Crater Detection, A New Tool for Mars Cartography and Chronology. Photogrammetric Engineering & Remote Sensing. **71(10)** (2005) 1205-1217

17. Kohonen, T.: Self-Organizing Maps. Springer-Verlag, Berlin Heidelberg New York (1995)
18. Koperski, K., Han, J., Stefanovic, N.: An Efficient two-Step Method for Classification of Spatial Data. Proc. Eight Symp. Spatial Data Handling. (1998) 45-55
19. Molloy, I., Stepinski, T.F.: Automatic Mapping of Valley Networks on Mars. Submitted to Computers & Geoscience (2006)
20. Mouginis-Mark, P.J., Garbeil, H., Boyce, J.M., Ui, C.S.E., Baloga, S.M.: Geometry of Martian Impact Craters: First Results from an Iterative Software Package. J. Geophys. Res. **109** (2004) E08996
21. Nock, R., Nielsen, F.: Stochastic Region Merging. IEEE Trans. Pattern Analysis and Machine Intelligence. **26(11)** (2004) 1452-1458
22. Rodionova, J.F., Dekchtyareva, K.I., Khramchikhin, A.A., Michael, G.G., Ajukov, S.V., Pugacheva, S.G., Shevchenko, V.V.: Morphological Catalogue Of The Craters Of Mars. ESA-ESTEC (2000)
23. Smith, D., Neumann, G., Arvidson, R.E., Guinness, E.A., Slavney, S.: Mars Global Surveyor Laser Altimeter Mission Experiment Gridded Data Record. NASA Planetary Data System, MGS-M-MOLA-5-MEGDR-L3-V1.0. (2003)
24. Soderblom, L.A., Condit, C.D. West, R.A., Herman, B.M., Kreidler, T.J.: Martian Planetwide Crater Distributions: Implications for Geologic History and Surface Processes. Icarus, **22** (1974) 239-263
25. Stepinski, T.F., Vilalta, R.: Digital Topography Models for Martian surfaces. IEEE Geoscience and Remote Sensing Letters. **2(3)** (2005) 260-264
26. Vincent, L., Soille, P.: Watersheds in Digital Spaces: An Efficient Algorithm Based on Immersion Simulations. IEEE Trans. Pattern Analysis and Machine Intelligence. **13(6)** (1991) 583-598
27. Vinogradova, T., Burl, M., Mjosness, E.: Training of a Crater Detection Algorithm for Mars Crater Imagery. Aerospace Conference Proc. 2002, IEEE. **7** (2002) 7-3201 - 7-3211
28. Wetzler, P.G., Enke, B., Merline, W.J., Chapman, C.R., Burl, M.C.: Learning to Detect Small Impact Craters. Seventh IEEE Workshops on Computer Vision (WACV/MOTION'05). **1** (2005) 178-184
29. Witten I. H., Frank E.: Data Mining: Practical Machine Learning Tools and Techniques. 2nd Edition, Morgan Kaufmann, San Francisco, (2005)

A Multilingual Named Entity Recognition System Using Boosting and C4.5 Decision Tree Learning Algorithms

György Szarvas[1], Richárd Farkas[2], and András Kocsor[2]

[1] University of Szeged, Department of Informatics,
6720 Szeged, Árpád tér 2., Hungary
[2] MTA-SZTE, Research Group on Artificial Intelligence,
6720 Szeged, Aradi Vértanúk tere 1., Hungary
{rfarkas, szarvas, kocsor}@inf.u-szeged.hu

Abstract. In this paper we introduce a multilingual Named Entity Recognition (NER) system that uses statistical modeling techniques. The system identifies and classifies NEs in the Hungarian and English languages by applying AdaBoostM1 and the C4.5 decision tree learning algorithm. We focused on building as large a feature set as possible, and used a split and recombine technique to fully exploit its potentials. This methodology provided an opportunity to train several independent decision tree classifiers based on different subsets of features and combine their decisions in a majority voting scheme. The corpus made for the CoNLL 2003 conference and a segment of Szeged Corpus was used for training and validation purposes. Both of them consist entirely of newswire articles. Our system remains portable across languages without requiring any major modification and slightly outperforms the best system of CoNLL 2003, and achieved a 94.77% F measure for Hungarian. The real value of our approach lies in its different basis compared to other top performing models for English, which makes our system extremely successful when used in combination with CoNLL modells.

Keywords: Named Entity Recognition, NER, Boosting, C4.5, decision tree, voting, machine learning.

1 Introduction

The identification and classification of proper nouns in plain text is of key importance in numerous natural language processing applications. In Information Extraction systems proper names generally carry important information about the text itself, and thus are targets for extraction and Machine Translation. These have to handle proper nouns and other sort of words in a different way due to the specific translation rules that apply to them. These two topics are in the focus of our research.

1.1 Related Work

Research and development efforts in the last few years have focused on other languages, domains or cross-language recognition. Hungarian NER fits into this trend quite well, due to the special agglutinative property of the language.

N. Lavrač, L. Todorovski, and K.P. Jantke (Eds.): DS 2006, LNAI 4265, pp. 267–278, 2006.
© Springer-Verlag Berlin Heidelberg 2006

Machine learning methods have been applied to the NER problem with remarkable success. The most frequently applied techniques were the Maximum Entropy Model, Hidden Markov Models (CoNLL-2003) and Support Vector Machines (JNLPBA-2004, [10]).

We use AdaBoostM1 and C4.5 learning techniques which have an inherently different theoretical background from the machine learning algorithms that have been used most frequently for NER (like Maximum Entropy Models, Support Vector Classifiers, etc.). The results of this paper prove that this can significantly improve classification accuracy in a model combination scheme. Another reason for using decision trees was that we needed a fast and efficient model to exploit the potentials of our large feature set.

There are some results on NER for the Hungarian language as well but all of them are based on expert rules [9], [12]. To our knowledge, no statistical models have yet been constructed for the Hungarian language.

1.2 Structure of the Paper

In the following section we will introduce the NER problem in general, along with the details of the English and Hungarian tasks performed and the evaluation methodology. In Section 3 we discuss the learning methods, the pre- and post-processing techniques we applied and the structure of our complex NER system. The experimental results are then presented in Section 4 along with a brief discussion, followed in Section 5 by some concluding remarks and suggestions for future work.

2 The NER Task

The identification of proper names can be regarded as a tagging problem where the aim is to assign the correct tag (label) to each token in a simple text. This classification determines whether the lexical unit in question is part of a proper noun phrase and if it is, which category it belongs to.

The NER task was introduced during the nineties as a part of the shared tasks in the Message Understanding Conferences (MUC) [4]. The goal of these conferences was the recognition of proper nouns (person, organization, location names), and other phrases denoting dates, time intervals, and measures in texts collected from English newspaper articles. The best systems [1] following the MUC task definition achieved outstanding accuracies (near 95% F measure).

Later, as a part of the Computational Natural Language Learning (CoNLL) conferences [15], a shared task dealt with the development of systems like this that work for multiple languages (first introduced in [5]) and were able to correctly identify a person, an organization and location names, along with other proper nouns treated as miscellaneous entities. The collection of texts consisted of newswire articles, in Spanish + Dutch and English + German, respectively. There are several differences between the CoNLL style task definition and the 1990s MUC approach that made NER a much harder problem:

- Multilinguality was introduced, thus systems had to perform well in more than one language without any major modification.

- The NE types that are very simple to identify (phrases denoting dates, time intervals, measures and so on.) were excluded from the CoNLL task. This way, systems were evaluated on the 4 most problematic classes out of the many used in MUCs.

- A more strict evaluation script was introduced that penalizes the misclassification of an inner part of a long phrase twice (one error for finding a wrong shorter phrase and another for the misclassified term).

These modifications made the NER task harder (the accuracy of the best performing systems [8] dropped below 89% for English) but more practical since real world applications like Information Extraction benefit from these types of NEs and by doing this (only whole phrases classified correctly contribute to other applications). In our studies we always followed the CoNLL style task definition and used the same evaluation script.

In accordance with the task definition of the CoNLL conferences we distinguish four classes of NEs, namely *person, location, organization names* and *miscellaneous entities*. This classification is not straightforward in many cases and a human annotator needs some background knowledge and additional information about the context to perform the task. Many proper nouns can denote entities of more than one class depending on the context, and occasionally a single phrase might fall into any of the four categories depending on the context (like "Ford", which can refer to a person, the company, an airport or the car type).

2.1 English NER

An NER system in English was trained and tested on a sub-corpus of the Reuters Corpus[1], consisting of newswire articles from 1996 provided by Reuters Inc. The data is available free of charge for research purposes and contains texts from diverse domains ranging from sports news to politics and the economy. The best result published in the CoNLL 2003 conference was an F measure of 88.76% obtained from the best individual model, and 90.3% for a hybrid model based on the majority voting of five participating systems.

2.2 Hungarian NER

To train and test our NER model on Hungarian texts, we decided to use a sub-corpus of the Szeged Treebank [6] which contains business news articles from 38 NewsML[2] topics ranging from acquisitions to stock market changes or the opening of new industrial plants. We annotated this collection of texts with NE labels that followed the current international standards as no other NE corpus of reasonable size is available for Hungarian. The data can be obtained free of charge for research purposes[3].

One major difference between Hungarian and English data is the domain specificity of the former corpus. The Hungarian texts we used consist of short newspaper articles from the domain of economy, and thus the *organization* class dominates the

---

[1] http://www.reuters.com/researchandstandards/
[2] See http://www.newsml.org/pages/docu_main.php for details.
[3] http://www.inf.u-szeged.hu/~hlt/index.html

other three in frequency. This difference undoubtedly makes NER an easier problem on the Hungarian text, while the special characteristics of the Hungarian language (compared to English) like agglutinativity or free word order usually makes NLP tasks in Hungarian very difficult. Thus it is hard to compare the results for Hungarian with other languages but achieving similar results in English (the language for which the best results have been reported so far) is still a remarkable feature.

The annotation procedure of the corpus consisted of several phases where two independent annotators tagged the data and discussed problematic cases later on. In the final phase all the entities that showed some kind of similarity to one that was judged inconsistent were collected together from the corpus for a review by the two annotators and the chief annotator. The resulting corpus had an inter-annotator agreement rate of 99.89% and 99.77% compared to the annotations made by the two linguists on their own [14].

This corpus then is completely equivalent to other corpuses used on the CoNLL-2002 and CoNLL-2003 conferences, both in format and annotation style (the same classes are labeled). We hope that this will make both cross-language comparison and the use of the corpus in developing NER systems more straightforward.

No independent results on Hungarian NER using this corpus have yet been published. The results here are compared to our previous results, which are the best that have been published so far [7].

2.3 Evaluation Methodology

To make our results easier to compare with those given in the literature, we employed the same evaluation script that was used during the CoNLL conference shared tasks for entity recognition[4]. This script calculates Precision, Recall and F value scores by analyzing the text at the phrase level. This way evaluation is very strict as it can penalize single mistakes in longer entity phrases two times.

It is worth mentioning that this kind of evaluation places a burden on the learning algorithms as they usually optimize their models based on a different accuracy measure. Fitting this evaluation into the learning phase is not straightforward because of some undesired properties of the formula that can adversely affect the optimization process.

3 Complex NER Model

We regard the NER problem as essentially a classification of separate tokens. We believe that this approach is competitive with the – theoretically more suitable – sequence tracking algorithms (like Hidden Markov Models, Maximum Entropy approaches or Conditional Random Fields) and we could choose a decision tree which requires less computation time and thus enables us for example to use an enormous feature set. Of course our model takes into account the relationship between consecutive words as well through a window with appropriate window size.

---

[4] The evaluation script can be downloaded from the CoNLL conference web site.

To solve classification problems effectively it is worth applying various types of classification methods, both separately[5] and in combination. The success of hybrid methods lies in tackling the problem from several angles, so algorithms of inherently different theoretical bases are good subjects for voting and for other combination schemes. Feature space construction and the proper pre-processing of data also have a marked impact on system performance. In our experiments we incorporated all these principles into a complex statistical NER model.

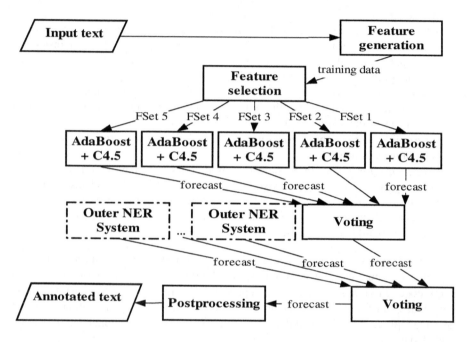

Fig. 1. Outline of the structure of our NER model. The result of our model working alone is discussed for English and Hungarian, along with the results of a voting system for English treated as a hybrid model. We used our model in combination with the two top performing CoNLL systems.

The building blocks of our system are shown in Figure 1. We expected our model would perform well in combination with other popular models (noted as "Outer NER System" in Figure 1) like the Maximum Entropy approach, Hidden Markov Model or Support Vector Classifiers. Our results on the English dataset where outputs of such systems were available justify this expectation.

3.1 Feature Set

Initial features. We employed a very rich feature set for our word-level classification model, describing the characteristics of the word itself along with its actual

---

[5] We investigated several algorithms but because of the lack of space we present here only the best performing one. For details of our past experiments, please see [7].

context (a moving window of size four). Our features fell into the following major categories:

- gazetteers of *unambiguous NEs* from the train data: we used the NE phrases which occur more than five in the train texts and got the same label more than 90 percent of the cases,
- *dictionaries* of first names, company types, sport teams, denominators of locations (mountains, city) and so on: we collected 12 English specific lists from the Internet and 4 additional to the Hungarian problem,
- *orthographical features*: capitalization, word length, common bit information about the word form (contains a digit or not, has uppercase character inside the word, and so on). We collected the most characteristic character level bi/trigrams from the train texts assigned to each NE class,
- *frequency information*: frequency of the token, the ratio of the token's capitalized and lowercase occurrences, the ratio of capitalized and sentence beginning frequencies of the token,
- *phrasal information*: chunk codes and forecasted class of few preceding words (we used online evaluation),
- *contextual information*: POS codes (we used codes generated by our POS tagger for Hungarian instead of the existing tags from the Szeged Treebank), sentence position, document zone (title or body), topic code, trigger words (the most frequent and unambiguous tokens in a window around the NEs) from the train text, is the word between quotes and so on.

The same features were used in the experiments on Hungarian texts and only the semantics of some feature varied (e.g. we used a different categorization of POS and chunk codes in Hungarian and the company type suffixes were different). Only the topic code and document zone features were omitted for Hungarian as all articles had the same topic (economics) and the titles of the articles were not capitalized like in the English dataset.

Feature set splitting and recombination. Using the above six groups of features we trained a decision tree for all possible subset of the groups (63 models), and of course not every subset describe the NER problem equally well. We used simple C4.5 trees here because their training are very fast and we assumed that the differences between single trees would not change significantly while boosting them. We evaluated these models on the CoNNL development set.

The 11 best performing models achieved very similar results; the others were far behind them. We decided to keep 5 of these models for CPU consumption reasons. We chose the five models – from the 11 - that showed the greatest average variety in features used (we did not omit a category because it achieved a slightly worse result).

We trained a classifier using each of these five feature sets (note that they are not disjunctive) and then recombined the resulting models in a voting scheme (which will be introduced later in detail). The same five sets of features (group of categories) were used in the experiments on Hungarian language.

3.2 Classifiers

Boosting [13] and C4.5 [11] are well known algorithms for those who are acquainted with pattern recognition. Boosting has been applied successfully to improve the performance of decision trees in several NLP tasks. A system that made use of AdaBoost and fixed depth decision trees [2] came first on the CoNLL-2002 conference shared task for Dutch and Spanish, but gave somewhat worse results for English and German (it was ranked fifth, and had an F measure of 85.0% for English) in 2003. We have not found any other competitive results for NER using decision tree classifiers and AdaBoost published so far.

As our results show, their combination can compete with state-of-the-art recognition systems solving the NER problem, as well as bring some improvement in classification accuracy and in preserving the superiority of decision tree learning when it comes to the CPU time used in training and evaluating a model. In our experiments we used the implementations available in the WEKA [16] library, an open-source data mining software written in Java.

Boosting was introduced by Shapire as a way of improving the performance of a weak learning algorithm. The algorithm generates a set of classifiers (of the same type) by applying bootstrapping on the original training data set and it makes a decision based on their votes. The final decision is made using a weighted voting schema for each classifier that is many times more accurate than the original model. 30 iterations of Boosting were performed on each model. Further iterations gave only slight improvements in the F measure (less than 0.05%), thus we decided to perform only 30 iterations in each experiment.

C4.5 is based on the well-known ID3 tree learning algorithm, which is able to learn pre-defined discrete classes from labeled examples. Classification is done by axis-parallel hyperplanes, and hence learning is very fast. This makes C4.5 a good subject for boosting. We built decision trees that had at least 5 instances per leaf, and used pruning with subtree raising and a confidence factor of 0.33. These parameters were determined after the preliminary testing of some parameter settings and evaluating the decision trees on the development phase test set. A more thorough analysis of learning parameters will be performed in the near future.

3.3 Combination

There are several well known meta-learning algorithms in the literature that can lead to a 'better' model (in terms of classification accuracy) than those serving as a basis for it, or can significantly decrease the CPU time of the learning phase without loss of accuracy. In our study we chose to concentrate on improving the accuracy of the system.

The decision function we used to integrate the five hypotheses (learnt on different subsets of features) was the following: *if any three of the five learners' outputs coincided we accepted it as a joint prediction, with a forecasted 'O' label referring to a non-named entity class otherwise.* This cautious voting scheme is beneficial to system performance as a high rate of disagreement often means a poor prediction rate. For a CoNLL type evaluation it is better to make such mistakes that classifies an NE as non-named entity than place an NE in a wrong entity class (the latter detrimentally affects precision and recall, while the former only affects the recall of the system).

3.4   Post-processing Data

Several simple post-processing methods can bring about some improvement to system accuracy. Take, for instance, full person names which consist of first names and family names, which are easier to recognize than 'standalone' family names which refer to a person (e.g. "John Nash" or "Nash"). Here if we recognize a full name and encounter the family name later in the document we simply overwrite its label with a person name. This is a reasonable assumption that holds true in most cases.

Certain types of NEs rarely follow each other without any punctuation marks so if our term level classification model produces such an output we overwrite all class labels of this sequence with the label assigned to its head.

Table 1. Improvements of the post processing steps based on the previous step (percentile)

	Family names	Rare sequence filter	Acronym
CoNLL Test	+0.63	+0.59	+0.70
CoNLL Develop	+0.25	+0.16	+0.47

Acronym words are often easier to disambiguate in their longer phrase form, so if we find both in the same document we change the prediction given for the acronym when it does not coincide with the encountered longer form.

These simple post processing heuristics do not involve any learning or adaptation, but have been simply evaluated on the development dataset and found to be useful for both English and Hungarian – although their improvement on the Hungarian NER system was only marginal. Similar and other simple post-processing steps were performed in several NER systems (for example in [3], which came second in the CoNLL-2003 conference).

4   Results and Discussion

In this section we give a summary of our results and discuss the similarities and differences between Hungarian and English NER.

Tables 2 and 3 give a summary of the accuracies of the system elements for English and Hungarian texts. The effect of each element (which was built on the previous one) can be followed from top to bottom. In the first row one can see the performance of the baseline algorithm which selects complete unambiguous named entities appearing in the training data. The subsequent rows contain the results of the original feature set, the worst and the best models built on the five previously chosen feature sets, while the fourth row gives the performance of their combination. Here the feature set splitting procedure brought a significant (15-30%) error reduction. Finally the effect of the post processing steps can be seen in the last row.

Table 4 summarizes the F measure classification for each NE class. For English, location and person classes achieve the best accuracy, while organization is somewhat worse, and the miscellaneous class is much harder to identify. Our results for Hungarian indicate that organization can achieve an F measure comparable to location

Table 2. F measures of the recognition process for English

	Develop	**Test**	**Error reduction (best)**
Baseline		59.61	
Full FS	87.17	84.81	
Five models	85.9-89.8	81.3-84.6	
Voting	91.40	86.90	
Postproc	92.28	89.02	2.32%[6] (6.08%)
Hybrid	94.72	91.41	11.44%

and person names (in the Hungarian data we had many more examples of organization names than those in the English corpus).

4.1 Results for English Texts

Our system got an F measure of 92.28% on the pre-defined development phase test set and 89.02% on the evaluation set (after a retraining which incorporated the development set into the training data) with the CoNLL evaluation script. This corresponds to a 2.32% error reduction relative to the best model known that was tested on the same data [8]. We should point out here that the system in [8] made use of the output of two externally trained NE taggers and thus the best standalone model in the system was [3]. When compared to it, it showed an error reduction of 6.08%.

Interestingly, we could improve both text sets at the same level (5 and 6.5 percentile), but while the feature set splitting procedure plays a key role in this improvement on the development set, post-processing helped the evaluation set more. This is because of the different characteristics of the sets.

Our algorithm was combined with the best two systems ([8], [3]) that were submitted to the CoNLL 2003 shared task[7], and performed significantly better than the best hybrid NER system reported in the shared task paper which employed the 5 best participating models (having a 91.41% F measure compared to 90.3%). This means a significant (11.44%) reduction in misclassified NEs. The successful applicability of our model in such a voting system is presumably due to ours having an inherently different theoretical background, which is usually beneficial to combination schemes. Our system uses Boosting and C4.5 decision tree learning, while the other two systems incorporate Robust Linear Classifier, Transformation-Based Learning, Maximum Entropy Classifier and Hidden Markov Model.

---

[6] The 4th row of Table 2 refers to the best individual system made by us and thus the error reduction was calculated against the best individual models, while the 5th row refers to the the hybrid model using our and two other CoNLL-2003 systems. The error reduction was calculated against the best hybrid system reported in the CoNLL-2003 shared task paper.

[7] Their output on the test set can be downloaded from the CoNLL homepage.

4.2 Results for Hungarian Texts

For Hungarian, the kind of results produced by inherently different but accurate systems are presently unavailable (thus the last voting step of Figure 1 is omitted in this case). However, our system gives fair results without the aid of the voting phase with outer systems. This is perhaps due to the domain specific nature of the input and makes NER a bit easier. The combined model which incorporated the predictions of the five AdaBoost+C4.5 models into a joint decision achieved an F measure of 94.77%.

Table 3. F measures of the recognition process for Hungarian

	Develop	Test	Error reduction (best)
Baseline		70.99	
Full FS	95.21	92.77	3.08%
Five models	90.3 – 94.7	88.1 – 93.7	0%- 15.55%
Voting	95.91	94.69	28.82%
Postproc	96.20	94.77	29.89%

These results are quite satisfactory if we take into account the fact that the results for English are by far the best known, and NLP tasks in Hungarian are many times more difficult to handle because Hungarian has many special (and from a statistical learning point of view, undesirable) characteristics.

Table 4. The per class F measures on the evaluation sets

	CoNLL individual	CoNLL hybrid	Hungarian
LOC	92.90	93.43	95.07
MISC	79.67	82.29	85.96
ORG	84.53	88.32	95.84
PER	93.55	96.27	94.67
overall	89.02	91.41	94.77

4.3 Discussion

Overall, then, we achieved some remarkably good results for NER; our systems can compete with the best known ones (and even perform slightly better on the CoNLL dataset). Being inherently different from those models that have been known to be successful in NER for English makes our system even more useful when it is combined with these competitive models in a decision committee. We should also mention here that our NER system remains portable across languages as long as language specific resources are available; and it can be applied successfully to languages with very different characteristics.

For English our standalone model using AdaBoost and C4.5 with majority voting slightly outperforms other systems described in literature, although we should say that this difference is not significant. In spite of this, in experiments our system achieved a significant increase in prediction accuracy in combination with other competitive models.

5   Conclusions and Future Work

Our first conclusion here is that the building and testing of new or less frequently applied algorithms is always worth doing, since they can have a positive effect when combined with popular models. We consider the fact that we managed to build a competitive model based on a different theoretical background as the main reason for the significant (11.44%) decrease in misclassified NE phrases compared to the best hybrid system known.

Second, having a rich feature representation of the problem (which permits a feature set split and recombine procedure) often turns out to be just as important as the choice of the learning method.

Thirdly, our results demonstrate that combining well-known *general* machine learning methods (C4.5, Boosting, Feature Selection, Majority Voting) and *problem-specific* techniques (large feature set, post processing) into a complex system works well for NE recognition. What is more, this works well for different languages without the need for modifying the model itself, hence this task can be solved efficiently and in way that is language independent.

There are of course many ways in which our NER system could be improved. Perhaps the two most obvious ones are to implement those more popular models that we make use of majority voting (which is beneficial for the Hungarian model) and also to enlarge the size and improve the quality of our training data (the English dataset may contain some annotation errors and inconsistencies). This is what we plan to do in the near future.

References

1. Daniel M. Bikel, Richard L. Schwartz, and Ralph M. Weischedel: An algorithm that learns what's in a name. Machine Learning, 34 -1-3 (1999) 211--231
2. Xavier Carreras, Lluís Márques and Lluís Padró: Named Entity Extraction using AdaBoost In: Proceedings of CoNLL-2002, Taipei, Taiwan, (2002) 167-170
3. Hai L. Chieu and Hwee T. Ng.: Named Entity Recognition with a Maximum Entropy Approach. Proceedings of CoNLL-2003 (2003) 160-163
4. Nancy Chinchor.: MUC-7 Named Entity Task Definition, in Proceedings of Seventh Message Understanding Conference (1998)
5. Silviu Cucerzan and Daniel Yarowsky: Language-independent named entity recognition combining morphological and contextual evidence. Proceedings of Joint SIGDAT Conf. on EMNLP/VLC (1999)

6. Dóra Csendes, János Csirik and Tibor Gyimóthy: The Szeged Corpus: A POS tagged and Syntactically Annotated Hungarian Natural Language Corpus. Proceedings of TSD 2004, vol. 3206 (2004) 41-49.
7. Farkas Richárd, Szarvas György, Kocsor András: Named Entity Recognition for Hungarian using various Machine Learning Algorithms accepted for publication in Acta Cybernetica (http://www.inf.u-szeged.hu/~rfarkas/ACTA2006_hun_namedentity.pdf)
8. Radu Florian, Abe Ittycheriah, Hongyan Jing and Tong Zhang: Named Entity Recognition through Classifier Combination. Proceedings of CoNLL-2003 (.2003) 168-171.
9. Kata Gábor, Enikő Héja, Ágnes Mészáros, Bálint Sass: Nyílt tokenosztályok reprezentációjának technológiája. IKTA-00037/2002, Budapest, Hungary (2002)
10. Jin-Dong Kim, Tomoko Ohta, Yoshimasa Tsuruoka, Yuka Tateisi and Nigel Collier: Introduction to the Bio-Entity Task at JNLPBA. Proceedings of the International Joint Workshop on Natural Language Processing in Biomedicine and its Applications (2004)
11. Ross Quinlan: C4.5: Programs for machine learning, Morgan Kaufmann (1993)
12. Gábor Prószéky: Syntax As Meta-Morphology. Proceedings of COLING-96, Vol.2 (1996) 1123–1126
13. Rob E. Shapire: The Strength of Weak Learnability. Machine Learnings, Vol. 5 (1990) 197-227
14. György Szarvas, Richárd Farkas, László Felföldi, András Kocsor, János Csirik: A highly accurate Named Entity corpus for Hungarian, Proceedings of International Conference on Language Resources and Evaluation (2006)
15. Erik F. Tjong Kim Sang, and Fien De Meulder: Introduction to the CoNLL-2003 Shared Task: Language-Independent Named Entity Recognition, Proceedings of CoNLL-2003 (2003)
16. Ian H. Witten and Eibe Frank: Data Mining: Practical machine learning tools and techniques, 2nd Edition, Morgan Kaufmann, San Francisco (2005)

Model-Based Estimation of Word Saliency in Text

Xin Wang and Ata Kabán

School of Computer Science, The University of Birmingham,
Birmingham, B15 2TT, UK
{X.C.Wang, A.Kaban}@cs.bham.ac.uk

Abstract. We investigate a generative latent variable model for model-based word saliency estimation for text modelling and classification. The estimation algorithm derived is able to infer the saliency of words with respect to the mixture modelling objective. We demonstrate experimental results showing that common stop-words as well as other corpus-specific common words are automatically down-weighted and this enhances our ability to capture the essential structure in the data, ignoring irrelevant details. As a classifier, our approach improves over the class prediction accuracy of the Naive Bayes classifier in all our experiments. Compared with a recent state of the art text classification method (Dirichlet Compound Multinomial model) we obtained improved results in two out of three benchmark text collections tested, and comparable results on one other data set.

1   Introduction

Information discovery from textual data has attracted numerous research efforts over the last few years. Indeed, a large part of communications now-days is computer-mediated and is being held in a textual form. Examples include email communication, digital repositories of literature of various kinds and much of the online resources. Text analysis techniques are therefore of interest for a number of diverse subjects.

Interestingly, it is apparent from the literature that relatively simple statistical approaches, ignoring much of the syntactical and grammatical complexity of the language are found to be surprisingly effective for text analysis, and are able to perform apparently difficult tasks such as topic discovery, text categorisation, information retrieval and machine translation, to name just a few. It appears as though the natural language contains so much redundancy that simplifying statistical models still capture useful information and are able to work effectively in principle.

It is common-sense however that word occurrences do not carry equal importance and the importance of words is context-dependent. Yet, many current approaches to text modelling make no attempt to take this into consideration. Although the problem of selection/weighting of salient features for supervised text categorisation problems has been studied extensively in the literature [10],

N. Lavrač, L. Todorovski, and K.P. Jantke (Eds.): DS 2006, LNAI 4265, pp. 279–290, 2006.
© Springer-Verlag Berlin Heidelberg 2006

a model based approach that could be used for structure discovery problems is still somewhat lacking. Model-based approaches exist for continuous valued data [3], however these are not directly applicable to discrete data such as text.

In this paper we overcome the deficiency of other text modeling approaches by considering word saliency as a context-dependent notion. We formulate and investigate a generative latent variable model, which includes word saliency estimation as integral part of a multinomial mixture-based model for text modelling. The obtained model can be used either for supervised classification or unsupervised class discovery (clustering). The unsupervised version of the problem is particularly challenging because there is no known target to guide the search. In this case we need to assess the relevance of the words without class labels, but with reference to our model of the structure inherent in the data.

In previous work [9], we have studied model-based feature weighting in a Bernoulli mixture framework. We found that, for documents, the absences of words are less salient than their presences. This has brought the suitability of the binary representation of text into question. Based on these and earlier results [4], a frequency based representation is arguably more appropriate and therefore in this paper we build on multinomial mixtures. We recognise a close relationship of our model with the Cluster Abstraction Model (CAM) [2], which however has not been previously used for word saliency estimation and its capabilities for down-weighting non-salient words have not been made explicit. The question of how salient is a particular word in comparison to other words has not been asked previously in a model-based manner and indeed in its basic form, CAM cannot answer this question. Making this capability of the model explicit is therefore a useful addition made in this paper, which could be used e.g. in text summarisation problems.

We demonstrate experimental results showing that a multinomial mixture-based model equipped with a feature saliency estimator is able to automatically down-weight common stop-words as well as other corpus-specific common words and this enhances our ability to capture the essential structure in the data, ignoring irrelevant details. As a classifier, our approach improves over the class prediction accuracy of the Naive Bayes classifier in all our experiments. Compared with a more recent state of the art text classifier, the Dirichlet Compound Multinomial (DCM) [2], we obtained improved results in two out of three benchmark text collections tested, and comparable results on one other data set.

2   The Model

Our method is based on the multinomial mixture model and a common, cluster-independent multinomial. In addition, a binary latent variable ϕ is introduced for each draw of a word from the dictionary, to indicate whether the cluster-specific or the cluster-independent multinomial will generate the next word. The cluster-specific multinomials generate salient words whereas the common multinomial generates common words. The process can be formulated as a generative model.

2.1   Notation

- Data
  Let $\mathcal{X} = (\boldsymbol{x}_1, \cdots, \boldsymbol{x}_N)$ denote N documents. Each document
  $\boldsymbol{x}_n = (x_1, \cdots, x_{L_n})$ contains L_n words from a T-size dictionary.
  $\mathcal{Y} = (\boldsymbol{y}_1, \cdots, \boldsymbol{y}_N)$ is the terms $\times$ documents matrix of $\mathcal{X}$ where each $\boldsymbol{y}_n$ is
  a T-dimensional histogram over the dictionary.
- Hidden variables
  Let $\mathcal{Z} = (z_1, \cdots, z_N)$ denote the class labels of the N documents. With each
  document, $\boldsymbol{\Phi}_n = (\phi_1, \cdots, \phi_{L_n})$ is the sequence of the binary indicators of
  the saliency for each word in $\boldsymbol{x}_n, n = 1, \cdots, N$.
- Parameters
  $\boldsymbol{\theta}_k$ is the k-th multinomial of the K components of the model, $\boldsymbol{\lambda}$ denotes
  the common multinomial component. The prior probability that a word is
  picked from cluster k of the K clusters is denoted by ρ_k, as opposed to $1 - \rho_k$,
  which is the probability with which the common component activates and
  generates words. $\alpha_1, \ldots, \alpha_K$ will denote the individual prior probabilities of
  these clusters. Let $\boldsymbol{\Theta} \equiv \{\boldsymbol{\alpha}, \boldsymbol{\theta}_1, \ldots, \boldsymbol{\theta}_K, \boldsymbol{\lambda}, \boldsymbol{\rho}\}$ denote the full parameter set.

2.2   The Generative Process

Assume a document (data sequence) $\boldsymbol{x} = (x_1, \cdots, x_L)$ is to be generated.

- As in a standard finite mixture, a component label $z = k$ is selected by sampling from a multinomial distribution with parameters $(\alpha_1, \ldots, \alpha_K)$; Then
  for each word $l = 1, \ldots, L$:
- Generate ϕ_l from flipping a biased coin, whose probability of getting a head
  is ρ_k.
- If $\phi_l = 1$, then use the cluster-specific multinomial distribution $\boldsymbol{\theta}_k$ to generate the word $x_l = t$, with probability θ_{tk};
- Else ($\phi_l = 0$), use the common multinomial distribution $\boldsymbol{\lambda}$ to generate the
  word $x_l = t$, with probability λ_t.

Model formulation. Following the above generative process, the joint probability of $\mathcal{X}$ and Φ, given the model parameters and under the i.i.d assumption
of document instances is:

$$P(\mathcal{X}, \Phi | \boldsymbol{\Theta}) = \prod_{n=1}^{N} \sum_{k=1}^{K} \alpha_k \prod_{l=1}^{L_n} \left[\rho_k P(x_{ln} | \boldsymbol{\theta}_k) \right]^{\phi_{ln}} \left[(1 - \rho_k) P(x_{ln} | \boldsymbol{\lambda}) \right]^{1 - \phi_{ln}}$$

$$= \prod_{n=1}^{N} \sum_{k=1}^{K} \alpha_k \prod_{l=1}^{L_n} \prod_{t=1}^{T} \left[\left[\rho_k P(x_{ln} = t | \boldsymbol{\theta}_k) \right]^{\phi_{ln}} \left[(1 - \rho_k) P(x_{ln} = t | \boldsymbol{\lambda}) \right]^{1 - \phi_{ln}} \right]^{\delta(x_l, t)}$$

$$= \prod_{n=1}^{N} \sum_{k=1}^{K} \alpha_k \prod_{l=1}^{L_n} \prod_{t=1}^{T} \left[\left[\rho_k \theta_{tk} \right]^{\phi_{ln}} \left[(1 - \rho_k) \lambda_t \right]^{1 - \phi_{ln}} \right]^{y_{tn}}$$

Therefore the marginal probability (likelihood function) by summing out the hidden variable ϕ_l is the following.

$$P(\mathcal{Y}|\boldsymbol{\Theta}) = \sum_{\Phi} P(\mathcal{X}, \boldsymbol{\Phi}|\boldsymbol{\Theta})$$

$$= \prod_{n=1}^{N} \sum_{k=1}^{K} \alpha_k \prod_{l=1}^{L_n} \prod_{t=1}^{T} \sum_{\phi_l=0}^{1} \left[\left[\rho_k \theta_{tk} \right]^{\phi_{ln}} \left[(1-\rho_k)\lambda_t \right]^{1-\phi_{ln}} \right]^{y_{tn}}$$

$$= \prod_{n=1}^{N} \sum_{k=1}^{K} \alpha_k \prod_{t=1}^{T} \left[\rho_k \theta_{tk} + (1-\rho_k)\lambda_t \right]^{y_t} \tag{1}$$

Model estimation. *E-step*: the class posteriors, that is the expected value of the latent variables (z_n) associated with each observation given the current parameter estimates are calculated

$$\gamma_{kn} \equiv P(z_{kn}=1|\boldsymbol{y}_n) \propto \alpha_k \prod_{t=1}^{T} \left[\rho_k \theta_{tk} + (1-\rho_k)\lambda_t \right]^{y_{tn}} \tag{2}$$

We note that the parameter estimation can be conveniently carried out without computing the posterior probabilities of the saliency variables, and so those will be computed after the parameter estimation is complete.

M-step: the parameters are re-estimated as follows,

$$\hat{\alpha}_k = \frac{\sum_n \gamma_{kn}}{\sum_{nk} \gamma_{kn}} = \frac{\sum_n \gamma_{kn}}{N} \propto \sum_n \gamma_{kn}$$

$$\hat{\theta}_{tk} \propto \frac{\rho_k \theta_{tk}}{\rho_k \theta_{tk} + (1-\rho_k)\lambda_t} \sum_{n=1}^{N} \gamma_{kn} y_{tn}$$

$$\hat{\lambda}_t \propto \sum_{k=1}^{K} \frac{(1-\rho_k)\lambda_t}{\rho_k \theta_{tk} + (1-\rho_k)\lambda_t} \sum_{n=1}^{N} \gamma_{kn} y_{tn}$$

$$\hat{\rho}_k = \frac{1}{\sum_{t=1}^{T} \sum_{n=1}^{N} \gamma_{kn} y_{tn}} \sum_{t=1}^{T} \frac{\rho_k \theta_{tk}}{\rho_k \theta_{tk} + (1-\rho_k)\lambda_t} \sum_{n=1}^{N} \gamma_{kn} y_{tn}$$

Scaling. It is to be noted that this algorithm can be efficiently implemented using sparse matrix manipulation routines. In particular, the data is typically very sparse – many entries y_{tn} are zero, since in most text documents the majority of the words of the overall dictionary are not used. Therefore the log of the E-step expression can be rewritten as a multiplication between a sparse and a dense matrix. Also each one of the M-step equations, where the data appears, can be re-arranged as matrix multiplications where one of the matrices (the data matrix) is sparse. Therefore the scaling per iteration is linear in the number of non-zero entries in the terms × documents matrix.

Inferring the probability that a word is salient. In this section we show that word saliency estimates can be computed from the presented model. This interpretation adds a new and useful functionality to the model and also helps us to better understand the working of the model. Despite the calculations are straightforward, this issue has not been addressed or noticed in any of the related previous works.

After the parameter estimation is completed, the expected saliency of a word t can be inferred as the following. The probability that an arbitrary occurrence of the term t is salient will be denoted $P(\phi = 1|t)$ and this probability is evaluated as

$$
P(\phi = 1|t) = \frac{\sum_{n=1}^{N} \sum_{l=1}^{n_l} P(\phi_l = 1|x_{ln} = t)}{\sum_{n=1}^{N} y_{tn}} = \frac{\sum_{n=1}^{N} y_{tn} P(\phi_l = 1|x_{ln} = t)}{\sum_{n=1}^{N} y_{tn}}
$$

$$
= \frac{1}{\sum_{n=1}^{N} y_{tn}} \sum_{n=1}^{N} y_{tn} \frac{\sum_{k=1}^{K} \gamma_{kn} \rho_k \theta_{tk}}{\sum_{k=1}^{K} \gamma_{kn} [\rho_k \theta_{tk} + (1 - \rho_k) \lambda_t]}
$$

$$
= \frac{1}{\sum_{n=1}^{N} y_{tn}} \sum_{n=1}^{N} y_{tn} \sum_{k=1}^{K} \gamma_{kn} \frac{\rho_k \theta_{tk}}{\rho_k \theta_{tk} + (1 - \rho_k) \lambda_t}
$$

where, according to Bayes rule,

$$
P(\phi_l = 1|x_{ln} = t) = \frac{\sum_{k=1}^{K} \gamma_{kn} \rho_k \theta_{tk}}{\sum_{k=1}^{K} \gamma_{kn} \rho_k \theta_{tk} + \sum_{k=1}^{K} (1 - \rho_k) \gamma_{kn} \lambda_t}
$$

The latter computes the probability that a specific occurrence of a word t is salient. Note that all words may have both salient and non-salient occurrences.

Note also that such saliency based ranking of the dictionary-words would not be possible to obtain from the model parameters alone. From the parameters θ_k and λ we can obtain a list of most probable words for each cluster and the list of the most probable common words respectively, similarly to CAM [2]. However these do not answer the question of how salient a word is and how it compares to other words in terms of saliency, simply because θ_{tk} is not comparable to λ_t. This is a very important issue in problems where one needs to know which handful of words are the most responsible for the inherent topical structure of a document collection, e.g. text summarisation problems. Therefore we regard this as a notable contribution of our approach, and it allows us to have a principled model-based estimation of word saliency as an integral part of a model-based cluster analysis.

In the sequel, we provide experimental evidence of the working of our approach, and its advantages over a mixture of multinomials that does not have an in-built word saliency estimator.

3   Experiments

3.1   Synthetic Data

In a first experiment, 300 data points have been generated over a 100-symbol dictionary, of which 60 are uninformative. The data is shown on the leftmost plot of Fig.1. To avoid local optima, the model estimation was repeated 40 times and the estimated model with the best in-sample log likelihood was selected. The average length of each sequence is 75 (Poisson parameter). There are four clusters mixed with one common component. The parameter ρ_k is set to 0.53, that is, when generating each word, the chance of using the cluster component is 0.53. The middle plot of Fig.1 shows the estimated feature saliences — indeed the 40

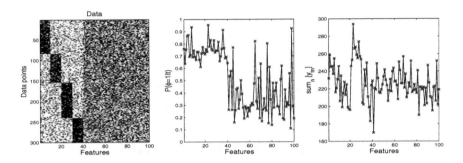

Fig. 1. Left: The data, with 4 clusters defined by 40 salient symbols (features) and having another 60 common features. Darker means higher frequency of occurrence, white stands for zero. Middle: Estimated saliency for each symbol. Right: Frequency of occurrences for each feature. We see, the frequency would have been misleading in this case, whereas the model-based approach identifies the salient features much more accurately. It is also obvious, that other possible feature weighting schemes that are unrelated to the essential structure of the data would also be fooled in some cases. E.g. if the common features are distributed more sparsely across the data set, then the tf-idf weighting would be misleading as well.

informative features, which define the four true clusters, are accurately identified. The rightmost plot depicts the frequency of occurrence for each symbol, to show that frequency counts would have been misleading for determining which are the important features in this case. Clearly, other possible feature weighting schemes that are unrelated to the essential structure of the data would also be fooled in some cases. E.g. if the common features are distributed more sparsely across the data set, then the tf-idf weighting would be misleading as well. Thus the principle behind the advantage of a model-based approach is now evident.

3.2   Real Data

Finding common terms in 10 Newsgroups data. We apply the model to text document data from 10 newsgroups of the 20 Newsgroups corpus[1]: alt.atheism, comp.graphics, comp.sys.ibm.pc.hardware, misc.forsale, rec.autos, rec.sport.baseball, sci.electronics, sci.med, sci.space and talk.politics.mideast. The data was processed using the Rainbow toolbox [2], without stop-word removal

---

[1] Available from http://www.cs.cmu.edu/~textlearning
[2] http://www.cs.cmu.edu/~mccallum/bow/rainbow/

but word stemming only. Rare words, with less than 5 occurrences were also removed in the preprocessing phase. The resulting data matrix is $22,945 \times 10,000$. Each class contains 1000 documents. We observed that the algorithm is sensitive to initialisation and to alleviate this problem of local optima, we initialise the common component to the sample mean of the data.

Fragments from the lower end of the word saliency ranking obtained are shown in Table 1 together with their actual estimated saliency probabilities. Clearly, most of them are common words, as checked against a standard stop-list [1]. However, notably, there are some corpus-specific common terms identified too, which don't fall into the scope of general common stop words, such as 'subject', 'question', 'article', 'write', 'information', 'people', 'world', etc. Therefore using a pre-defined stop-word list would not be able to eliminate these.

Table 1. Fragments from the lower end of the estimated word saliency ranking list in 10 Newsgroups, together with each words estimated saliency estimates. Some of the down-weighted (low saliency) words are indeed in the stop-words list, others (in the leftmost column) are common words that are specific to this corpus and would therefore not be detected, removed or down-weighted without a model-based approach.

0.0068	the	0.0741	you			0.2655	article
0.0147	to	0.0761	as			0.2660	ask
0.0313	in	0.0778	that			0.2691	read
0.0489	it	0.0790	have			0.2694	once
0.0500	this	0.0845	do	...		0.2699	post
0.0563	be	0.0853	all			0.2707	why
0.0579	and	0.0853	so			0.2710	news
0.0623	some	0.0855	of			0.2771	writes
0.0644	are	0.0893	about			0.2797	wrote
0.0667	is	0.0894	an			0.2999	information

To further illustrate our method at work, we pick an article at random from the talk.politics.mideast group, which was not used for model training. The following paragraph shows the text with the words made with different fonts, according to their saliences as follows. The underlined words are those which are estimated to be the most salient, with saliency ≥ 0.8 and ≤ 1.0. (Note that, the words "prerequisites", "co-operation", "reaffirm" were removed in preprocessing due to rareness, therefore they don't have any saliency specified.) The words with saliency between 0.41 and 0.8 are in normal font, and the least salient words, with saliency ≤ 0.4 are in grey.

"(The participating States) recognize that pluralistic democracy and the rule of law are essential for ensuring respect for all human rights and fundamental freedoms. . . They therefore welcome the commitment expressed by all participating States to the ideals of democracy and political pluralism. . . The participating States express their conviction that full respect for human rights and fundamental freedoms and the development of societies based on pluralistic democracy. . . are prerequisites for progress in setting up the lasting order of peace, security, justice, and co-operation. . . They therefore reaffirm their commitment to implement fully all provisions of the Final Act and of the other CSCE documents relating to the human dimension. . . In order to strengthen respect for, and enjoyment of, human rights and fundamental freedoms, to develop human contacts and to resolve issues of a related humanitarian character, the participating States agree on the following".

As a final illustrative experiment, we look at the induced geometry of the word features. To this end, we train the model on 2 newsgroups: `sci.space` and `talk.politics.mideast`. For showing corpus-specific common terms, the words on the stop-words list are removed in this experiment. Then each word t is visualised in 3D by its coordinates $\lambda_t, \theta_{t,1}$ and $\theta_{t,2}$. So the number of points on the plot will equal the dictionary size (8,824 words in this case). This plot is seen on Fig. 2 as follows. Each point has a colour between red and green. The proportion of the red and the green component, for each word, is given by its word saliency estimate $P(\phi = 1|t)$. Pure red would stand for a saliency value of 1. Pure green would stand for a saliency value of 0. Intermediate colors signify the probability of saliency. For some of the points further away from the centre, we also show the actual word content. As we can see, the salient words for the two classes are distributed along two of the axes and are well separated from each other: 'space', 'nasa', 'earth', 'launch' clearly are salient, coming from the newsgroup on `sci.space`. Similarly, 'israel', 'armenian', 'jews', 'turkish' are salient, coming from the group `talk.politics.mideast`. In turn, the common, unsalient words lie along the third axis: 'people', 'write', 'article', 'time' are all common words that are specific to newsgroup messages. In other corpora and other contexts they may well be salient — therefore they are not on a general-purpose stop-words list and would be difficult to appropriately deal with in a non-model-based approach.

Text classification. Although our algorithm has been derived from an unsupervised model formulation, it can also be used in supervised classification if class labels are available. To be comparable with other algorithms, we carry out experiments on three standard corpora: the 'industry sector', '20 Newsgroups' and 'Reuters-21578' document collections[3]. The same experimental settings and criteria are used as in [2]. Documents are preprocessed and count vectors are extracted using the Rainbow toolbox. The 500 most common words are removed from the vocabulary in this experiment, in order to enable a direct comparison with recent results from the literature. The characteristics of the three document collections employed are given in Table 2.

Table 3 shows the classification results, that we obtained together with a result taken from [2] in the last row of the table, for comparison. For the latter, the readers can refer to [2] for more details. It should be pointed out, that for single-labelled datasets, such as the 'Industry sector' and '20 newsgroups', the

---

[3] Downloaded from http://www2.imm.dtu.dk/~rem/index.php?page=data

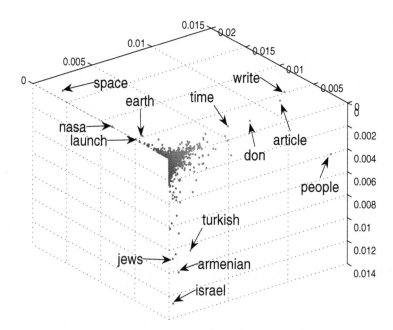

Fig. 2. The 3D view of the words in the vocabulary, as obtained from the two newsgroups `sci.space` and `talk.politics.mideast`. The colour combination visualises the estimated word saliences: redder stands for more salient, green stands for less salient. We see the salient words for the two classes are distributed along two of the main axes and are well separated from each other. The common, unsalient words lie along the third axis: 'people', 'write', 'article', 'time' are all common words specific to newsgroup messages.

precision is used as a measure of accuracy of the classification. For multi-labelled data in turn, such as the 'Reuters-21578', precision and recall are combined by computing the "break-even" point, which is defined using either micro or macro averaging [7]:

$$\mathrm{BE}_{micro} = \frac{\sum_k \mathrm{TP}_k}{\sum_k (\mathrm{TP}_k + \mathrm{FP}_k)}; \qquad \mathrm{BE}_{macro} = \frac{1}{K} \sum_k \frac{\mathrm{TP}_k}{\mathrm{TP}_k + \mathrm{FP}_k}$$

K is the number of document classes. TP_k and FP_k are the number of true positives and false positives for class k respectively.

The classification results of our model are significantly superior to those obtained with a multinomial mixture, in all cases, as tested using the nonparametric Wilcoxon rank sum test at the 5% level.

Finally, we compare our results with a recently proposed text classifier, the Dirichlet Compound Multinomial (DCM) [2]. The DCM was proposed as an alternative to the multinomial for modeling of text documents, with one additional degree of freedom, which allows it to capture the burstiness phenomenon

Table 2. Characteristics of the '20 newsgroups', 'Industry' and 'Reuters' data collections

Dataset	Vocabulary Size	Total nr of docs	Nr of Classes	Avg length of docs	Multiple labels	Train/Test splitting
Industry Sector	55,055	9,555	104	606	N	50/50
20 Newsgroups	61,298	18,828	20	116	N	80/20
Reuters-21578	15,996	21,578	90	70	Y	7,770/3,019

Table 3. Classification results for the '20 newsgroups', 'Industry sector' and 'Reuters-21578' data sets. Standard deviations are not given in the fourth column since there is a single standard training set/test set split for the Reuters-21578 corpus.

METHOD	20 Newsgroups PRECISION $\pm\sigma$	Industry Sector PRECISION $\pm\sigma$	Reuters-21578 MACRO BE	MICRO BE
Mixture model	0.866±0.005	0.8085± 0.006	0.2487	0.6428
Our model	0.888±0.003	**0.877±0.003**	**0.4513**	**0.7644**
DCM	**0.890±0.005**	0.806±0.006	0.359	0.740

of words. It was found to be superior to the multinomial on text modeling, and promising classification improvements have been reported [2].

Interestingly, when compared to DCM, we find our proposed model performs comparably on the full 20 Newsgroups corpus, and significantly better on the other two datasets. The latter two data sets are more sparse and we conjecture this may be a reason for the success of our method.

These results are very promising because they are obtained from different considerations than those of [2]. Therefore the possibility of combining our feature saliency modelling approach with their Dirichlet-compound-multinomial building blocks is a potentially fruitful line of further research, which may bring further improvements. It is also worth mentioning that Madsen *et. al* report even better results obtained via a heuristically modified version of multinomial, such as a log transformation of the data, complement modeling approaches. Such approaches modify the input data and distribution parameters, and therefore don't give probability distributions properly normalised [2]. All such heuristics could be included in our framework too.

3.3 Discussion

We are currently trying to analyse the reasons behind the success of our model against multinomial mixture and the implications of having included a word saliency estimator as an integral part of the mixture. Intuitively, it may appear that assuming two bags of words — a bag of topic-bearing, class-specific words and a bag of common words — for generating each document offers more flexibility than assuming one bag only. However, this is not the only reason, since a mixture of multinomials is just another multinomial. It is important, that one of these bags is constrained to be the same for all the topical clusters. The cluster-specific multinomial is then 'relieved' from having to represent both the common words and the content-bearing words, whose distributions, as observed in [2], are fundamentally different.

It is interesting to view our approach from the algorithmic point of view and follow up the effects of the parameters ρ_k and λ in terms of a 'shrinkage' of the multinomial mean parameters towards a common distribution. The probabilities ρ_k control the extent of this shrinkage — if ρ_k is small, it is more probable to have non-salient words in average, therefore the multinomial mean, $\rho_k \theta_{tk} + (1 - \rho_k)\lambda_t$ parameters are 'shrunken' closer to the common component. Shrinkage methods have widely been used for data denoising e.g. in the wavelet literature, for continuous valued data. Little work has been devoted to shrinkage estimators for discrete data, and text in particular [6] and even less to analysing shrinkage effects where these are somewhat implicit.

One may also wonder why we bother devising new classification methods that bring some improvements over Naive Bayes, when Support Vector Machines (SVM) [8] are so effective and successful for text classification [5]. The main reason for this is that SVMs are 'black-box' type methods in the sense that they do not provide explanatory information regarding the text being classified. One often wants to quickly understand or summarise a text corpus, to find out what it is about and which is the most important topical information that it contains. Tools for automating this are very useful. Ideally, we should aim for methods that exhibit both excellent class prediction accuracy as well as intuitive explanatory ability for human interpretation. However achieving this is not trivial and there is a lot more to be desired. In this work we hope to have made a small step in this direction.

4   Conclusions

We proposed a generative latent variable model for feature saliency estimation based on multinomial mixture. This provides a computationally efficient algorithm that can be used in both unsupervised and supervised classification problems. The model is able to infer the saliency of words in a model-based principled way. Experimental results have shown that, common stop-words as well as other corpus-specific common words are automatically down-weighted. As a classifier, our approach improves over the class prediction accuracy of the Naive Bayes classifier in all our experiments. Compared with DCM, we obtained improved results in two out of three benchmark text collections tested, and comparable results on one other data set.

References

1. W. Nelson Francis and Henry Kucera. Frequency analysis of English usage, 1982.
2. Rasmus E. Madsen, David Kauchak, and Charles Elkan. Modeling word burstiness using the Dirichlet distribution. In *ICML '05: Proceedings of the 22nd international conference on Machine learning*, pages 545–552, New York, NY, USA, 2005. ACM Press.

3. Mario A. T. Figueiredo Martin H. C. Law and Fellow-Anil K. Jain. Simultaneous feature selection and clustering using mixture models. *IEEE Trans. Pattern Anal. Mach. Intell.*, 26(9):1154–1166, 2004.

4. A. McCallum and K. Nigam. A comparison of event models for Naive Bayes text classification. In *In AAAI-98 Workshop on Learning for Text Categorization*, 1998.

5. Thorsten Joachims. Text Categorization with Support Vector Machines: Learning with Many Relevant Features. Proceedings of the European Conference on Machine Learning, Springer, 1998.

6. Andrew McCallum, Ronald Rosenfeld, Tom M. Mitchell, and Andrew Y. Ng. Improving text classification by shrinkage in a hierarchy of classes. In *ICML '98: Proceedings of the Fifteenth International Conference on Machine Learning*, pages 359–367, San Francisco, CA, USA, 1998. Morgan Kaufmann Publishers Inc.

7. Fabrizio Sebastiani. Machine learning in automated text categorization. *ACM Comput. Surv.*, 34(1):1–47, 2002.

8. Vladimir N. Vapnik. The Nature of Statistical Learning Theory. Springer, 1995.

9. Xin Wang and Ata Kabán. Finding uninformative features in binary data. In *Proceedings of IDEAL05*, pages 40–47, 2005.

10. Yiming Yang and Jan O. Pedersen. A comparative study on feature selection in text categorization. In Douglas H. Fisher, editor, *Proceedings of ICML-97, 14th International Conference on Machine Learning*, pages 412–420, Nashville, US, 1997. Morgan Kaufmann Publishers, San Francisco, US.

Learning Bayesian Network Equivalence Classes from Incomplete Data

Hanen Borchani, Nahla Ben Amor, and Khaled Mellouli

LARODEC, Institut Supérieur de Gestion de Tunis, 41 Avenue de la liberté,
2000 Le Bardo, Tunisie
hanene.borchani@gmail.com, nahla.benamor@gmx.fr,
khaled.mellouli@ihec.rnu.tn

Abstract. This paper proposes a new method, named Greedy Equivalence Search-Expectation Maximization (GES-EM), for learning Bayesian networks from incomplete data. Our method extends the recently proposed GES algorithm to deal with incomplete data. Evaluation of generated networks was done using expected Bayesian Information Criterion (BIC) scoring function. Experimental results show that GES-EM algorithm yields more accurate structures than the standard Alternating Model Selection-Expectation Maximization (AMS-EM) algorithm.

1 Introduction

Bayesian networks (BNs) are a popular tool for representing uncertainty in artificial intelligence [8,12]. They have been implemented in many expert systems and real applications in areas such as medical diagnosis and fraud detection.

This paper proposes a new approach for learning BNs from incomplete data, that is, in the presence of missing values or hidden variables. It takes advantage of the equivalence class search space via extending the Greedy Equivalence Search (GES) algorithm [3] to deal with incomplete data. In order to estimate missing values, we adopt the commonly used Expectation Maximization (EM) algorithm [4]. Thus, this novel extension of GES, which we call GES-EM algorithm, deals with model selection as well as parameter estimation from incomplete data.

This paper is organized as follows: Section 2 recalls basics of BNs. Section 3 presents our new approach for learning BN equivalence classes from incomplete data. Finally, Section 4 describes experimental study comparing the proposed GES-EM algorithm and the standard AMS-EM algorithm proposed by [7].

2 Bayesian Networks

The following syntactical conventions are used: $\mathbf{U} = \{X_1, ..., X_n\}$ denotes a finite set of n discrete random variables. A variable is denoted by an upper case letter and its state by a lower-case letter. A set of variables is denoted by a bold-face capitalized letter and the corresponding boldface lower-case letter denotes the state for each variable in the set. Calligraphic letters denotes statistical models.

N. Lavrač, L. Todorovski, and K.P. Jantke (Eds.): DS 2006, LNAI 4265, pp. 291–295, 2006.
© Springer-Verlag Berlin Heidelberg 2006

A Bayesian network $\mathcal{B}$ is a pair $(\mathcal{G}, \Theta)$. The structure $\mathcal{G}$ is a directed acyclic graph (DAG). The vertices represent the variables and the directed edges correspond to dependence relationships among those variables. The set of parameters $\Theta = \{\Theta_1, \Theta_2, ..., \Theta_n\}$, such that $\Theta_i = P(X_i \mid \mathbf{Pa}(X_i))$ denotes the conditional probability distribution of each node X_i given its parents, denoted by $\mathbf{Pa}(X_i)$. A BN encodes a unique joint probability distribution over $\mathbf{U}$ expressed by:

$$P(X_1, ..., X_n) = \prod_{i=1}^{n} P(X_i \mid \mathbf{Pa}(X_i)) \ . \tag{1}$$

Definition 1. *Two DAGs $\mathcal{G}$ and $\mathcal{G}'$ are equivalent if for every Bayesian network $\mathcal{B} = (\mathcal{G}, \Theta)$, there exists a Bayesian network $\mathcal{B}' = (\mathcal{G}', \Theta')$ such that $\mathcal{B}$ and $\mathcal{B}'$ define the same probability distribution, and vice versa.*

Theorem 1. *Two DAGs are equivalent if and only if they have the same skeletons and the same v-structures [16].*

Each set of equivalent $DAGs$ defines an equivalence class of BNs, denoted by $\mathcal{E}$, which can be represented via a unique completed partially directed acyclic graph $(CPDAG)$ [3]. Also, every DAG $\mathcal{G}$ in an equivalence class $\mathcal{E}$ is a *consistent extension* of the $CPDAG$ representation for that class [3].

3   GES-EM Algorithm

The few proposed approaches dealing with BN learning from incomplete data make use of search algorithms operating either in the BN search space [7,14] or the tree search space [6,10]. Although, searching over the tree search space fades away exponential computation demands, it can unfortunately return bad quality networks. In the other hand, the BN search space is, first, super exponential, and second allows redundant representations for the same state, which may affect considerably the accuracy of the learned networks [1].

Thus, to learn more accurate networks from incomplete data, we propose a new approach that makes use of the equivalence class search space, having better advantages than the traditional used ones [1,3]. The main idea is to alternate between EM iterations and structure search ones using the GES algorithm [3].

3.1   Evaluation Function

To evaluate learned structures, we have used the expected Bayesian Information Criterion (BIC) scoring function defined by [7]. Let $\mathbf{O}$ denotes the set of observable variables and $\mathbf{M} = \mathbf{U} - \mathbf{O}$ denotes the set of variables partially or completely unobserved. Based on missing at random (MAR) assumption [13], the expected BIC score of any neighbor $\mathcal{G}^i$ is deduced by taking expectation over all possible values of $\mathbf{m}$ given the observed variables $\mathbf{O}$ and the current state $(\mathcal{G}, \Theta)$ [7]:

$$Q(\mathcal{G}^i, \Theta^i : \mathcal{G}, \Theta) = E_{P(\mathbf{m}\mid\mathbf{O}:\mathcal{G},\Theta)}[logP(\mathbf{O}, \mathbf{m} \mid \mathcal{G}^i, \Theta^i)] - \frac{logN}{2}Dim(\mathcal{G}^i) \ . \tag{2}$$

Proposition 1. *Let $\mathcal{E}$ be an equivalence class, then $\forall\ \mathcal{G}^i, \mathcal{G}^j \in \mathcal{E}$ s.t. $i \neq j$,*

$$Q(\mathcal{G}^i, \boldsymbol{\Theta^i} : \mathcal{G}, \boldsymbol{\Theta}) = Q(\mathcal{G}^j, \boldsymbol{\Theta^j} : \mathcal{G}, \boldsymbol{\Theta})\ . \tag{3}$$

From Proposition 1, it follows that all *DAGs* contained within the same equivalence class $\mathcal{E}$ have the same expected BIC scores. Hence, the equivalence class $\mathcal{E}$ can be evaluated using any *DAG* member.

3.2   GES-EM Algorithm

The GES-EM algorithm starts with an initial equivalence class $\mathcal{E}^0$, represented via the empty structure $\mathcal{G}^0$, and the randomly initialized parameter set $\boldsymbol{\Theta}^0$, then, it executes a single edge insertion phase followed by a single edge deletion one.

The first phase, starts by running EM algorithm [4] to convergence in order to find improved parameter values $\boldsymbol{\Theta}'^0$ for the current structure. Next, it applies all valid *Insert* operations to $\mathcal{E}^0$ in order to obtain the set of its neighbors.

Each resulting neighbor is not necessarily completed. Therefore, we should convert each *PDAG* neighbor to its potential corresponding consistent extension $\mathcal{G}$ by applying the conversion procedure *PDAG-to-DAG* proposed by [5]. Firstly, this conversion allows us to check the validity of the applied *Insert* operations. In addition, using Proposition 1, the expected BIC score can only be computed for consistent extension relative to each equivalence class in order to evaluate it.

If the higher obtained expected score is lower than the one found in the previous iteration, then a local maximum is reached and the second phase starts from the previous best state. Otherwise, the best equivalence class is generated, via converting the best selected consistent extension $\mathcal{G}$ to *CPDAG*. This is performed via the conversion algorithm *DAG-to-CPDAG* as described in [3]. The resulting class becomes the current state and the search pursue until no consistent extension is generated or a local maximum is reached.

The second phase of GES-EM algorithm is quite analogous to the first one, except that it replaces *Insert* operations by *Delete* ones, in order to determine the neighbors of the current equivalence class.

To show that our algorithm is useful, we need to prove that it, indeed, improves the real score in each iteration. This result can be obtained using the following theorem given by Friedman in [7]:

Theorem 2. *If $Q(\mathcal{G}^i, \boldsymbol{\Theta^i} : \mathcal{G}, \boldsymbol{\Theta}) > Q(\mathcal{G}, \boldsymbol{\Theta} : \mathcal{G}, \boldsymbol{\Theta})$, then $Score(\mathcal{G}^i, \boldsymbol{\Theta^i}) > Score(\mathcal{G}, \boldsymbol{\Theta})$.*

Indeed, the GES-EM algorithm selects at each iteration, during the first and the second phase, a state having a higher expected score than the previous one. This means that, at each iteration, it improves the real score until it converges.

4   Experimental Results

This section reports experimental results in order to evaluate the proposed GES-EM algorithm by comparing it to the standard AMS-EM algorithm [7].

For the experiments, we considered two well-known BNs: Asia [9] with 8 nodes and 8 arcs, and Insurance [2] with 27 nodes and 52 arcs. We randomly sampled data sets of different sizes and then randomly removed values from each one to get incomplete data sets with various missing value rates (%mv). For each data set, we run the GES-EM and AMS-EM five times starting from an empty network and setting the maximum number of iterations to 30, for experiments carried out over Asia, and to 80, for those carried out over Insurance.

The results, summarized in Table 1, convey two important evaluation criteria: the accuracy or quality of the learned Asia and Insurance networks, measured using the mean expected BIC scores of the maxima reached by each algorithm, and the performance of the two algorithms provided in terms of iteration number that each algorithm performs before converging to a local maximum.

Table 1. Mean expected BIC scores and iteration number given in brackets

Asia network					
%mv	Algo.	1000	5000	10000	20000
10	GES-EM	-2437.93(9)	-11475.45(11)	-23572.63(11)	-45176.62(11)
	AMS-EM	-2435.77(13)	-11503.66(11)	-23594.55(12)	-45185.64(13)
20	GES-EM	-2477.66(11)	-12007.68(11)	-23875.51(11)	-45928.97(18)
	AMS-EM	-2489.37(30)	-12015.98(15)	-23936.06(13)	-45900.38(17)
30	GES-EM	-2531.91(13)	-12300.97(13)	-23783.22(20)	-47624.75(21)
	AMS-EM	-2530.40(11)	-12301.90(30)	-25497.92(30)	-47672.79(30)
Insurance network					
%mv	Algo.	250	500	1000	5000
10	GES-EM	-4770.38(25)	-8699.02(33)	-16943.55(53)	-82177.03(49)
	AMS-EM	-4865.05(80)	-9510.93(80)	-17734.69(80)	-82344.53(80)
20	GES-EM	-4876.67(41)	-9583.65(72)	-18516.03(66)	-88565.87(73)
	AMS-EM	-4983.33(80)	-9797.15(80)	-19297.02(80)	-88878.72(80)
30	GES-EM	-5296.69(57)	-10041.22(95)	-19234.23(114)	-91532.09(89)
	AMS-EM	-5322.12(80)	-10489.54(80)	-21075.80(80)	-91993.44(80)

The obtained mean expected BIC scores show that the GES-EM returns more often higher quality networks than AMS-EM algorithm. This is performed in the most cases with less iteration number. Moreover, we note that our algorithm, contrary to AMS-EM algorithm, always, escapes bad local maxima.

The obtained iteration number also indicates that our algorithm is faster than AMS-EM algorithm, this is confirmed by Table 2 which gives the execution time ratios, which expressed how many times longer GES-EM took than AMS-EM.

Table 2. Execution time ratio (GES-EM /AMS-EM)

	Asia				Insurance			
%mv	1000	5000	10000	20000	250	500	1000	5000
10	0.614	0.938	0.774	0.978	0.447	0.561	0.754	0.897
20	0.292	0.658	0.726	1.050	0.722	1.060	0.736	0.954
30	1.076	0.553	0.681	0.748	0.894	1.034	1.153	1.287

5 Conclusion

This paper proposes a novel approach for learning BN equivalence classes from incomplete data. The main idea of our approach is to use the equivalence class search space instead of the traditional used ones, i.e. BN search space or tree search space. We proved theoretically and experimentally the convergence of our algorithm. Experimental results showed that GES-EM algorithm yields, in less execution time, to more accurate structures than the AMS-EM algorithm [7].

The GES-EM algorithm ensures the learning of equivalence classes incorporating some hidden variables, under the condition that the number and the location of these variables are known. A further work will be to extend it to discover hidden variables.

Another interesting issue is incremental learning of BNs from incomplete data. Few works deal with this problem and existing ones have been mostly restricted to updating the parameters assuming a fixed structure [11,15]. Thus, we aim to extend the GES-EM algorithm to deal with incremental learning.

References

1. S. A. Andersson, D. Madigan, and M. D. Perlman. A characterization of Markov equivalence classes for acyclic digraphs. Annals of Statistics, 25:505-541, 1997.
2. J. Binder, D. Koller, S. Russell and K. Kanazawa. Adaptive probabilistic networks with hidden variables. Machine Learning, 29:213-244, 1997.
3. D. M. Chickering. Optimal Structure Identification With Greedy Search. Journal of Machine Learning Research, pages 507-554, 2002.
4. A. Dempster, N. Laird, and D. Rubin. Maximum likelihood from incompete data via the EM algorithm. Journal of the Royal Statistical Society, B39:1-38, 1977.
5. D. Dor and M. Tarsi. A simple algorithm to construct a consistent extension of a partially oriented graph. Tech. Rep. R-185, Cognitive Systems Laboratory, 1992.
6. O. François and P. Leray. Bayesian network structural learning and incomplete data. International and Interdisciplinary conference on AKRR, Espo, Finland, 2005.
7. N. Friedman. Learning belief networks in the presence of missing values and hidden variables. In 14th Conference of Machine Learning, 1997.
8. F. V. Jensen. An Introduction to Bayesian Networks. UCL Press, London, 1996.
9. S. L. Lauritzen and D. J. Spiegelhalter. Local computations with probabilities on graphical structures and their application to expert systems. Journal of the Royal Statistical Society, 50(2):157-224, 1988.
10. M. Meila and M. I. Jordan. Estimating dependency structure as a hidden variable. In NIPS 10, 1998.
11. R. M. Neal, G. E. Hinton. A View of the EM algorithm that justifies incremental sparse and other variants. Learning in Graphical Models, 1998.
12. J. Pearl. Probabilistic Reasoning in Intelligent Systems: Networks of Plausible inference. Morgan Kaufmann Publishers, San Francisco (California), 1988.
13. D. Rubin. Inference and missing data. Biometrika, 63:581-592, 1976.
14. M. Singh. Learning Bayesian networks from incomplete data. In AAAI97, 1997.
15. D. J. Spiegelhalter and S. L. Lauritzen. Sequential updating of conditional probabilities on directed graphical structures. Networks, 20:579-605, 1990.
16. T. Verma and J. Pearl. Equivalence and synthesis of causal models. In Proceedings of the Sixth Conference on UAI, Morgan Kaufmann Publishers, 1990.

Interesting Patterns Extraction Using Prior Knowledge

Laurent Brisson

Laboratoire I3S - Université de Nice, 06903 Sophia-Antipolis, France
`brisson@i3s.unice.fr`

Abstract. One important challenge in data mining is to extract interesting knowledge and useful information for expert users. Since data mining algorithms extracts a huge quantity of patterns it is therefore necessary to filter out those patterns using various measures. This paper presents IMAK, a part-way interestingness measure between objective and subjective measure, which evaluates patterns considering expert knowledge. Our main contribution is to improve interesting patterns extraction using relationships defined into an ontology.

1 Introduction

In most data mining projects, prior knowledge is implicit or is not organized as a structured conceptual system. We use ExCIS framework [1] which is dedicated to data mining situations where the expert knowledge is crucial for the interpretation of mined patterns. In this framework the extraction process makes use of a well-formed conceptual information system (CIS) for improving the quality of mined knowledge. A CIS is defined by Stumme [8] as a relationnal database together with conceptual hierarchies.

Numerous works focused on indexes that measure the interestingness of a mined pattern [3]. They generally distinguished objective and subjective interest. Silberschatz and Tuzhilin [6] proposed a method to define unexpectedness and actionability via belief systems while Liu [3] developed a method that use user expectations. In this paper we present an interestingness measure called IMAK, which evaluates extracted patterns according prior knowledge. The novelty of this approach lies in the use of a Conceptual Information System in order to extract rules easily comparable with knowledge. This ontology based approach for unexpected and actionable patterns extraction differs from works on interestingness measures.

The paper is organized as follows. In section 2, we study related works. Section 3 focus on interesting patterns extraction. Section 4 concludes the paper.

2 Interestingness Measures

Among all indexes that measure the interestingness of a mined pattern there are measures of objective interestingness such as confidence, coverage, lift, success

N. Lavrač, L. Todorovski, and K.P. Jantke (Eds.): DS 2006, LNAI 4265, pp. 296–300, 2006.

rate while unexpectedness and actionability are proposed for subjective criteria. In this section we presents only subjective interestingness measures.

2.1   What Makes Patterns Interesting?

Silberschatz [7] presents a classification of measures of interestingness and identifies two major reasons why a pattern is interesting from the subjective (user-oriented) point of view:

- Unexpectedness: a pattern is interesting if it is surprising to the user
- Actionnability: a pattern is interesting if the user can do something with it to his or her advantage

Therefore a pattern can be said to be interesting if it is both unexpected and actionable. This is clearly a highly subjective view of the patterns as actionability is dependent not only on the problem domain but also on the user's objectives at a given point in time [4]. According to the actionability criteria, a model is interesting if the user can start some action depending on it [7]. On the other hand, unexpected models are considered interesting since they contradict user expectations which depend on his beliefs.

2.2   Belief System [6]

Silberschatz and Tuzhilin proposed a method to define unexpectedness via belief systems. In this approach, there are two kinds of beliefs: soft beliefs that the user is willing to change if new patterns are discovered and hard beliefs which are constraints that cannot be changed with new discovered knowledge. Consequently this approach assumes that we can believe in certain statements only partially and some degree or confidence factor is assigned to each belief. A pattern is said to be interesting relatively to some belief system if it "affects" this system, and the more it "affects" it, the more interesting it is.

2.3   User Expectations [3]

User expectations is a method developed by Liu. User had to specify a set of patterns according to his previous knowledge and intuitive feelings. Patterns had to be expressed in the same way that mined patterns. Then Liu defined a fuzzy algorithm which matches these patterns. In order to find actionable patterns, the user has to specify all actions that he can take. Then, for each action he specifies the situation under which he is likely to run the action. Finally, the system matches each discovered pattern against the patterns specified by the user using a fuzzy matching technique.

3   Interesting Patterns Extraction

3.1   Knowledge Properties

We chose to express knowledge like "if ... then ..." rules in order to simplify comparison with extracted association rules. Each knowledge has some essential properties to select the most interesting association rules:

- Confidence level: 5 different values are available to describe knowledge confidence according a domain expert. These values are range of confidence value: 0-20%, 20-40%, 40-60%, 60-80% and 80-100%. We call confidence the probability the consequence of a knowledge occurs when the condition holds.
- Certainty:
  - Triviality: cannot be contradicted
  - Standard knowledge: domain knowledge usually true
  - Hypothesis: knowledge the user want to check

Since our project deals with data from the "family" branch of the French national health care system (CAF), our examples are related to CAF domain. Let's consider the following knowledge:

KNOWLEDGE 1
Objective='To be paid' ∧ Allowance='Housing Allowance' ∧ Distance='0km' → Contact='At the agency'

- Confidence level: 60-80%
- Certainty: Hypothesis

3.2   IMAK : An Interestingness Measure According Knowledge

We propose an interestingness measure $IMAK$ which considers actionnalibity (using certainty knowledge property) and unexpectedness (using generalization relationships between ontology concepts). Although unexpected patterns are interesting it's necessary to consider actionable expected patterns. In our approach we deal with actionnability using knowledge certainty property:

- If a pattern match a trivial knowledge it isn't actionable since actions concerning trivial knowledge are most likely known
- Since user knowledge define his main points of interest, a pattern matching standard knowledge could be actionable
- If a pattern match a hypothesis, it is highly actionable

IMAK only considers confidence as objective interestingness measure. Consequently it can't be applied on rules with lift ≤ 1. It makes no sense to compare with knowledge rules whose antecedent and consequent aren't positively correlated.

Our measure describes four levels of interest:

- none: uninteresting information
- low: confirmation of standard knowledge
- medium: new information about a standard knowledge / confirmation of a hypothesis
- high: new information about a hypothesis

As you can see in table 1, IMAK value increases when a pattern matches a hypothesis and decreases when it matches a triviality. Furthermore IMAK value increases when a pattern is more general than a knowledge or when its confidence

Table 1. IMAK values

Pattern is ⟍ Knowledge Certainty	Triviality	Standard knowledge	Hypothesis
Case 1. Pattern with better confidence level than knowledge			
more general	medium	high	high
similar	none	low	medium
more specific	none	medium	high
Case 2. Pattern and knowledge with similar confidence level			
more general	low	medium	high
similar	none	low	medium
more specific	none	low	medium
Case 3. Pattern with lesser confidence level than knowledge			
more general	none	none	low
similar	none	low	medium
more specific	none	none	low

level is the best. Generalization level of a rule compared to a knowledge is defined with the help of the embedded ontology in ExCIS framework.

3.3 Experimental Results

Let's consider the knowledge rule 1, and the two following extracted rules:

EXTRACTED RULE 1
Objective='To be paid' ∧ Allowance='Housing Allowance' → Contact='At the agency' [confidence=20%]

EXTRACTED RULE 2
Objective='To be paid' ∧ Allowance='Housing Allowance'
∧ Distance='LessThan30km' → Contact='At the agency' [confidence=95%]

Rule 1 is a generalization of the knowledge. But its confidence is lesser than knowledge confidence level. Consequently *IMAK* value is "low" since the knowledge is a "hypothesis" (see table 1 column 3 line 7).

Rule 2 is also a generalisation of the knowledge. Its confidence is better than than knowledge confidence level. Consequently *IMAK* value is "high" since the knowledge is a "hypothesis" (see table 1 column 3 line 1). Now let's consider the rule:

EXTRACTED RULE 3
Objective='To be paid' ∧ Allowance='Student Housing Allowance'
∧ Distance='0km' → Contact='At the agency' [confidence=75%]

Rule 3 is more specific than knowledge and its confidence is similar. Consequently *IMAK* value is "medium" since the knowledge is a "hypothesis" (ref table 1 column 3 line 6).

We apply IMAK measure on 5000 rules extracted by several runs of CLOSE algorithm [5]. CAF experts couldn't deal with such a number of rules. However after having defined their knowledge we could present them a hundred of interesting rules classified into few categories.

4   Conclusion

We presented IMAK, an interestingness measure, which evaluates extracted patterns according to prior knowledge. Some works on subjective interestingness measures [2,3,6] use templates or beliefs in order to express knowledge. Our contribution is to improve interesting patterns extraction using relationships defined into an ontology [1]. IMAK measure doesn't make syntaxic matching but uses semantic relationships between concepts, analyzes rules cover, compares confidence level and takes into account the knowledge certainty. Consequently it is part-way between objective and subjective measure. In future works we plan to compute IMAK using ontology relationships which aren't generalization/specialization relationships and to evaluate our measure on a less subjective application domain.

References

1. L. Brisson, M. Collard and N. Pasquier. *Improving the Knowledge Discovery Process Using Ontologies*. Proceedings of Mining Complex Data workshop in ICDM Conference, November 2005
2. M. Klemettinen, H. Mannila, P. Ronkainen, H. Toivonen and A. Verkamo. Finding interesting rules from large sets of discovered association rules In CIKM-94, 401 – 407, November 1994.
3. B. Liu, W. Hsu, L.-F. Mun and H.-Y. Lee. *Finding Interesting Patterns using User Expectations*. Knowledge and Data Engineering, 11(6):817-832, 1999.
4. K. Mcgarry *A Survey of Interestingness Measures for Knowledge Discovery* The knowledge engineering review, vol. 00:0, 1-24, 2005.
5. Pasquier N., Taouil R., Bastide Y., Stumme G. and Lakhal L. *Generating a Condensed Representation for Association Rules*. Journal of Intelligent Information Systems, Kerschberg L., Ras Z. and Zemankova M. editors, Kluwer Academic Publishers
6. A. Silberschatz and A. Tuzhilin. *On Subjective Measures of Interestingness in Knowledge Discovery*. Proceedings 1st KDD conference, pp. 275-281, august 1995.
7. A. Silberschatz and A. Tuzhilin. *What Makes Patterns Interesting in Knowledge Discovery Systems*. IEEE Transaction On Knowledge And Data Engineering, 8(6):970-974, december 1996.
8. G. Stumme. *Conceptual On-Line Analytical Processing*. K. Tanaka, S. Ghandeharizadeh and Y. Kambayashi editors. Information Organization and Databases, chpt. 14, Kluwer Academic Publishers, pp. 191-203, 2000.

Visual Interactive Subgroup Discovery with Numerical Properties of Interest*

Alípio M. Jorge[1], Fernando Pereira[1], and Paulo J. Azevedo[2]

[1] LIACC, Faculty of Economics, University of Porto, Portugal
amjorge@liacc.up.pt
[2] Departamento de Informática, University of Minho, Portugal
pja@di.uminho.pt

Abstract. We propose an approach to subgroup discovery using distribution rules (a kind of association rules with a probability distribution on the consequent) for numerical properties of interest. The objective interest of the subgroups is measured through statistical goodness of fit tests. Their subjective interest can be assessed by the data analyst through a visual interactive subgroup browsing procedure.

1 Subgroup Discovery

Subgroup discovery is an undirected data mining task, first identified by Klösgen [3], and meanwhile studied by others. A subgroup is a subset of a population having interesting values w.r.t. a property of interest. For example, if the average level cholesterol for all the patients of an hospital is 190, we may find interesting that people who smoke and drink have a cholesterol of around 250. In this case, we have a property of interest (the level of cholesterol) and a subgroup of patients with a precise description. This subgroup can be regarded as relevant or interesting due to the fact that the mean of the property of interest is significantly different from a value of reference, such as the mean of the whole population.

Definition - Given a population of individuals U and a criterion of interest, a *subgroup* $G \subseteq U$ is a subset of individuals that satisfies the criterion. Each subgroup has a description, (a set of conditions) satisfyed by all its members. ⋄

We will define the interest of a subgroup w.r.t. a chosen property of interest (P.O.I.). We assume that the P.O.I. is one and is numerical, although in general we may consider other types of variables. In our work, the notion of interest of a subgroup is not limited to the value of particular measures such as mean. Instead, we compare the observed distribution of the values of the P.O.I. with the distribution of the whole population.

Definition - Let y be a numerical property of interest, and G a subgroup with description $desc_G$. The *distribution of the P.O.I.* y for the individuals $x \in G$ is approximated by the observed $\Pr(y|desc_G)$ and is denoted by $D_{y|desc_G}$.   ⋄

The *a priori* distribution of the P.O.I. is the one for the whole population. Subgroup discovery methods work typically with categorical properties of interest.

---

* Supported by POSI/SRI/40949/2000/ Modal Project (Fundação Ciência e Tecnologia), FEDER e Programa de Financiamento Plurianual de Unidades de I & D.

N. Lavrač, L. Todorovski, and K.P. Jantke (Eds.): DS 2006, LNAI 4265, pp. 301–305, 2006.

Our method constructs the subgroups from discovered distribution rules [2], a kind of association rules with a statistical distribution on the consequent.

Example - Suppose we have clinical data describing habits of patients and their level of cholesterol. The distribution rule $smoke \wedge young \rightarrow chol = \{180/2,$ $193/4, 205/3, 230/1\}$ represents the information that, of the young smokers on the data set, 2 have a *cholesterol* of 180, 4 of 193, 3 of 205 and 1 of 230. This information can be represented graphically as a frequency polygon. The attribute *chol* is the property of interest.   ◇

The objective interest of a subgroup is given by the unexpectedness of its distribution for the property of interest, which can be measured with existing statistical goodness of fit tests. We will define the interest of a subgroup as the deviation of the distribution of the property of interest with respect to the *a priori* distribution. In this sense, the interest of a subgroup is akin to the interest of an association rule as measured by lift, conviction, or χ^2 [4]. However, in the current approach, we take into account the distribution of the possible values of the property of interest, instead of only one such value.

Definition - The *interest of a subgroup* G is given by the dissimilarity between the distribution of the property of interest for the subgroup $D_{y|desc_G}$ and a reference distribution $D_{y|ref}$.   ◇

The reference is typically the *a priori* distribution. The degree of similarity can be measured using statistical goodness of fit tests. In this work we have used Kolmogorov-Smirnov. Given a dataset S, the task of subgroup discovery consists in finding all the distribution rules $A \rightarrow y = D_{y|A}$, where A has a support above a determined mininum σ_{min} and $D_{y|A}$ is statistically significantly different from the *a priori* distribution D_y (the p-value of the K-S test is low).

2   The Visual Interactive Process

Given a population and a criterion of interest, the number of interesting subgroups/distribution rules can be very large. As in the discovery of association rules, for the data analyst to explore the discovered patterns it is useful that a post processing rule browsing environment exists [1].

We propose a visual interactive subgroup discovery procedure that graphically displays the distribution of each subgroup and allows the navigation by the data analyst in a chosen continuous space of subgroups. To represent the continuous space of subgroups we propose an x-y plot, where the coordinates x and y represent statistical measures of the distribution of the property of interest. A simple example is a mean-variance plot. Other subgroup spaces such as median-mode, skewness-kurtosis and mean-kurtosis have also been considered. Skewness and kurtosis are well known distribution shape measures, median, mode and mean are location measures and variance is a spread measure.

Given a two-dimensional plot, each subgroup is represented as a point. This plot will serve as a browsing device (Fig. 1). The data analyst can click on one of the points of that space and visualize the distributions (as frequency polygons) and definitions of the corresponding sugroups. In this phase the selected subgroup is also visually and statistically compared to a reference group. This process is iterated and interesting subgroups found can be saved.

The skewness-kurtosis space gives the data analyst an overall picture of the shapes of the distributions of the subgroups. The mean-kurtosis gives an idea of the location of the distributions as well as of how their shapes are more or less flat. The mean-standard deviation space identifies the subgroups that have their mean below and above the whole population. The mode-median space depicts the location of the distributions of the property of interest for the subgroups, both through mode and median.

2.1   Studying Algal Blooms

This subgroup discovery approach is being applied to study algae population dynamics in a river which serves as an urban water supply resource. The quantity and diversity of the algae are important for the quality of the water, which makes this an economically and socially critical eco-system. Blooms of these algae may reduce the life conditions in a river and cause massive deaths of fish, thus degrading water quality. The state of rivers is affected by toxic waste from industrial activity, farming land run-off and sewage water treatment [5]. Being able to understand and predict these blooms is therefore very important. This problem has been studied in the MODAL project (modys.niaad.liacc.up.pt/projects/modal) in collaboration with the local water distribution company.

The data were collected from 1998 to 2003. All attributes are continuous and are divided in three groups: *phytoplancton*, *chemical and physical* properties, and *microbiological* parameters. The phytoplancton attributes record the quantity of 7 micro-algae species, the chemical and physical attributes record the levels of various algae nutrients and other environmental parameters, the microbiological attributes record the quantities of some bacteria relevant for water quality.

The original data were pre-processed, so that each record stores the phytoplancton observed in one particular sample, and also each of the other descriptive attributes, aggregating the values observed between two samples of phytoplancton. Aggregating functions were maximum, minimum and median. Attributes with the values of the previous sample of the 7 phytoplancton species were also added, as well as two summary attributes measuring the Diversity and the Density of the algae. These attributes are important since a bloom of one of the species is characterized by a low diversity of species and a high density of algae. Three other attributes were added: Normalized Density and Normalized Diversity (normalized versions of Density and Diversity); and BLOOM.N calculated as the difference between normalized density and normalized diversity. High values of BLOOM.N indicate high possibility of a bloom. After pre-processing we have 72 input variables, 7+5 target variables and 131 cases. In the examples, variable names appear in Portuguese. The results were analysed by a biologist.

We have conducted several studies with different P.O.I., using a minimum support of 0.05. Here we provide a summary of results [1]. With DIVERSIDADE.N

---

[1] Please consult www.liacc.up.pt/~amjorge/docs/VSG-TR-06.pdf for more.

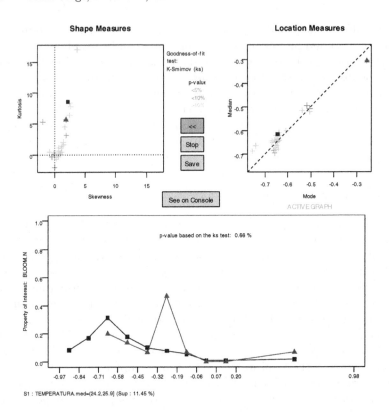

Fig. 1. Screen of the protoype showing two navigation plots (top), and the selected subgroup, where the P.O.I. BLOOM.N is affected by relatively high temperature (frequency polygon in triangles), and compared with the *a priori* distribution (in squares)

(normalized diversity) as the P.O.I., the median values of the microbiology and physical-chemical parameters as explanatory variables, we obtain 98 subgroups. Low diversity is a necessary condition for a bloom to occur. On the mode-median plot we can click on the subgroups with lowest mode and median. One of the rules obtained indicates that, for relatively low values of oxygen and low-medium values of iron there is a relatively high probability (when compared to the whole population) that DIVERSITY is low (between 0.3 and 0.4) or very low (below 0.1). While oxygen is necessary for phytoplancton primary production, the low quantity of one nutrient (iron) may reduce the phytoplancton to the species that live well under those conditions. This situation may lead to a bloom of one of the species. High values of BLOOM.N may indicate algae blooms (high density of micro-algae and low diversity). With this P.O.I., 27 subgroups were found. One of the rules relates relatively high temperatures (around 25 degrees Celsius) with a distribution of bloom values shifted right. This is a well known effect of high temperatures (Fig. 1).

Distribution rule generation is very fast (less than 5 seconds), and moving from subgroup to subgroup is made easy by the graphical interface. The mode-median plot is a very useful browsing device for these data. Looking for extremely skewed distributions of the target variables is facilitated with this subgroup space. By having immediate acces to all the generated subgroups, the data analyst can compare nearby subgroups and examine their descriptions.

The display of the distribution provides information that may be hidden by a summary measure such as mean. A distribution curve with two modes, for example, may indicate that a particular subgroup has two possible outcomes. If one of those outcomes is critical, than the antecedent of the subgroup may become an alarm trigger for water monitoring. This also implies that not only extreme values of median or mode indicate potentially interesting subgroups.

3   Conclusions

We have presented a visual interactive subgroup discovery approach for numerical properties of interest. Subgroups are discovered as distribution rules (DR) with sufficient support and having a distribution for the property of interest distinct from the whole population. The similarity between distributions is measured as the Kolmogorov-Smirnov statistical test's p-value. A large set of subgroups is presented to a data analyst as a two dimensional plot, corresponding to a space of subgroups. Each point on the plot is a different subgroup. The data analyst can inspect each of the subgroups by clicking on the respective point. Each subgroup is displayed with its definition, support, and the distribution of the P.O.I. The approach is being used in a project for monitoring the quality of water in a river. In this application, the properties of interest are the ones related with the control of algal blooms, which affect the quality of water.

Acknowledgments. We thank Luís Torgo and Rita Ribeiro for making available the problem and data, and also the biologist Dr. Catarina Magalhães, .

References

1. A. Jorge, Poças, and P. J. Azevedo. Post-processing operators for browsing large sets of association rules. In S. Lange, K. Satoh, and C. H. Smith, editors, *Proceedings of Discovery Science, DS 02, Luebeck, Germany*, number 2534 in Lecture Notes in Computer Science, pages 414–421. Springer-Verlag, 2002.
2. A. M. Jorge, P. J. Azevedo, and F. Pereira. Distribution rules with numerical properties of interest. In *Proceedings of Principles of Data Mining and Knowledge Discovery (PKDD-06)*, LNAI. Springer-Verlag, 2006.
3. W. Klösgen. Explora: A multipattern and multistrategy discovery assistant. In U. Fayyad, G. Piatetsky-Shapiro, P. Smyth, and R. Uthurusamy, editors, *Advances in Knowledge Discovery and Data Mining*. AAAI Press, Menlo Park, CA, 1996.
4. B. Liu, W. Hsu, and Y. Ma. Pruning and summarizing the discovered associations. In *KDD '99: Proceedings of the fifth ACM SIGKDD international conference on Knowledge discovery and data mining*, pages 125–134, New York, NY, USA, 1999. ACM Press.
5. R. P. Ribeiro and L. Torgo. Predicting harmful algae blooms. In M. e. a. Pires, editor, *Proceedings of Portuguese AI Conference (EPIA'03)*, volume 2902 of *LNAI*, pages 308–312. Springer-Verlag, 2003.

Contextual Ontological Concepts Extraction

Lobna Karoui, Nacéra Bennacer, and Marie-Aude Aufaure

Ecole Supérieure d'Electricité
Plateau de Moulon 3 rue Joliot Curie
91192 Gif-sur-Yvette cedex, France
{Lobna.Karoui, Nacera.Bennacer, Marie-Aude.Aufaure}@Supelec.fr

Abstract. Ontologies provide a common layer which plays a major role in supporting information exchange and sharing. In this paper, we focus on the ontological concept extraction process from HTML documents. We propose an unsupervised hierarchical clustering algorithm namely "Contextual Ontological Concept Extraction" (COCE) which is an incremental use of a partitioning algorithm and is guided by a structural context. This context exploits the html structure and the location of words to select the semantically closer cooccurrents for each word and to improve the words weighting. Guided by this context definition, we perform an incremental clustering that refines the words' context of each cluster to obtain semantic extracted concepts. The COCE algorithm offers the choice between either an automatic execution or an interactive one. We experiment the COCE algorithm on French documents related to the tourism. Our results show how the execution of our context-based algorithm improves the relevance of the clusters' conceptual quality.

1 Introduction

Ontologies provide a common layer which plays a major role in supporting infor-mation exchange and sharing by extending syntactic interoperability to semantic interoperability in the semantic Web. Ontology learning systems as defined in [1, 2, 3, 4, 5] have different purposes. Faure and Nédellec [1] define a system called ASIUM where a cooperative conceptual clustering is applied to technical texts using a syntactic parser to produce an acyclic conceptual graph of clusters. Basic clusters are formed by words that occur with the same verb after the same preposition. TEXT-TO-ONTO [2] is an ontology learning environment that extracts concepts, taxonomic and non taxonomic relations. WebOntEx [3] is based on HTML tags, lemmatization tags and conceptual tags. It uses WorldNet and a logic inductive programming method to build an ontology. Based on the HTML regularities, OntoMiner [4] analyzes sets of domain specific web sites and generates a taxonomy of particular concepts and their instances. OntoLearn [5] extracts domain terminology from Web documents by using a linguistic processor and a syntactic parser.

In this paper, we focus on the ontological concept extraction process from HTML documents. In order to improve it, we propose an unsupervised hierarchical clustering algorithm namely "Contextual Ontological Concept Extraction" (COCE) which is based on an incremental use of a partitioning algorithm and is guided by a structural context.

N. Lavrač, L. Todorovski, and K.P. Jantke (Eds.): DS 2006, LNAI 4265, pp. 306–310, 2006.
© Springer-Verlag Berlin Heidelberg 2006

The context, based on the html structure and the location of words in the documents, guides the clustering algorithm to refine the context of each word by improving the semantically closer cooccurrents selection for each word, the word weighting and the words pair's similarity. By performing an incremental process, this algorithm refines the context of each word cluster and improves the conceptual quality of the resulting clusters and, consequently, the extracted concepts. The COCE algorithm offers the choice between either an automatic execution or an interactive user's processing. We experiment the contextual clustering algorithm on French html corpus related to the tourism domain. The results show that our context-based clustering method and the successive refinements of clusters improve the relevance of the extracted concepts. In the following section, we define and experiment this algorithm.

2   The Contextual Ontological Concept Extraction Algorithm

In this section, we explain the context definition. Then, we present the algorithm.

- Context Definition

Our challenge is to answer correctly to the following question: how can we give a weighting that effectively illustrates the importance of a term in its domain and characterizes its relations with other words?

Let us go back to our corpus and the various analyses performed on the HTML documents. We note that there exist relations between the existing HTML elements (<h1> → <p>). We also note that key tags [7] such as <keywords> and <title> are related to other existing HTML elements. For example: <TITLE_URL> (header of a hyperlink) → <H1> (headings of a part of document). The first group of links is physical because it depends of the structure of the HTML document. But the second group shows a logical link that is not always visible (elements are not necessary consecutives). In order to represent links between tags, we define two new concepts namely "contextual hierarchy" based on HTML elements and "link co-occurrence". A contextual hierarchy (C.H) is a tag hierarchy. It illustrates possible relations existing within HTML documents and between them. In our study, the context of a word can vary; so the process is performed according to the contextual hierarchy. By complaining with this structure, we can define a link between terms, if they appear in the same block-level tag (<p>, <td>, etc.). In this case, the link is a "neighbourhood co-occurrence" and the context is limited to the tag. However, if they appear in different tags that are related by a physical or a logical link defined in the contextual hierarchy, we define the concept of "co-occurrence link". In this case, the context is the association of the two related tags (for example: <H1> + <p> or <TITLE> + <H1>, etc.). So, a co-occurrence link is a co-occurrence which context is not limited to one unit (phrase or document), but is a generic context that will be instantiated according to the location of the term in the document. The generic context represents the term's adaptability in the corpus. This contextual model respects the word location in order to take a multitude of term situations into account when computing its weighting by applying the "Equivalence Index" [6].

- Algorithmic Principles

In the following, we present an unsupervised hierarchical clustering algorithm namely "Contextual Ontological Concept Extraction" (COCE) (Fig.1) to extract ontological concepts from HTML documents. It is based on an incremental use of a partitioning algorithm and is driven by the structural contexts in which words occur. COCE proceeds in an incremental manner. It computes the occurrences of each word and selects their semantically closer cooccurrents according to the context. Then, it divides the clusters obtained at each step in order to refine the context of each cluster. So, the algorithm refines at the same time the context of each word and the context of each cluster. Also, it offers to the user the possibility to choose either a complete automatic execution or an interactive one. If he decides the automatic execution, he should either define some parameters or choose the default ones resulting from our empirical experiments. These parameters are: the highest number of words per cluster P, the accepted margin M representing an additional number of words in a resulting cluster accepted by the user and the similarity measure S. If he prefers to evaluate the intermediate word clusters, he chooses the interactive execution. In this case, the algorithm allows him to analyse the word cluster at the end of each intermediate clustering in order to define the value of k' (Fig.1-Eq1) and to decide either he prefers to continue within an interactive execution or to run an automatic one. In the interactive execution, the process is longer than the automatic one but it offers an opportunity to the user to obtain better hierarchical word clusters. As input to this algorithm, the user, firstly, should choose the input file of dataset F. Secondly, he defines the number of clusters K, and chooses an automatic or an interactive execution. An automatic execution is defined in four steps:

Step 0 and 1: These steps concern the application of the structural context which is deduced from the various analyses [7], computed from the dataset F and stored in a matrix and the execution of the partitioning algorithm to obtain the first k clusters.

```
 0: Algorithm Contextual-Ontological-Concepts-Extraction (In: F, K, P, M, S, Out: WC)
 1: Apply our Context definition and compute the occurrences of the population {/*Step 0*/}
 2: Dᵢ ← Φ {/* Step 1*/}  {/* Di is the word distribution into the various clusters Ci with Di ={C1, C2,, Ci}*/}
 3: Choose sporadically the K first centers
 4: Assign each word to the cluster that has the closest centroid
 5: Recalculate the positions of the centroids
 6: if (the positions of the centroids did not change) then
 7:    go to the step 10
 8: else
 9:    go to the  step 4.
10: Dᵢ ← Dᵢ U {C₁, C₂, C₃, ..., Cₖ}   {/* Cᵢ is a word cluster belonging to Dᵢ As Cᵢ ∈ Dᵢ */}
11: For all Cᵢ ∈ Dᵢ do {/* Step 2*/}
12:    if (Word-Number (Cᵢ) ≤ P) then
13:      WC ← WC U {Cᵢ}   {/* WC is the set of word clusters WC ={C₁, C₂ , C_T} */}
14:      Dᵢ ← Dᵢ \ {Cᵢ}
15:    else
16:      Dᵢ ← Dᵢ \ {Cᵢ}
17:      Unbolt the words Wᵢ belonging to the cluster Cᵢ   {/* Wᵢ is each word belonging to Ci So Ci = {W 1, W 2 , ..., W i} */}
18:      Compute the value of K' { /* Eq1:  K' = a *ln (Word-Number (Ci) * b)*/}
19:      Go to the steps 3, 4, 5 and 6
20: For all Cᵢ ∈ WC do {/* Step 3*/}
21:    if (Word-Number (Cᵢ) = 1 and Wᵢ ∈ Cᵢ) then   {/* Word-Number (Ci) is the number of word into the cluster Ci */}
22:      Calculate the position of Wᵢ to the existing centroids of the clusters Cᵢ ∈ WC
23:      if (Word-Number (C₀₋c) > P+M) then    {/* The cluster that has the closest centroid to the target word Wᵢ */}
24:        Choose another cluster C₀₋c that has the following closest centroid
25:        Go to the step 20
26:    else
27:        WC ← WC \ {C₀₋c, Cᵢ}
28:        Assign Wᵢ to the cluster C₀₋c
29:        WC ← WC U {C₀₋c}
30: Return (WC)
31: End
```

Fig. 1. The Contextual Ontological Concepts Extraction Algorithm

Step 2: This step finds clusters respecting the criterion P. For each intermediate execution of kmeans, we should define the value K' representing the number of clusters which should be obtained after the division. In an automatic execution of the COCE algorithm, this value is not defined by the user but computed by the system. We implement a proportional function (established on several experiments and the experts' knowledge) that automatically defines the value of K' computed by the formula (Eq1) in Fig1.

Step 3: When applying the division process, we can obtain clusters with only one word. Our idea is to automatically associate each word alone into a cluster resulting from step 2. Another problem appears when the algorithm affects too many words to the same cluster. In this case, and for a few clusters, we can obtain clusters with a great number of words. If a word is assigned to a cluster already containing P+M words, the algorithm will choose the cluster which is the closest centroid to the target word.

Choosing an interactive process implies the user intervention after step 1 and during step 2 and 3. In step 2, if a cluster is divided, the algorithm allows to refine the context of each cluster by taking into account only the associated attributes of its belonging words. By applying this method, the similarity computed better represents the association degree between each two words. By applying the step 3, it avoids the cases of having only one word in a cluster and those containing the majority of the single words.

- Experimental Evaluation of the COCE algorithm

In our experiments, we begin the process with key and title tags to give an outline of the domain's semantic information. These terms are those chosen by the web site designer as keyword, glossary, etc. In this section, we evaluate the COCE algorithm results by comparing them to those obtained by the Kmeans algorithm. We chose the Euclidian distance as a similarity measure. Our dataset is composed of 872 words. The Kmeans algorithm distributes them in 156 clusters while the COCE algorithm, experimented with the values 20, 10 and 22 for k, P and M respectively (more significant results), gives 162 clusters. In [8], some criteria for the statistical evaluation of unsupervised learners have been defined that we could not use them since the cluster homogeneity does not imply that the words in the cluster are semantically close or that the associated label satisfies the domain expert. In our case, we propose a manually domain experts' evaluation. We present the results to two domain experts. As a first step, each of them individually evaluates and labels manually the word clusters. Then, they work together in order to discuss about the results, their labels proposition and to give us only one evaluation and labeling for which they agree. To evaluate the word clusters' conceptual quality, we defined four criteria and compared our algorithm to the Kmeans one. Concerning the word distribution, we have for kmeans and COCE algorithms respectively 13% and 3.66% of our initial words that are grouped together. For the semantic interpretation, the domain expert notes that there are three types of word clusters. Advisable clusters are those for which the expert is able to associate a label. Improper clusters are clusters where the expert finds a difficulty to give a label. Unknown clusters are clusters where the expert could not find any semantic interpretation. Thanks to the P and the M values, in each word cluster, the percentage of noisy elements decreases a lot. As a consequence, the percentage of unknown and improper clusters is reduced. Moreover, we obtain 68.52% advisable clusters with the COCE algorithm which is more important than only 53.2% with the Kmeans one. Concerning the extracted concepts, we take into account only the advisable clusters and we obtain respectively 86.18% and 86.61% with the kmeans algorithm and the COCE one. Another element which affects the concept's quality is the level of generality for a concept. We obtain respectively with the kmeans algorithm and the COCE one 78.31% and 85.58% general concepts. Also, we remark that the clusters obtained with the COCE algorithm are more enhanced than with the Kmeans one. In order to keep our method functional, we should have a minimum structure in our documents. However, the absence of some HTML tags does not affect our method's functioning since it is incremental and searches the word's cooccurrents in other lower levels of the C.H.

3 Conclusion

As known, knowledge acquisition is a difficult task. In this paper, we focus on acquiring ontological domain concepts by exploiting HTML structure, relation between html tags and word location. Our structural context definition improves the word weighting, the selection of the semantically closer cooccurrents and the resulting word clusters, since it refines the context of each word. Then, we defined a hierarchical clustering method namely "Contextual Ontological Concept Extraction" algorithm guided by a structural contextual definition that refine the words' context of each cluster. The COCE algorithm improves the relevance of the conceptual quality of the ontological extracted concepts and offers the choice between either an automatic execution or an interactive one. In our research perspectives, we will define and experiment a linguistic context combined with the structural one and applied them to words belonging to other HTML tags. We will also propose a new evaluation method.

References

1. Faure, D., Nedellec, C. and Rouveirol, C. (1998). Acquisition of semantic knowledge uing machine learning methods: the system ASIUM. Technical report number ICS-TR-88-16, inference and learning group, University of Paris-sud.
2. Meadche, A. and Staab S. : "Ontology learning for the semantic Web, IEEE journal on Intelligent Systems, Vol. 16, No. 2, 72-79, 2001.
3. Han, H. and Elmasri, R.: "Architecture of WebOntEx: A system for automatic extraction of ontologies from the Web". Submitted to WCM2000.
4. Davulcu, H., Vadrevu, S. and Nagarajan, S. : OntoMiner: Boostrapping ontologies from overlapping domain specific web sites. In AAAI'98/IAAI'98 Proceedings of the 15th National Conference on Artificial Intelligence, 1998.
5. Navigli, R. and Velardi, P.: Learning domain ontologies from document warehousees and dedicated web sites. In AAAI'98/IAAI'98 Proceedings of the 15th National Conference on Artificial Intelligence, 1998.
6. Michelet, B. :L'analyse des associations. Thèse de doctorat, Université de Paris VII, UFR de Chimie, Paris, 26 Octobre 1988.
7. Karoui, L. and Bennacer, N. "A framework for retrieving conceptual knowledge from Web pages" Semantic Web Applications and Perspectives SWAP, Italy 2005.
8. Vazirgiannis, M., Halkidi, M. and Gunopoulos, D.: uncertaintly handling and quality assessmen in data mining. Springer, 2003.

Experiences from a Socio-economic Application of Induction Trees

Fabio B. Losa[1], Pau Origoni[1], and Gilbert Ritschard[2]

[1] Statistical Office of Ticino Canton, CH-6500 Bellinzona, Switzerland
`fabio.losa@ti.ch`
[2] Dept of Econometrics, University of Geneva, CH-1211 Geneva 4, Switzerland
`gilbert.ritschard@metri.unige.ch`

Abstract. This paper presents a full scaled application of induction trees for non-classificatory purposes. The grown trees are used for highlighting regional differences in the women's labor participation, by using data from the Swiss Population Census. Hence, the focus is on their descriptive rather than predictive power. Trees grown by language regions exhibit fundamental cultural differences supporting the hypothesis of cultural models in female participation. The explanatory power of the induced trees is measured with deviance based fit measures.

1 Introduction

Induced decision trees have become popular supervised classification tools since [1]. Though their primary purpose is to predict and to classify, trees can be used for many other relevant purposes: as exploratory methods for partitioning and identifying local structures in datasets, as well as alternatives to statistical descriptive methods like linear or logistic regression, discriminant analysis, and other mathematical modeling approaches [5].

This contribution demonstrates such a *non-classificatory* use of classification trees by presenting a full scaled application on female labor market data from the Swiss 2000 Population Census (SPC). The use of trees for our analysis was dictated by our primary interest in discovering the interaction effects of predictors of the women's labor participation. The practical experiment presented is original in at least two respects: 1) the use of trees for microeconomic analysis, which does not appear to be a common domain of application; 2) the use of induction trees for a complete population census dataset.

Note that since our goal is not to extract classification rules, but to understand — from a cross-cultural perspective — the forces that drive women's participation behavior, we do not rely on the usual misclassification rates for validating the trees. Rather, we consider some deviance based fit criteria [7, 8] similar to those used with logistic regression.

Before presenting our experiment, let us shortly recall the principle of classification trees. They are grown by seeking, through successive splits of the learning data set, some optimal partition of the predictor space for predicting the outcome

N. Lavrač, L. Todorovski, and K.P. Jantke (Eds.): DS 2006, LNAI 4265, pp. 311–315, 2006.

class. Each split is done according to the values of one predictor. The process is greedy. At the first step, it tries all predictors to find the "best" split. Then, the process is repeated at each new node until some stopping rule is reached. For our application, we used CART [1] that builds only binary trees by choosing at each step the split that maximizes the gain in purity measured by the Gini index. CART uses relatively loose stopping rules, but proceeds to a pruning round after the preliminary growing phase. We chose CART for our analysis, despite the gain in purity seems less appropriate for a non-classificatory purpose than, for example, the strength of association criterion used by CHAID [2]. Indeed, the great readability of the binary CART trees was decisive when compared with the n-ary CHAID trees that had, even at the first level, a much too high number of nodes to allow for any useful interpretation.

2   The Applied Study

We begin by setting the applied research framework, then we sketch our global analysis procedure and, finally, we present selected findings.

Female labor market participation reveals significant differences across countries. In Europe, scholars often identify at least two general models: a Mediterranean one (Italy, Greece, Portugal, etc.) versus a model typical for Central and Northern Europe [6]. The first is represented by an inverse L-shaped curve of the *activity* or *participation rate* by age, where after a short period of high rate (at entry in the labor market) the proportion of women working or seeking work begins to steadily decline up to retirement. The same graph depicts a M-shaped curve in Central and Northern European countries, characterized by high participation at entry, followed by a temporary decline during the period of motherhood and childbearing, and a subsequent comeback to work, up to a certain age where the process of definite exit starts.

In this respect, Switzerland is an interesting case. Firstly, Switzerland is a country placed in a nutshell across the Alps, which naturally divides Southern Europe from Central and Northern Europe. Secondly, there are three main languages, spoken by people living in three geographically distinct regions: French in the western part on the border with France, German in the northern and eastern parts on the border with Germany and Austria, and Italian south of the Alps in a region leading to Italy. The existence of three regions, with highly distinctive historical, social and cultural backgrounds and characters, and the fact that the Italian-speaking part is divided from the other two by the Alps highlight the very specific particularity of this country for a cross-cultural analysis of the female participation in the labor market. Moreover, the fact that the comparative analysis is performed amongst regions of the same country guarantees, despite differences stemming from the Swiss federal system, a higher degree of comparability on a large series of institutional, political and other factors than one would get with cross-country studies.

The idea of the research project was to verify the existence of differing cultural models of female labor market participation, by analysing activity rates and hours worked per week — in terms of proportions of full-timers and part-timers — across the three linguistic regions in Switzerland, by using the SPC 2000 data.

To shortly describe the data, we can say that the Federal Statistical Office made us available a clean census dataset covering the about 7 millions inhabitants of Switzerland. For our study, only the about 3.5 millions women were indeed of interest. In the preprocessing step we disregarded young ($< 20, 23\%$) and elderly ($> 61, 18\%$) women, as well as non Swiss women not born in Switzerland (1.6%), i.e. about 43% of the women. This left us with about 2 millions cases. Finally, we dropped about 350000 cases with missing values, and hence included 1667494 cases into the analysis.

The Empirical Research Design. The research procedure used classification trees at two different stages, with differing but complementary purposes. A tree was first grown in what we refer to as the *preliminary step*. Its main goal was to find a sound partition of the analysed population into a limited number of homogeneous groups — homogeneous female labor supply behavior in terms of activity and choice between full-time and part-time employment — over which a tailored analysis could be performed. This first step was run on the whole Swiss female population of age 20 to 61, using their *labor market status*[1] as outcome variable, and general socio-demographic characteristics (civil status, mother/non mother, ...) as predictive attributes. From this, a robust partition in three groups was chosen: the *non-mothers*, the *married or widowed mothers*, and the *divorced or single mothers*. The first group is composed by 609,861 women (36.6%), the second one by 903,527 (54.2%) and the third one by 154,106 (9.2%).

The second application of classification trees took place in the analysis of cross-cultural female labor supply behavior for each selected group. Here again the outcome variable was the *labor market status* of the women. A much broader series of predictive variables was retained however: age, profession, educational level, number of kids, age of last-born kid, type of household, etc. Classification trees have been produced separately for each region and then compared in order to analyse cultural patterns in the participation behavior of the three main language regions in Switzerland.

It is worth mentioning here that the final trees retained are simplified versions of those that resulted from the stopping and pruning criteria. They were selected on the basis of comprehensibility and stability factors. We checked for instance that the splits retained stayed the same when removing randomly 5% of the cases from the learning data set.

Results. In order to identify cultural models of female labor supply, three trees (one per region) were generated for each group. These — in combination with the results of the traditional bivariate analyses — were compared and thoroughly analysed in terms of structure and results. We give hereafter a very brief overview

---

[1] Labor market status is a categorical variable with four values: full-time active (at least 90% of standard hours worked per week), long part-time active (50% to 90%), short part-time active (less than 50%) and non active, where active means working or seeking for a job.

of the main results for the third group, i.e. divorced or single mothers. For details interested readers may consult the research report [3, 4].

A first obvious outcome is that opting for inactivity is much more frequent in the Italian speaking region. Italian speaking mothers without high education who have a last-born child less than 4 are most often inactive. For the other regions, part time activity is typical for women with a child less than 14, who either have at most a medium education level or work in fields such as health, education or sciences. A single overall splitting threshold at 54 for the age of the mothers in the Italian speaking region tends to confirm that this region conforms to the reversed L-shape of the Mediterranean model. In comparison, for the French speaking region, the major difference regarding the age of the mothers concerns only those whose last born child is more than 14. Among them, those who have a low or medium education level stop working at 60.

3 Explanatory Power of Our Non-classificatory Trees

Table 1 reports some of the quality figures we have computed for each of the three regional trees for divorced or single mothers: CHI for the Italian speaking, CHF for the French speaking and CHG for the German speaking region. The figures reported are q the number of leaves of the tree, c^* the number of different observed profiles in terms of the retained predictors, $D(m_0|m)$ the deviance between the induced tree and the root node, which is a likelihood ratio Chi-square measuring the improvement in explanatory power of the tree over the root node, d and sig, respectively the degrees of freedom and the significance probability of the Chi-square, Theil's uncertainty u, i.e. the proportion of reduction in Shannon's entropy over the root node, and $\sqrt{u}$ its square root, which can be interpreted as the part of the distance to perfect association covered by the tree. For technical details and justifications on the measures considered see [7].

The deviances $D(m_0|m)$ are all very large for their degrees of freedom. This tells us that the grown trees make much better than the root node and, hence, clearly provide statistically significant explanations. The Theil uncertainty coefficient u seems to exhibit a low proportion of gain in uncertainty. However, looking at its square root, we see that we have covered about 25% of the distance to perfect association. Furthermore, the values obtained should be compared with the maximal values that can be achieved with the attributes considered. These are about .5, i.e. only about twice the values obtained for the trees. Thus, with the grown trees that define a partition into q classes only instead of c^* for the finest partition, we are about half the way from the finest partition.

Table 1. Trees quality measures

	q	c^*	n	$D(m_0|m)$	d	sig.	u	$\sqrt{u}$
CHI	12	263	5770	822.2	33	.00	.056	.237
CHF	10	644	35239	4293.3	27	.00	.052	.227
CHG	11	684	99641	16258.6	30	.00	.064	.253

4   Conclusion

The experiment reported demonstrates the great potential of classification trees as an analytical tool for investigating socio-economic issues. Especially interesting is the visual tree outcome. For our study, this synthetic view of the relatively complex mechanisms that steer the way women decide about their participation in the labor market provided valuable insight into the studied issue. It allowed us to highlight cultural differences in the interaction effects of attributes like age of last-born child, number of children, profession and education level that would have been hard to uncover through regression analysis, for example.

It is worth mentioning that generating reasonably sized trees is essential when the purpose is to describe and understand underlying phenomenon. Indeed, complex trees with many levels and hundred of leaves, even with excellent classification performance in generalization, would be too confusing to be helpful. Furthermore, in a socio-economic framework, like that considered here, the tree should make sense from the social and economic standpoint. The tree outcomes should therefore be confronted with other bivariate analyses and modeling approaches. Our experience benefited a great deal from this interplay.

References

[1] Breiman, L., Friedman, J.H., Olshen, R.A., Stone, C.J.: *Classification And Regression Trees*. Chapman and Hall, New York (1984)

[2] Kass, G.V.: An exploratory technique for investigating large quantities of categorical data. *Applied Statistics* **29** (1980) 119–127

[3] Losa, F.B., Origoni, P.: Partecipazione e non partecipazione femminile al mercato del lavoro. Modelli socioculturali a confronto. Il caso della Svizzera italiana nel contesto nazionale. Aspetti statistici, Ufficio cantonale di statistica, Bellinzona (2004)

[4] Losa, F.B., Origoni, P.: The socio-cultural dimension of women's labour force participation choices in Switzerland. *International Labour Review* **144** (2005) 473–494

[5] Murthy, S.K.: Automatic construction of decision trees from data: A multi-disciplinary survey. *Data Mining and Knowledge Discovery* **2** (1998) 345–389

[6] Reyneri, E.: *Sociologia del mercato del lavoro*. Il Mulino, Bologna (1996)

[7] Ritschard, G.: Computing and using the deviance with classification trees. In Rizzi, A., Vichi, M., eds.: *COMPSTAT 2006 - Proceedings in Computational Statistics*. Springer, Berlin (2006) forthcoming

[8] Ritschard, G., Zighed, D.A.: Goodness-of-fit measures for induction trees. In Zhong, N., Ras, Z., Tsumo, S., Suzuki, E., eds.: *Foundations of Intelligent Systems, ISMIS03*. Volume LNAI 2871. Springer, Berlin (2003) 57–64

Interpreting Microarray Experiments Via Co-expressed Gene Groups Analysis (CGGA)

Ricardo Martinez[1], Nicolas Pasquier[1],
Claude Pasquier[2], and Lucero Lopez-Perez[3]

[1] Laboratoire I3S, 2000, route des lucioles,
06903 Sophia-Antipolis cedex, France
{rmartine, pasquier}@i3s.unice.fr
[2] Laboratoire Biologie Virtuelle, Centre de Biochimie, Parc Valrose,
06108 Nice cedex 2, France
claude.pasquier@unice.fr
[3] INRIA Sophia Antipolis, 2004, route des Lucioles,
06903 Sophia-Antipolis cedex, France
lucero.lopez@gmail.com

Abstract. Microarray technology produces vast amounts of data by measuring simultaneously the expression levels of thousands of genes under hundreds of biological conditions. Nowadays, one of the principal challenges in bioinformatics is the interpretation of huge data using different sources of information.

We propose a novel data analysis method named CGGA (Co-expressed Gene Groups Analysis) that automatically finds groups of genes that are functionally enriched, i.e. have the same functional annotations, and are co-expressed.

CGGA automatically integrates the information of microarrays, i.e. gene expression profiles, with the functional annotations of the genes obtained by the genome-wide information sources such as Gene Ontology (GO)[1].

By applying CGGA to well-known microarray experiments, we have identified the principal functionally enriched and co-expressed gene groups, and we have shown that this approach enhances and accelerates the interpretation of DNA microarray experiments.[2]

1 Introduction

One of the main challenges in microarray data analysis is to highlight the principal functional gene groups using distinct sources of genomic information. These sources of information, constantly growing by an ever-increasingly volume of genomic data, are: semantic (taxonomies, thesaurus and ontologies), literature and bibliographic databases (articles, on-line libraries, etc.), experience databases (ArrayExpress, GEO, etc.) and nomenclature databases (HUGO: human, Flybase: fruit fly, SGD: yeast...).

---

[1] Gene Ontology project: http://www.geneontology.org/
[2] CGGA program is available at http://www.i3s.unice.fr/~rmartine/CGGA.

N. Lavrač, L. Todorovski, and K.P. Jantke (Eds.): DS 2006, LNAI 4265, pp. 316–320, 2006.
© Springer-Verlag Berlin Heidelberg 2006

Actually, one of the major goals in bioinformatics is the automatic integration of biological knowledge from distinct sources of information with genomic data [1]. A first assessment of the methods developed to answer this challenge was proposed by Chuaqui [3]. We target here the enrichment of two recently developed research axes, *sequential* and *a priori*, that exploit multiple sources of annotations such as GO.

The sequential axis methods build co-expressed gene clusters (groups of genes with a similar expression profiles). Then they detect co-annotated gene subsets (sharing the same annotation). Afterwards, the statistical significance of these co-annotated gene subsets is tested. Among the methods in this axis let us quote *Onto Express* [5], *Quality Tool* [6], *EASE* [7], *THEA* [11] and *Graph Modeling* [15].

The a priori axis methods first finds functionally enriched groups (FEG), i.e. groups of co-annotated genes by function. Then they integrate the information contained in the profiles of expression. Later on, the statistical significance of the FEG is tested by an *enriched score* [10], a *pc-value* [2], or a *z-score* test [8].

Our approach, called CGGA (Co-expressed Gene Groups Analysis), is inspired by the a priori axis: the FEG are initially formed from the Gene Ontology, next a function, which synthesizes the information contained in the expression data, is applied in order to obtain an arranged gene list. In this list, the genes are sorted by decreasing expression variability. The statistical significance of the FEG obtained is then tested using a similar hypothesis proof as presented in *Onto Express* [5]. Finally, we obtain co-expressed and statistically significant FEG.

This article is organized in the following way: in section 2 we describe the validation data as well as the tools used: databases, ontologies and statistical packages; our algorithm CGGA is described in section 3; the results obtained are presented in section 4 and the last section presents our conclusions.

2 Data and Methods

2.1 Dataset

In order to evaluate our approach, the CGGA algorithm was applied to the DeRisi dataset which is one of the most studied in this field [4]. DeRisi experience measures the variations in gene expression profiles during the cellular process of diauxic shift for the yeast *Saccharomyces Cerevisiae*. This process corresponds to the transition from fermentation to respiration that occurs when fermenting yeast cells, inoculated into a glucose-rich medium, turn to the utilization of the ethanol (aerobic respiration).

2.2 Ontology and Functionally Enriched Groups (FEG)

In order to fully exploit Gene Ontology (GO) we have generated: SGOD database. Our database contains all GO annotations for every yeast gene using Saccharomyces Genome Database (SGD)[3] nomenclature. We have stored all the functional annotations of each gene and his parents preserving the hierarchical structure of GO. Queries carried out on the SGOD database have built the whole set of the FEGs.

---

[3] Saccharomyces Genome Database: http://www.yeastgenome.org/

2.3 Expression Profile Measure of the Genes

In order to incorporate the expression profile of the genes, we have used a measurement of their variability of expression, *f-score* [13], which is more robust than other measurements such as *anova, fold change* or *t-student* statistics [13].

This measurement enables us to build a list of genes, *g-rank*, ordered by decreasing expression variability. We have used the SAM program [16] to calculate the *f-score* associated with each gene.

3   Co-expressed Gene Groups Analysis (CGGA)

The CGGA is based on the idea that any resembling change (co-expression) of a gene subset belonging to an FEG is physiologically relevant. We say that two genes are co-expressed if they are close in the sense of the metric given by the expression variability (*f-score*). The CGGA algorithm computes a *pc-value* for each FEG that estimates its coherence (according to the *g-rank*) and thus to detect the statistically significant groups.

3.1   CGGA Algorithm

The CGGA algorithm first builds the *g-rank* list from the expression levels and the FEG from the SGOD. For each FEG of n genes, the algorithm determines the $n(n+1)/2$ gene subsets that we want to test for co-expression. For each subset we compute the *pc-value* corresponding to the test described below.

Let H_0 be the hypothesis that x genes from one of the subsets were related by chance. The probability that H_0 is true follows from the hypergeometric distribution[4]:

$$p(X = x \mid N, \mathrm{R}_{g(x)}, n) = \frac{\binom{\mathrm{R}_{g(x)}}{x}\binom{N - \mathrm{R}_{g(x)}}{n - x}}{\binom{N}{n}} \quad \text{where} \quad p(X = 0 \mid N, \mathrm{R}_{g(x)}, n) = 0, \quad (1)$$

with: N: total number of genes in the dataset, n: number of genes in the FEG, x: position of the gene in the FEG (previously ordered by rank), $r_{g(x)}$: absolute rank of the gene of position x in the *g-rank* list and $\mathrm{R}_{g(x)}$: number of ranks between the gene of position x from its FEG predecessor. $\mathrm{R}_{g(x)}$ is calculated from the absolute ranks $r_{g(x)}$ according to the formula: $\mathrm{R}_{g(x)} = r_{g(x)} - r_{g(x-1)} + 1$ where $\mathrm{R}_{g(0)} = r_{g(0)} = 1$.

The *pc-value* corresponding to this hypothesis test is [5]:

$$pc - value(x) = 1 - \sum_{k=1}^{x} p(X = k \mid N, \mathrm{R}_{g(k)}, n). \quad (2)$$

In order to accept or reject H_0 we will use the following significance threshold: *p–value* = *Min* $\{N^{-1}, |\Omega|^{-1}\}$, where $|\Omega|$ is the cardinality of the set of functional annotations. So, for each FEG, if *pc–value(x)* < *p–value* then H_0 is rejected, i.e. the FEG is statistically significant.

---

[4] For more details on the computation of this probability, refer to [17].

4 Results

In order to evaluate our method, we compared the results obtained by DeRisi [4], IGA [2] and CGGA. The results obtained using CGGA for the over-expressed and under-expressed genes are presented in Table 1. As expected, all groups identified as significantly co-expressed by the DeRisi method have also been identified by the CGGA. The groups identified by CGGA and DeRisi are in **bold**, the ones identified only by CGGA are in *italics*, and the only group identified also by IGA is underlined.

Table 1. Over-expressed FEGs obtained by CGGA with a *p-value* = 6.88E-04

Functionally Enriched GO Group	*n* genes	*x* over-exp. genes	*pc-value*
proton-transporting ATP synthase complex	2	2	4.38E-06
invasive growth (sensu Saccharomyces)	5	3	6.13E-06
signal transduction filamentous growth	2	2	8.77E-06
respiratory chain complex II	4	4	3.75E-05
succinate dehydrogenase activity	4	4	3.75E-05
mitochondrial electron transport	4	4	3.75E-05
aerobic respiration	36	10	3.30E-05
tricarboxylic acid cycle	14	5	5.09E-05
tricarboxylic acid cycle	14	5	6.54E-05
gluconeogenesis	12	2	9.64E-05
response to oxidative stress	10	3	1.55E-06
filamentous growth	8	4	9.06E-05
<u>*vacuolar protein catabolism*</u>	4	2	2.63E-05
respiratory chain complex IV	8	2	4.05E-04
cytochrome-c oxidase activity	8	2	4.05E-04

In the case of over-expressed genes (Table 1), CGGA found seven of the nine groups obtained manually by DeRisi [4]. The two annotated groups "glycogen metabolism" and "glycogen synthase" have not been identified by CGGA because they are expressed only at the initial phase of the process. However CGGA identified eight other statistically significant and coherent groups. Only one of these eight other groups has also been identified by IGA and none of them by DeRisi. Similar results, available at CGGA web page, were obtained for the under-expressed FEGs.

5 Conclusion

The CGGA algorithm presented in this article makes it possible to automatically identify groups of significantly co-expressed and functionally enriched genes without any prior knowledge of the expected outcome. CGGA can be used as a fast and efficient tool for exploiting every source of biological annotation and different measure of gene variability.

In contrast to sequential approaches such as [5]-[7], [11], and [15], CGGA analyze all the possible subsets of each FEG and does not depend on the availability of fixed lists of expressed genes. Thus, it can be used to increase the sensitivity of gene detection, especially when dealing with very noisy datasets. CGGA can even produce statistically significant results without any experimental replication. It does not need that all genes in a significant and co-expressed group change, so it is therefore robust against imperfect class assignments, which can be derived from public sources (wrong annotations in ontologies) or automated processes (spelling or naming errors).

The automated functional annotation provided by our algorithm reduces the complexity of microarray analysis results and enables the integration of different sources of genomic information such as ontologies.

CGGA can be used as a tool for platform-independent validation of a microarray experiment and its comparison with the huge number of existing experimental databases and the documentation databases. Results show the interest of our approach and make it possible to identify relevant information on the analyzed biological processes.

In order to identify heterogeneous groups of genes expressed only in certain phases of the process, we plan to integrate the information concerning the metabolic pathway ontologies for future work.

References

1. Attwood T. and Miller C.J.: Which craft is best in bioinformatics? Computer Chemistry, Vol. 25. (2001) 329-339.
2. Breitling R., Amtmann A., Herzyk P.: IGA: A simple tool to enhance sensitivity and facilitate interpretation of microarray experiments. BMC Bioinformatics, Vol. 5. (2004) 34.
3. Chuaqui R.: Post-analysis follow-up and validation of microarray experiments. Nature Genetics, Vol. 32. (2002) 509-514.
4. DeRisi J., Iyer L. and Brown V.: Exploring the metabolic and genetic control of gene expression on a genomic scale. Science, Vol. 278. (1997) 680-686.
5. Draghici S. et al.: Global functional profiling of gene expression. Genomics. (2003). 81:1-7.
6. Gibbons D., Roth F., et al.: Judging the quality of gene expression-Based Clustering Methods Using Gene Annotation. Genome Research, Vol. 12. (2002)1574-1581.
7. Hosack D., Dennis G., et al.: Identifying biological themes within lists of genes with EASE. Genome Biology, Vol. 4. (2003) R70.
8. Kim S., Volsky D. et al.: PAGE: Parametric Analysis of Gene Set Enrichment. BMC Bioinformatics, Vol. 6. (2005) 144.
9. Masys D., et al.: Use of keyword hierarchies to interpret gene expressions patterns. BMC Bioinformatics, Vol. 17. (2001) 319-326.
10. Mootha V., et al.: PGC-1 α-reponsive genes involved in oxidative phosphorylation are coordinately downregulated in human diabetes. Nature Genetics, Vol. 34(3). (2003) 267-273.
11. Pasquier C., Girardot F., Jevardat K., Christen R.: THEA : Ontology-driven analysis of microarray data. Bioinformatics, Vol. 20(16). (2004).
12. Quackenbush J.: Microarray data normalization and transformation. Nature Genetics, Vol. 32 (suppl.). (2002) 496-501.
13. Riva A., Carpentier A., Torresani B., Henaut A.: Comments on selected fundamental aspects of microarray analysis. Computational Bio. and Chem., Vol. 29. (2005) 319-336.
14. Robinson M., et al.: FunSpec: a web based cluster interpreter for yeast. BMC Bioinformatics, Vol. 3. (2002) 35.
15. Sung G., Jung U., Yang K.: A graph theoretic modeling on GO space for biological interpretation of gene clusters. BMC Bioinformatics, Vol. 3. (2004) 381-386.
16. Tusher V., Tibshirani R., Chu G., et al.: Significance analysis of microarrays applied to the ionizing radiation response. Proc. Nat. Acad. Sci. USA, Vol. 98 (9). (2001) 5116-21.
17. Martinez R., et al.: CGGA: An automatic tool for the interpretation of gene expression experiments. Accepted (to appear) on the Journal of Integrative Bioinformatics. 2006.

Symmetric Item Set Mining
Based on Zero-Suppressed BDDs

Shin-ichi Minato

Graduate School of Information Science and Technology,
Hokkaido University
Sapporo, 060-0814 Japan

Abstract. In this paper, we propose a method for discovering hidden information from large-scale item set data based on the symmetry of items. Symmetry is a fundamental concept in the theory of Boolean functions, and there have been developed fast symmetry checking methods based on BDDs (Binary Decision Diagrams). Here we discuss the property of symmetric items in data mining problems, and describe an efficient algorithm based on ZBDDs (Zero-suppressed BDDs). The experimental results show that our ZBDD-based symmetry checking method is efficiently applicable to the practical size of benchmark databases.

1 Introduction

Frequent item set mining is one of the fundamental techniques for knowledge discovery. Since the introduction by Agrawal et al.[1], the frequent item set mining and association rule analysis have been received much attentions from many researchers, and a number of papers have been published about the new algorithms or improvements for solving such mining problems[3].

After generating frequent item set data, we sometimes faced with the problem that the results of item sets are too large and complicated to retrieve useful information. Therefore, it is important for practical data mining to extract the key structures from the item set data. Closed/maximal item set mining[10] is one of the useful method to find important item sets. Disjoint decomposition of item set data[8] is another powerful method for extracting hidden structures from frequent item sets.

In this paper, we propose one more interesting method for finding hidden structure from large-scale item set data. Our method is based on the symmetry of items. It means that the exchange of a pair of symmetric items has completely no effect for the database information. This is a very strict property and it will be a useful association rule for the database analysis.

The symmetry of variables is a fundamental concept in the theory of Boolean functions, and the method of symmetry checking has been studied for long time in VLSI logic design area. There are some state-of-the-art algorithms[9,5] using BDDs (Binary Decision Diagrams)[2] to solve such a problem. The BDD-based techniques can be applied to data mining area. Recently, we found that ZBDDs

N. Lavrač, L. Todorovski, and K.P. Jantke (Eds.): DS 2006, LNAI 4265, pp. 321–326, 2006.

(Zero-suppressed BDDs)[6] are very suitable for representing large-scale item set data used in transaction database analysis[7].

In this paper, we discuss the property of symmetric items in transaction database, and then present an efficient algorithm to find all symmetric item sets using ZBDDs. We also show some experimental results for conventional benchmark data. Our method will be useful for extracting hidden information from a given database.

2   Symmetry of Variables in Boolean Functions

The symmetry is a fundamental concept in the theory of Boolean functions. A symmetric Boolean function means that any exchange of input variables has no effect for the output value. In other words, the output value is decided only by the total number of true assignments in the n-input variables. The parity check functions and threshold functions are typical examples of symmetric functions.

When the function is not completely symmetric, we sometimes find partial groups of symmetric variables. If a pair of variables are exchangeable without any output change, we call them symmetric variables in the function. An obvious property holds that if the pairs (a, b) and (a, c) are both symmetric, then any pair of (a, b, c) is symmetric.

As finding symmetric variables leads to compact logic circuits, it has been studied for long time in VLSI logic design area. In order to check the symmetry of the two variables v_1 and v_2 in the function F, we first extract four sub-functions: F_{00}, F_{01}, F_{10}, and F_{11} by assigning all combinations of constant values 0/1 into v_1 and v_2, and then compare of F_{01} and F_{10}. If the two are equivalent, we can see the two variables are symmetric. In principle, we need $n(n-1)/2$ times of symmetry checks for all possible variable pairs. There have been proposed some state-of-the-art algorithms[9,5] using BDDs (Binary Decision Diagrams)[2] to solve such a problem efficiently.

3   Symmetric Item Sets in Transaction Databases

3.1   Combinatorial Item Sets and Boolean Functions

A combinatorial item set consists of the elements each of which is a combination of a number of items. There are 2^n combinations chosen from n items, so we have 2^{2^n} variations of combinatorial item sets. For example, for a domain of five items a, b, c, d, and e, we can show examples of combinatorial item sets as: $\{ab, e\}, \{abc, cde, bd, acde, e\}, \{1, cd\}, 0$. Here "1" denotes a combination of null items, and 0 means an empty set. Combinatorial item sets are one of the basic data structure for various problems in computer science, including data mining.

A combinatorial item set can be mapped into Boolean space of n input variables. For example, Fig. 1 shows a truth table of Boolean function: $F = (a\, b\, \bar{c}) \vee (\bar{b}\, c)$, but also represents a combinatorial item set $S = \{ab, ac, c\}$. Using BDDs for the corresponding Boolean functions, we can implicitly represent

a	b	c	F
0	0	0	0
1	0	0	0
0	1	0	0
1	1	0	1	→ ab
0	0	1	1	→ c
1	0	1	1	→ ac
0	1	1	0
1	1	1	0

As a Boolean function:
  F(a,b,c) = (a b ~c) V (~b c)
As a combinatorial item set:
  S(a,b,c) = {ab, ac, c}

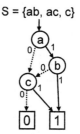

S = {ab, ac, c}

Fig. 1. A Boolean function and a combinatorial item set

Fig. 2. An example of ZBDD

and manipulate combinatorial item set. In addition, we can enjoy more efficient manipulation using "Zero-suppressed BDDs" (ZBDD)[6], which are special type of BDDs optimized for handling combinatorial item sets. An example of ZBDD is shown in Fig. 2.

The detailed techniques of ZBDD manipulation are described in the articles[6]. A typical ZBDD package supports cofactoring operations to traverse 0-edge or 1-edge, and binary operations between two combinatorial item sets, such as union, intersection, and difference. The computation time for each operation is almost linear to the number of ZBDD nodes related to the operation.

3.2 Symmetric Items in Combinatorial Item Sets

Here we discuss the symmetry of items in a combinatorial item set. For example we consider the following combinatorial item set: $S = \{abc, acd, ad, bcd, bd, c, cd\}$. In this case, the item a and b are symmetric but the other pairs of variables are not symmetric. The symmetry can be confirmed as follows. First we classify the combinations into four categories: (1) both a and b included, (2) only a is included, (3) only b is included, and (4) neither included. Namely, it can be written as: $S = abS_{11} \cup aS_{10} \cup bS_{01} \cup S_{00}$. Then, we can determine the symmetry of a and b by comparing S_{10} and S_{01}. If the two subsets are equivalent, a and b are exchangeable. For the above example, $S_{11} = \{c\}$, $S_{10} = \{cd, d\}$, $S_{01} = \{cd, d\}$, and $S_{00} = \{c, cd\}$. We can see a and b are symmetric as $S_{10} = S_{01}$.

Even if we do not know the actual meaning of the item a and b in the original database, we can expect that a and b would have somehow strong relationship if the symmetric property holds. It is a kind of hidden information. It would be a useful and interesting task to find all possible symmetric item sets from the given databases. This method can be used not only for original database but also for frequent item set data to find some relationships between the items.

3.3 ZBDD-Based Algorithm for Finding Symmetric Item Sets

As shown in article[7], the ZBDD-based data structure is quite effective (exponentially in extreme cases) for handling transaction databases, especially when

```
SymChk(S, v₁, v₂) /* Assume v₁ higher than v₂ in the ZBDD ordering.*/
{
    if (S = 0 or S = 1) return 1 ;
    r ← Cache(S, v₁, v₂) ;
    if (r exists) return r ;
    t₁ ← S.top ; /* Top item in S */
    if (t₁ higher than v₁)
        (S₁, S₀) ← (Cofactors of S by t₁) ;
        r ← SymChk(S₁, v₁, v₂) && SymChk(S₀, v₁, v₂) ;
    else
        (S₁, S₀) ← (Cofactors of S by v₁) ;
        t₂ ← Max(S₁.top, S₀.top) ; /* Top item in S₁, S₀ */
        if (t₂ higher than v₂)
            (S₁₁, S₁₀) ← (Cofactors of S₁ by t₂) ;
            (S₀₁, S₀₀) ← (Cofactors of S₀ by t₂) ;
            r ← SymChk((t₂S₁₁ ∪ S₁₀), t₂, v₂) && SymChk((t₂S₀₁ ∪ S₀₀), t₂, v₂) ;
        else
            (S₁₁, S₁₀) ← (Cofactors of S₁ by v₂) ;
            (S₀₁, S₀₀) ← (Cofactors of S₀ by v₂) ;
            r ← (S₁₀ = S₀₁)? 1 : 0 ;
        endif
    endif
    Cache(S, v₁, v₂) ← r ;
    return r ;
}
```

Fig. 3. Sketch of the symmetry checking algorithm

the item sets include many similar partial combinations. Now we show an efficient algorithm of finding symmetric item sets based on ZBDD operations.

First we explain the *cofactor* operation on ZBDDs. $\text{Cofactor}(S, v)$ classifies a combinatorial item set S into the two subsets, one of which includes the item v and the other does not. Namely, it extracts S_1 and S_0 such that $S = vS_1 \cup S_0$. If the item v is the top (highest ordered) item in the ZBDD, then S_1 and S_0 are the two sub-graphs pointed by 1-edge and 0-edge of the top decision node, and the cofactor operation can be done in a constant time. Therefore, if the item v_1 and v_2 are the first and second top items in the ZBDD, the symmetry checking is quite easy because S_{10} (subset with v_1 but not v_2) and S_{01} (subset with v_2 but not v_1) can be extracted and compared in a constant time.

When v_1 and v_2 are not in the highest oreder in the ZBDD, we may use recursive expansion for the symmetry checking. We get the top item t in the ZBDD S, and extract S_1 and S_0 as the cofactors of S by t. We then recursively check the symmetry of (v_1, v_2) for each subset S_1 and S_0, and if they are symmetric for the both, we can see they are symmetric for S.

This procedure may require an exponential number of recursive calls in terms of the number of items higher than v_1, v_2 in the ZBDD, however, we do not have to execute the procedure twice for the same ZBDD node because the results will be the same. Therefore, the number of recursive calls is bounded by the ZBDD size, by using a hash-based cache to save the result of procedure for each ZBDD node. In addition, if we found the two items are asymmetric either for S_1 or S_0, we may immediately quit the procedure and conclude they are asymmetric for S. The detailed algorithm is shown with a pseudo code in Fig. 3. Repeating this procedure for all item pairs in S, we can extract all possible symmetric item sets.

Table 1. Experimental result

Data name	#Item	#Record	#Tuple	ZBDD nodes	Time(sec) for ZBDD gen.	Sym. pairs	Time(sec) for sym.chk.
mushroom	119	8,124	8,124	8,006	1.1	19	0.6
T10I4D100K	870	100,000	89,135	547,777	59.2	0	61.7
pumsb	2,113	49,046	48,474	1,749,775	166.7	90	1,152.0
BMS-Web-View-1	497	59,602	18,473	42,629	24.9	6	30.2
accidents	468	340,183	339,898	3,876,468	127.5	11	18.0

Minimum support = 500: (Total patterns: 1,442,504,   ZBDD nodes: 4,011)

(x32 x7 x31) (x60 x64) x119 x48 x102 x91 x58 x80 x101 x95 x66 x61 x29 x17 x78
x68 x69 x77 x45 x117 x116 x56 x6 x111 x11 x44 x110 x43 x42 x94 x53 x37 x28
x24 x16 x10 x41 x15 x114 x99 x55 x39 x14 x2 x107 x98 x93 x90 x86 x85 x76 x67
x63 x59 x54 x52 x38 x36 x34 x23 x13 x9 x3 x1

Minimum support = 200: (Total patterns: 18,094,822,   ZBDD nodes: 12,340)

(x80 x71 x79 x70) (x32 x31) (x78 x68) (x27 x26) x35 x7 x119 x48 x112 x102 x91
x95 x66 x61 x29 x17 x46 x69 x77 x45 x60 x117 x65 x116 x56 x6 x111 x64 x11
x58 x101 x44 x110 x43 x42 x109 x94 x53 x37 x28 x24 x16 x10 x115 x41 x15 x4
x114 x108 x99 x55 x39 x14 x2 x113 x107 x98 x93 x90 x86 x85 x76 x67 x63 x59
x54 x52 x40 x38 x36 x34 x25 x23 x13 x9 x3 x1

Fig. 4. Results for the all frequent patterns in "mushroom" data

The total computation time is $O(n^2|G|)$, where n is number of items and $|G|$ is the ZBDD size for S. The time will be shorter in practice, when the most of item pairs are asymmetric, and/or the hash-based cache hits very well in repeating the procedure.

4   Experimental Results

We implemented our symmetric checking algorithm. The program is based on our own ZBDD package, and additional 70 lines of C++ code for the symmetry checking algorithm. We used a Pentium-4 PC, 800MHz, 1GB of main memory, with SuSE Linux 9. We can manipulate up to 20,000,000 nodes of ZBDDs in this PC.

For evaluating the performance, we applied our method to the practical databases chosen from FIMI2003 benchmark set[4]. We first constructed a ZBDD for the set of all tuples in the database, and then apply our symmetry checking algorithm for the ZBDD. The results are shown in Table 1. "Sym. pairs" shows the number of symmetric pairs we found. Our result demonstrates that we succeeded in extracting all symmetric item sets for a practical size of databases within a feasible computation time. We can see that no symmetric pairs are found in "T10I4D100K." It is reasonable because this data is randomly generated and there is no strong relationship between any pair of items.

We also conducted another experiment of symmetry checking for the set of frequent patterns in "mushroom." Figure 4 shows the results with the minimum support = 500 and 200. While the data includes a huge number of patterns, ZBDDs are in a feasible size, and our method quickly found the symmetric item groups (shown by parentheses) in a few seconds of computation time.

5   Conclusion

In this paper, we presented an efficient method for extracting all symmetric item sets in transaction databases. The experimental results show that our method is very powerful and will be useful for discovering hidden interesting information in the given data. Now we are going to apply our method to actual real-life data mining problems.

As our future work, we are considering more efficient algorithm to be applied for more larger ZBDDs, and it would also be interesting to develop "approximately" symmetry checking method which allows some errors or noise in the data.

References

1. R. Agrawal, T. Imielinski, and A. N. Swami, Mining Association rules between sets of items in large databases, In P. Buneman and S. Jajodia, edtors, *Proc. of the 1993 ACM SIGMOD International Conference on Management of Data*, Vol. 22(2) of SIGMOD Record, pp. 207–216, ACM Press, 1993.
2. Bryant, R. E., Graph-based algorithms for Boolean function manipulation, IEEE Trans. Comput., C-35, 8 (1986), 677–691.
3. B. Goethals, "Survey on Frequent Pattern Mining", Manuscript, 2003. http://www.cs.helsinki.fi/ u/goethals/publications/survey.ps
4. B. Goethals, M. Javeed Zaki (Eds.), Frequent Itemset Mining Dataset Repository, Frequent Itemset Mining Implementations (FIMI'03), 2003. http://fimi.cs.helsinki.fi/data/
5. N. Kettle and A. King: "An Anytime Symmetry Detection Algorithm for ROB-DDs," In Proc. IEEE/ACM 11th Asia and South Pacific Design Automation Conference (ASPDAC-2006), pp. 243–248, Jan. 2006.
6. Minato, S., Zero-suppressed BDDs for set manipulation in combinatorial problems, In Proc. 30th ACM/IEEE Design Automation Conf. (DAC-93), (1993), 272–277.
7. S. Minato and H. Arimura: "Efficient Combinatorial Item Set Analysis Based on Zero-Suppressed BDDs", IEEE/IEICE/IPSJ International Workshop on Challenges in Web Information Retrieval and Integration (WIRI-2005), pp. 3–10, Apr., 2005.
8. S. Minato: "Finding Simple Disjoint Decompositions in Frequent Itemset Data Using Zero-suppressed BDD," In Proc. of IEEE ICDM 2005 workshop on Computational Intelligence in Data Mining, pp. 3-11, ISBN-0-9738918-5-8, Nov. 2005.
9. A. Mishchenko, "Fast Computation of Symmetries in Boolean Functions," IEEE Trans. Computer-Aided Design, Vol. 22, No. 11, pp. 1588–1593, 2003.
10. T. Uno, T. Asai, Y. Uchida and H. Arimura, "An Efficient Algorithm for Enumerating Closed Patterns in Transaction Databases," In Proc. of the 8th International Conference on Discovery Science 2004 (DS-2004), 2004.

Mathematical Models of Category-Based Induction

Mizuho Mishima[1] and Makoto Kikuchi[2]

[1] Department of Information and Media Science,
Kobe University, Kobe 657-8501, Japan
mizuho@kurt.scitec.kobe-u.ac.jp
[2] Department of Computer and Systems Engineering,
Kobe University, Kobe 657-8501, Japan
mkikuchi@kobe-u.ac.jp
http://kurt.scitec.kobe-u.ac.jp/~kikuchi/

Abstract. Category-based induction is a kind of inductive reasoning in which the premise and the conclusion of the argument is in the form *all the member of a category have the property*. Rips and Osherson et al. investigated the argument strength of category-based induction, and Lopez et al. showed that there are differences of the acceptability of category-based induction between infants and growing-ups. There are two problems in their analysis. One is the ambiguity of the difference between categories and individuals, and the other is the reason of the changes of the acceptability in developmental process of logical inference. In this paper we give mathematical models category-based induction and, based on the models, propose a hypothesis which explains the reason of the problems.

1 Introduction

Induction is an inference from the observed facts that *several samples from a certain set has the property* P to the conclusion that *every element of the set has the property* P. The elements of the set are usually regarded as *individuals*. We call induction in this style as *individual-based induction* (IBI). The elements of the set are taken from some categories in some cases of inductive arguments, and such induction is called *category-based induction* (CBI). The concept of CBI was introduced by Rips [5] and investigated closely by Osherson et al [4]. Osherson et al. analyzed the strength of category-based arguments, which includes category-based induction. They classified category-based arguments according to the factors which affect inferences. They obtained 13 types, and 4 types correspond to induction. Lopez et al. [2] showed that adults can understand every 4 kinds, but infants cannot understand and do not manifest one of them.

There are two problems in these arguments. One is the difference between individuals and categories. The concepts of an individual and a category are quite different, but it is unclear where the characteristics of category-based arguments are used in Oshersons' discussions. The other is discussions about the changes

N. Lavrač, L. Todorovski, and K.P. Jantke (Eds.): DS 2006, LNAI 4265, pp. 327–331, 2006.

of the accessible types of category-based induction in the developmental process of logical reasoning. Medin [3] and Keil [1] considered that this fact is caused by the imperfect scientific knowledge of children. But we cannot explain the universality of the age of the changes of the acceptability by the differences of scientific knowledge. In this paper, we introduce two naive models of CBI, and propose another mathematical model of CBI. By using this model, we result in a hypothesis that the developmental change comes from the ability of logical operations of infants.

2   Category-Based Model

IBI is formalized as below. When some individuals in a category have a property, it is induced that all elements in the category has the property. For example, when *every dog we know is white*, we conclude that *every dog is white*. This form of an inference can be denoted as follows:

$$\frac{S \subseteq P \quad S \subseteq U}{U \subseteq P}$$

where U is the universal set of objects in a discussion, such as the set of all dog in the example. On the other hand, CBI is a reasoning of the following form: when some categories have a property, the superordinate category also has the property. For example, from the fact that *every lion barks* and *every tiger barks*, we conclude that *every mammal barks*. This argument can be formalized as follow. Suppose S_1 to be category, Lion, S_2 to be category, Tiger, U to be a set of all mammal and P to be a set of animals to bark.

$$\frac{(S_1 \subseteq P)\&(S_2 \subseteq P) \quad (S_1 \subseteq U)\&(S_2 \subseteq U)}{U \subseteq P}$$

In general, CBI is an argument of the following form: Suppose $S_1, S_2, \ldots, S_n (n \geq 1)$ are categories in premise of a CBI. Then, CBI is

$$\frac{(S_1 \subseteq P)\&(S_2 \subseteq P)\& \cdots \&(S_n \subseteq P) \quad (S_1 \subseteq U)\&(S_2 \subseteq U)\& \cdots \&(S_n \subseteq U)}{U \subseteq P}$$

This formulation of CBI will be called the *category-based model* of CBI in this paper.

CBI was proposed by Rips [5], and formalized by Osherson et al. [4]. Many of CBI researches have been based on their studies. Osherson introduced and formalized various sorts of category-based arguments and classified them into 13 types. Among them, 4 types called Premise Typicality, Premise Diversity, Conclusion Specificity, and Premise Monotonicity correspond to induction. In these types, categories in premise are basic level and the category in conclusion is superordinate, where these notions are introduced by Rosch [7]. Premise Typicality means that the more typical premise categories are, the stronger the argument is thought. Premise Diversity is a phenomenon that the more dissimilar premise categories are, the stronger the argument is seemed. Conclusion Specificity phenomenon is that the more conclusion category is constrained and concrete, the stronger the argument is felt. Premise Monotonicity phenomenon is that the more similar premise categories are given, the stronger the argument is regarded. Osherson et al. defined the argument strength by using the similarity degrees between categories in premise and elements of the category in conclusion, and the similarity degrees are obtained by psychological experiences. The arguments strength can be obtained also from experiences directly.

3   Individual-Based Model

Category-based model of CBI defined in the previous section is natural, but it does not capture some properties of CBI. In fact, the model does not corresponds to arguments by Osherson et al [4]. That is, what is important their arguments is the relationship between categories which are represented by two variables SIM and COV. These variables do not related the fact that a category is a collection of objects, and the elements of the premise categories are not referred at all. In this sense, CBI in their discussion should be regarded as IBI rather than a category-based argument. Hence we propose another model, which is suitable for Osherson's argument.

Let T_0 be a set of objects which can be observed directly. All elements of T_0 satisfing a property is the subset of T_0. T_1 is a set of subsets of T_0, that is, a subset of the power set of T_0. Similarly, a subset of the power set of T_1 is T_2. An element included in T_i is called a concept of i-rank and Let $T = (T_0, T_1, T_2)$. T is the structure of IBI. The higher the rank of a concept is, the more highly abstracted the concept is. In this structure T, IBI is described as follows. The universal set U in the conclusion is an element of T_1, the set of samples S in premise is a subset of U, and P is a predicate on $T_0 \cup T_1 \cup T_2$. Then, the premise of an induction is *all element in S is P*, and the conclusion is *all element in U is P*. In this structure, CBI is very similar to IBI. That is, the universal set U in the conclusion is an element of T_2, the set of samples S is a subset of U, and P is a predicate on $T_0 \cup T_1 \cup T_2$. The rank of the elements in U express the difference between IBI and CBI, it is 0 for IBI and 1 for CBI, and they have the same form.

In the structure T, elements of T_1 have two aspects of an individual and a category. That is to say, when one refers to the relation between T_0 and T_1, an element of T_1 is a *category* on T_0. When one thinks about the relation between T_1 and T_2, an element of T_1 is an *individual* for an element of T_2. When we consider CBI as inferences from knowledge on T_1 to knowledge on T_2, this inference must depend on the construction of the knowledge on T_1. But this knowledge on T_1 is thought as a result of some inference from knowledge on T_0, like induction from knowledge on T_0. We call this inference *pre-inference* and we call the CBI in this model *post-inference*. In the pre-inference, the premise categories function as category, and in the post-inference, the premise categories work as individual. In this way, the two aspects of premise category must be incorporated in whole inference system including pre-inference and post-inference. For this dual aspects, it is not easy to connect pre-inference and post-inference without using the concept of *reification* of a category to an individual.

4   Extended Category-Based Model

We shall propose an alternative mathematical model of CBI, called an *extended category-based model*, which is a unification of the former two models. This model contains two inferences which correspond to the pre-inference and the post-inference in the last model, although the correspondence is not exact since the intermediate proposition, a proposition about T_1 in the last model, is not the same. This model has a merit that we can connect the two inferences formally and naturally without using the concept of reification.

We started by assuming that we have the universal set U in the category-based model. Similarly, we assume U in an extended category-based model. Since we only assume the set U and everything in the inferences are categories, this model is a generalization of a category-based model. There are two steps of inferences about CBI in this model. The first-step is inferences from $(S_1 \subseteq P)\&(S_2 \subseteq P)\&\cdots\&(S_n \subseteq P)$ to $S \subseteq P$, and from $(S_1 \subseteq U)\&(S_2 \subseteq U)\&\cdots\&(S_n \subseteq U)$ to $S \subseteq U$. That is, the premise categories are combined as a single category by a set operation $S = S_1 \cup S_2 \cup \cdots \cup S_n$. This new category plays the role of the premise category in the second-step. In this step, we infer from two assumptions that $S \subseteq P$ and $S \subseteq U$ to $U \subseteq P$. We remark that this second-step is an IBI. Then, CBI is an inference which is a union of these two steps of inferences.

$$\frac{\dfrac{(S_1 \subseteq P)\&(S_2 \subseteq P)\&\cdots\&(S_n \subseteq P)}{S \subseteq P} \qquad \dfrac{(S_1 \subseteq U)\&(S_2 \subseteq U)\&\cdots\&(S_n \subseteq U)}{S \subseteq U}}{U \subseteq P}$$

The first-step of inferences of this model corresponds to the pre-inference in the individual-based model since the premises of the inference are assertions that some of the individuals satisfy the property in the argument, and the second-step corresponds to the post-inference, i.e. CBI itself, in the individual-based model since the conclusion is the same. Since the intermediate assertions, conclusion of the first-step inferences and the conclusion of the pre-inference, are different, these inferences do not correspond exactly. However, the set S plays the role of reification in the individual-based model. That is, it connects the two steps of inferences and we do not use reification in this model.

In a practical case, the construction of the set S is not easy, and we propose the assumption that this difficulty causes the difference of the understanding of the types of induction according to age.

Main thesis: The difficulty of category-based induction corresponds to the difficulty of construction of the structure of logical operations.

Lopez et al. [2] showed adults could understand every 4 types of induction, but infants could understand only two phenomenon. Medin [3] and Keil [1] considered that this fact was caused by the imperfect scientific knowledge of children. We think that this fact is not related to the scientific knowledge, but it is related to the difficulty of the construction of the set S. In the three types which can be understood by children, we do not need to construct the set S. For example, in the type of Premise Typicality, the premise category can be equal to the set S and we do not need to the set theoretic operation.

5 Conclusion

An object has two aspects: a category, and an individual. Both aspects are implicitly referred in category-based arguments. We gave a mathematical model of CBI by using notions of set theory. In this model, the two aspects can be treated simultaneously. Based on this model, we proposed a thesis with which some phenomenon about CBI can be explained. This correctness of thesis should be verified by psychological experiments, and this is one of the further problem of our research.

Whether an object is a category or an individual depends on the viewpoint of an observer, and the difference between a category and an individual comes from the difference of the degree of abstraction. An object can be seen as a category if there are other objects which belong to the object, and it can be seen as an individual if there is a category to which the object belongs. Our models partly express this phenomenon of the duality of an object. We need to expand our model to investigate the dual nature, and such investigations lead to further understanding of our category-based arguments and categorization.

References

1. Keil, F. Concepts, Kinds, and cognitive Development, Cambridge University Press, 1989.
2. Lopez, A., Gelman, S. A., Gutheil, G., Smith, E. E.: The development of category-based induction. Child Development, **63**, (1992) 1070–1090.
3. Medin, D L., Wattenmaker, W. D., Hampson, S. E. Family Resemblance, conceptual cohesiveness, and category construction Cognitive Psychology, **19**, (1987) 242–279.
4. Osherson, D., Smith, E. E., Wilkie, O., Lopez, A., Sharfir, E. Category-based induction. Psychological Review, **97** (2), (1990) 185–200.
5. Rips, L. Inductive Judgments about natural categories. Joural of Verbal Learning and Verbal Behavior, **14**, (1975) 665–681.
6. Ross B. H., Gelman, S. A., Rosengren K. S. Children's category-based inferences affect classification British Journal of Developmental Psychology, **23**, (2005) 1–24.
7. Rosch, E. Pronciples of categorization. in: Cognition and categorization (E. Rosch and B.L. Barbara ed.), Lawrence Erlbaum Associates, 1978, 27–48.

Automatic Construction of Static Evaluation Functions for Computer Game Players

Makoto Miwa, Daisaku Yokoyama, and Takashi Chikayama

Graduate School of Frontier Sciences, the University of Tokyo, Chiba, Japan
{miwa, yokoyama, chikayama}@logos.k.u-tokyo.ac.jp
http://www.logos.ic.i.u-tokyo.ac.jp/~miwa/

Abstract. Constructing evaluation functions with high accuracy is one of the critical factors in computer game players. This construction is usually done by hand, and deep knowledge of the game and much time to tune them are needed for the construction. To avoid these difficulties, automatic construction of the functions is useful. In this paper, we propose a new method to generate features for evaluation functions automatically based on game records. Evaluation features are built on simple features based on their frequency and mutual information. As an evaluation, we constructed evaluation functions for mate problems in shogi. The evaluation function automatically generated with several thousand evaluation features showed the accuracy of 74% in classifying positions into mate and non-mate.

1 Introduction

Static evaluation functions are one of the most important components of game-playing programs. They evaluate a position and assign it a heuristic value which indicates some measure of advantage. Most common static evaluation functions are expressed as a linear combination of evaluation features and output a scalar value called an evaluation value.

Developers usually construct and tune the static evaluation functions of difficult games, such as chess, shogi, and go by hand. They extract important evaluation features from a position and assign weights to the features. Expert knowledge of the target game and considerable efforts are required to extract appropriate features and assign weights to them.

Automatic construction of static evaluation functions is one approach to cut the cost of developers and avoid the local optimum problem. Using game records, several studies have been made on automatic construction of the static evaluation functions of such games as backgammon [1] and Othello [2,3]. These studies have made great successes on these games, but they have not been applied to more complicated games.

In this paper, we show a new method of automatic construction of static evaluation functions based on game records. This method treats two-class problems which include *win* or *lose* and *mated* or *not mated* and is widely applicable to many types of games.

N. Lavrač, L. Todorovski, and K.P. Jantke (Eds.): DS 2006, LNAI 4265, pp. 332–336, 2006.
© Springer-Verlag Berlin Heidelberg 2006

2   Proposal

Our method constructs static evaluation functions using training positions obtained from game records. We represent the functions as a linear combination of evaluation features. To weight the features, we use a Naïve Bayesian classifier [4] which is one of fast classification algorithms and the output is a continuous-valued probability estimate. Temporal difference [5] or ordinal correlation [6] might have been chosen instead.

2.1   Base Features

To deal with input positions, some expressions are required to represent them. Our method uses base boolean features selected by hand to represent them. It can deal with the continuous valued features by using discretization methods [7] which transform the features into ordered discrete ones. The set of the base features should be chosen so that all possible positions that can appear in the target game uniquely with least redundancy. Atomic features like 'x is on A1' in tic-tac-toe are selected as base features. The base features are not necessarily simple; introducing higher level features, such as 'x has made a line' in tic-tac-toe may reduce the cost to generate evaluation features.

2.2   Generation of Evaluation Features

In our method, an evaluation feature is made by a conjunction of base features called a *pattern*. Other combinations such as disjunctions, NOTs, and arithmetical operations are not considered to reduce the computational cost. Note that disjunctions can be expressed in the top-level linear combination of evaluation features to a certain degree and negations of features can be included in base features. We select evaluation features from all possible patterns based on frequency and conditional mutual information.

We first extract frequent patterns from training positions, because patterns which rarely appear in training positions often cause overfitting. By this, infrequent important patterns may be overlooked, but we can not tell them important without enough occurrences. A frequent pattern is defined as "a pattern which appears in training positions more than α times", where α is called a minimum support. On that selection, we select closed pattern only to eliminate fully dependent patterns. The definition of a closed pattern is defined as "maximal element of patterns which appear in exactly the same training positions". We use LCM (Linear time Closed set Miner) [8] for the extraction.

Next, we use CMIM (Conditional Mutual Information Maximization) [4] to select important features from the closed frequent patterns. CMIM selects a pattern with the largest conditional mutual information with already selected patterns. When there are many closed frequent patterns, k patterns with the largest mutual information are chosen from each of the training positions and the set of patterns used is the union of them. By this, every training position has at least k selected patterns. With large k, preselection in each position will not change the set of selected patterns much.

3   Experimental Settings and Results

We have applied this method to a mate problem in shogi and tried to automatically construct an evaluation function which returns a prediction value how likely the enemy king is mated.

Shogi is a game similar to chess. The goal of its player is to checkmate the opponent's king, as is the same for chess. The most distinctive rule in shogi is that a captured opponent piece can be dropped on an empty square and used as a part of the allies. Mate problem [9] is deciding whether the attacker can check mate the opponent's king by a sequence of checks.

3.1   Experimental Settings

We use 41,224 base features: positions and effects of pieces. "Effects of a piece" are places where the piece can move to. As captured pieces in hand can be placed on any empty squares, "in hand" is a possible position. We express the state of a piece with its kind, position, and the side it belongs to. An effect of a piece is expressed with its side, start point, end point, and the side of the piece on the end point. When the end point is empty, the side is *empty*.

The data set consists of 80,000 training positions and 9,768 test positions. These positions are created from ending positions of 9,144 game records between players who have a rating of over 2,200 in Shogi Club 24 [1]. We label these positions based on the result of the mate search algorithm PDS (Proof-number and Disproof-number Search) [9]. A position is labeled as positive if PDS proves the position mated in searching 5,000,000 nodes. The position is labeled as negative otherwise.

The experiments are conducted with three machines: two 3.06GHz Intel Xeons and 2GB RAM (Xeon), four 1.4GHz AMD Opterons and 4GB RAM (Opteron1.4G), and two 2.4GHz AMD Opterons and 8GB RAM (Opteron2.4G). Programs are written in C++.

3.2   Experiment Results

The number of obtained closed frequent patterns along with the minimum support is displayed in figure 1. With larger minimum support, the number of closed frequent patterns decreased rapidly. Extraction of the patterns was fast and took about 38 minutes for minimum support 2% using Xeon. Figure 2 shows the number of frequent patterns and closed frequent patterns along with the length of the pattern for minimum support 2%. The extraction of closed patterns eliminated 88% of redundant patterns from frequent patterns.

Figure 3 shows the result of evaluation features extracted from 10,000 training positions using minimum support 5%. This figure shows the effect of pre-selection of top k patterns for each training position.

Selection using large k does not affect the final selection of the evaluation features much. Half of the evaluation features are covered even if k is as small

---

[1] http://www.shogidojo.com/

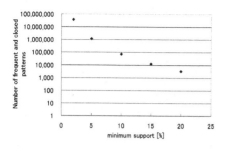

Fig. 1. Number of closed frequent patterns along the minimum support

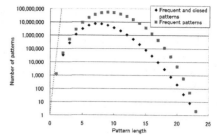

Fig. 2. Number of frequent patterns and closed frequent patterns along with the length of the pattern for minimum support 2%

as one fifth of the number of the evaluation features. We selected evaluation features when k =100, 200, 500, 1,000. It took about 5 days using Opteron2.4G when k was 10,000. With k is 500 or greater, the number of selected patterns were too large to be treated by CMIM. We divided the training set to perform CMIM selection in memory.

We show the classification accuracy of the obtained evaluation functions in figure 4 obtained with 5-fold cross validation. The classification accuracy did not necessarily increase with the number of evaluation features or k and was affected by the division of the training set. The number of evaluation features is needed to be small for fast evaluation, and we can generate a useful evaluation function by selecting an evaluation function with a small number of evaluation features and high accuracy. The highest classification accuracy was 74.0% when k was 200 and the number of evaluation features was 2,000. For comparison, we have also tried other classifiers. This classifier was better than three classifiers using base features in accuracy. 3-layer neural network did not converge and their accuracy varied from 40% to 70%. Naïve Bayesian classified all positions mated. Linear discriminant analysis converged and its accuracy was 68%. We

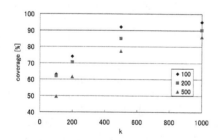

Fig. 3. Coverage of evaluation features using CMIM only by extracted evaluation features from selected top k patterns

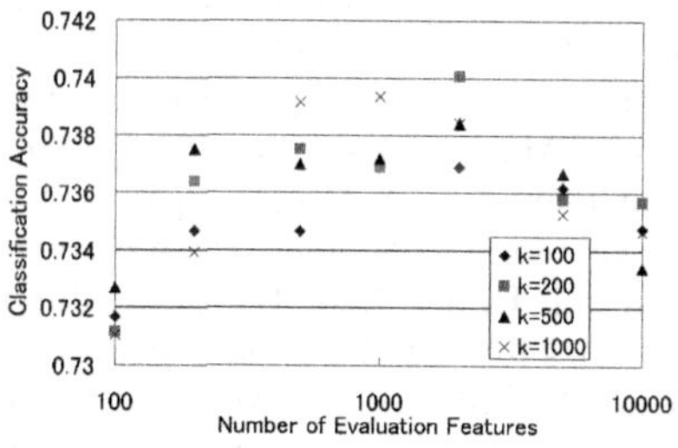

Fig. 4. Classification accuracy of obtained evaluation functions

also tried to use Support Vector Machine to learn from base features, but its learning did not end in a week.

This accuracy is about 7% less than the accuracy of an evaluation function made from evaluation features selected by hand. This is because the minimum support was too large and the evaluation features by hand included features which can not be constructed from base features with only the combination methods used such as the number of pieces and the number of escape ways.

4   Conclusion and Future Work

In this paper, we proposed a new method to construct static evaluation functions automatically using game records. This method constructs evaluation functions from base features by selections based on frequency and conditional mutual information. We applied this method to the mate problem in shogi, and generated evaluation functions which predict the probability of Several thousands of evaluation features were generated from 41,224 base features, and we obtained an evaluation function with 74.0% in classification accuracy.

We plan to apply this method to other games to check its applicability and compare with other methods such as GLEM [2] or Kaneko's methods [3]. We also would like to improve the way of selecting evaluation features. GLEM's *pattern* [2] applied to Othello is useful for this purpose.

References

1. Tesauro, G.: TD-Gammon, a self-teaching backgammon program, achieves master-level play. Neural Comput. **6**(2) (1994) 215–219
2. Buro, M.: From simple features to sophisticated evaluation functions. In van den Herik, H.J., Iida, H., eds.: Proc. of the First International Conference on Computers and Games. Volume 1558., Tsukuba, Japan, Springer-Verlag (1998) 126–145
3. Kaneko, T., Yamaguchi, K., Kawai, S.: Automated Identification of Patterns in Evaluation Functions. In van den Herik, H.J., Iida, H., Heinz, E.A., eds.: Advances in Computer Games 10, Kluwer Academic Publishers (2004) 279–298
4. Fleuret, F.: Fast Binary Feature Selection with Conditional Mutual Information. In: JMLR Vol.5. (2004) 1531–1555
5. Baxter, J., Tridgell, A., Weaver, L.: Learning to Play Chess Using Temporal Differences. Machine Learning **40**(3) (2000) 243–263
6. D. Gomboc, T. A. Marsland, M.B.: Evaluation fuction tuning via ordinal correlation. In van den Herik, Iida, H., ed.: Advances in Computer Games, Kluwer (2003) 1–18
7. Fayyad, U.M., Keki, B.: Multi-Interval Discretization of Continuous-Valued Attributes for Classification Learning. In: IJCAI-93. (1993) 1022–1027
8. Uno, T., Kiyomi, M., Arimura, H.: Lcm ver.3: collaboration of array, bitmap and prefix tree for frequent itemset mining. In: Proc. of the 1st international workshop on open source data mining, New York, NY, USA, ACM Press (2005) 77–86
9. Sakuta, M., Iida, H.: And/or-tree search algorithms in shogi mating search. ICGA Journal **24**(4) (2001) 218–229

Databases Reduction Simultaneously by Ordered Projection

Isabel Nepomuceno[1], Juan A. Nepomuceno[1],
Roberto Ruiz[1], and Jesús S. Aguilar–Ruiz[2]

[1] Department of Computer Science, University of Sevilla, Sevilla, Spain
{isabel, janepo, rruiz}@lsi.us.es
[2] Area of Computer Science, University Pablo de Olavide, Sevilla, Spain
jsagurui@upo.es

Abstract. In this paper, a new algorithm *Database Reduction Simultaneously by Ordered Projections* (RESOP) is introduced. This algorithm reduces databases in two directions: editing examples and feature selection simultaneously. Ordered projections techniques have been used to design RESOP taking advantage of symmetrical ideas for two different task. Experimental results have been made with UCI Repository databases and the performance for the latter application of classification techniques has been satisfactory.

1 Introduction

Nowadays the huge amount of information produced in different disciplines implies that manual analysis of data is not possible. Knowledge discovery in databases (KDD) deal with the problem of reducing and analyzing data with the use of automated analysis techniques. Data mining process is the previous process of extracting trends or patterns from data in order to transform data in useful and understandable information.

Data mining algorithms must work with databases with thousands of attributes and thousands of examples in order to extract trends or pattern from data. Databases preprocessing techniques are used to reduce the number of examples or attributes as a way of decreasing the size of the database with which we are working. There are two different types of preprocessing techniques, **editing**: reduction of the number of examples by eliminating some of them or finding representatives patterns or calculating prototypes and, secondly, **feature selection**: eliminating non-relevant attributes. Today's standard technology motivates more powerful methods which embed two different tasks at the same time.

In this paper, we propose an algorithm to embed horizontal and vertical database reduction simultaneously, that is, editing and feature selection at the same time. In section two the algorithm is presented. Section three shows experimental results carry out with the method proposed. Finally, conclusions are presented.

N. Lavrač, L. Todorovski, and K.P. Jantke (Eds.): DS 2006, LNAI 4265, pp. 337–341, 2006.

2   RESOP Algorithm

RESOP, *Reduction Database Simultaneously by Ordered Projection*, carries out editing of examples and features selection simultaneously using ideas of two algorithms based on ordered projections: EPO, see [1], and SOAP, see [2]. Techniques based on ordered projections use the idea of projecting every example over each attribute in order to create a partition in subsequence. Each subsequence is composed with examples of the same class. The aim is to built a partition of the space of examples in order to evaluate what example can be eliminated and what attribute is more relevant.

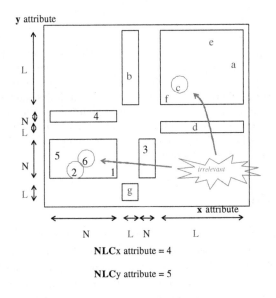

Fig. 1. RESOP algorithm applied over a two-dimensional database. Examples c, 6, 2 are eliminated and x attribute with NLC=4 is better to classify than y attribute with NLC=5.

RESOP idea is illustrated in Figure 1. A database with two attributes, x and y, and examples with two possible labels, letters or numbers, is given. Every example is projected over each axis: the valor over the corresponding attribute for each example is considered. We obtain in x axis the sequence $\{[5, 2, 6, 4, 1], [g, b], [3], [f, c, d, e, a]\}$ with label $\{N, L, N, L\}$. In y axis we obtain the sequence $\{[g], [1, 2, 6, 5, 3], [d], [4], [f, c, b, a, e]\}$. First attribute, x, has 4 label changes (NLC=4) $\{N, L, N, L\}$ and the second one, y, NLC=5, $\{L, N, L, N, L\}$. Firstly for vertical reduction, best attributes in order to classify are attributes with the lowest NLC. The generate ranking of examples is first attribute x and y the second one. Secondly, in order to horizontal reduction, examples which will be eliminated are examples which are not necessary to define the regions (i.e., to define classification rules): 2, 6 and c.

The algorithm has two different parts to handle: continuous and discrete attributes. See algorithm 1. Algorithm 2 is necessary to calculate the Number of Label Changes (NLC) for each attribute. λ parameter is used in order to relax the condition of editing for databases with a huge number of attributes and few examples, see [3].

3   Experimental Methodology

In this section, we present experiments carried out and the methodology used. The behavior of our algorithm (editing and selection of attributes) has been studied in several data sets available from UCI Repository [1], see [4].

Table 1. Comparison of the percentage of correctly classified instances for C4.5, IB1 and NB algorithms over the different medium size databases obtained with RESOP. PR is the percentage of data retention after the reduction. *na* is not available data after reduction process.

Data	Size	λ	PR	IBK Original ER	IBK RESOP ER	J48 Original ER	J48 RESOP ER	NB Original ER	NB RESOP ER
ads	3279 × 1558	1	7.89	90.95	89.15	93.52	93.09	93.86	92.54
		0.95	na	na	na	na	na	na	na
		0.85	na	na	na	na	na	na	na
hypothyroid	3772 × 29	1	16.11	91.07	91.7	99.44	92.29	95.49	92.23
		0.95	9.41	91.07	87.54	99.44	89.63	95.49	89.66
		0.85	na	na	na	na	na	na	na
isolet	1559 × 617	1	50.00	84.22	68.36	74.67	63.43	81.58	76.04
		0.95	50.00	84.22	68.36	74.67	63.43	81.58	76.04
		0.85	50.00	84.22	68.36	74.67	63.43	81.58	76.04
letter	20000 × 16	1	49.18	95.53	86.51	87.54	81.45	64.08	52.87
		0.95	47.73	95.53	86.48	87.54	81.16	64.08	51.73
		0.85	43.34	95.53	86.34	87.54	81.41	64.08	50.21
mushroom	8124 × 22	1	14.47	99.79	95.09	99.8	94.74	0	82.92
		0.95	na	na	na	na	na	na	na
		0.85	na	na	na	na	na	na	na
musk2	6598 × 166	1	46.04	62.77	59.1	72.18	68.5	37.29	74.89
		0.95	15.98	62.77	52.87	72.18	66.82	37.29	72.69
		0.85	1.73	62.77	15.99	72.18	18.46	37.29	21.88
splice-2	3190 × 60	1	44.52	65.71	52.32	91.22	47.08	93.76	55.33
		0.95	na	na	na	na	na	na	na
		0.85	na	na	na	na	na	na	na

Data sets used are partitioned using ten-fold-cross-validation procedure. The algorithm runs for each data set. The original and the reduced data set are used as a training set in a classification algorithm and the percentage of correctly classified instances is measured. C4.5, IB1 and NB are used as classification algorithms. The purpose is to study the relevance of our method and the greater percent of correctly classified instances when the classification method is applied on the reduced data set. The reduction method algorithm is executed taken into account that a ranking of attributes is produced selecting the best 50% of this attributes and removing the remainder.

---

[1] http://www1.ics.uci.edu/ mlearn/MLRepository.html

In order to proof the goodness of our approach, in Table 1 results of the classification using C4.5, IB1 and NB techniques are shown. In this table we modify λ parameter. This parameter allow us to control the level of reduction of examples. The main objective is to compare the performance of our reduction method when the λ parameter adjust the number of instances to delete. The ER, error rate, for the original database and the different reduced databases is presented. ER is the percent of correctly classified instances produced when classifiers algorithms are applied. Finally PR is the percentage of data retention after the reduction. We must consider how the error changes when the database is reduced considerably. Our aim is to keep or increase the ER value after reduction database.

4   Conclusions

In this paper a new technique for reducing databases in two directions simultaneously is presented. On the one hand removing examples (editing examples) or vertical reduction, and on the other hand removing attributes (feature selection) or horizontal reduction. The method is based on using techniques of ordered projection, see [1,2], in order to reduce simultaneously examples and attributes.

Result obtained are satisfactory in order to evaluate the goodness of the proposal. Take into account today's standard technology, the method is very interesting from data mining techniques application point of view.

Future works will focus on making a comparison with order similar methods which edits examples and selects attributes, and studding the behavior of the algorithm with huge databases as for example microarrays.

References

1. Riquelme, José C.; Aguilar-Ruiz, Jesús S.; Toro, Miguel: Finding representative patterns with ordered projections Pattern Recognition 36 (2003), pp. 1009-1018.
2. Ruiz, R.; Riquelme, Jose C.; Aguilar-Ruiz, Jesus S.: NLC: A Measure Based on Projections 14th International Conference on Database and Expert Systems Applications, DEXA 2003Lecture Notes in Computer Science, Springer-VerlagPrague, Czech Republic, 1-5 September, (2003).
3. Jesús S. Aguilar-Ruiz, Juan A. Nepomuceno, Norberto Díaz-Díaz, Isabel A. Nepomuceno-Chamorro: A Measure for Data Set Editing by Ordered Projections. IEA/AIE (2006), pp. 1339-1348.
4. Blake, C.; Merz, E.K.: UCI repository of machine learning databases. (1998).

Algorithm 1 RESOP - database Reduction Simultaneously by Ordered Projections

---

INPUT D: data base
OUTPUT D: data base reduced, k parameter
begin
  for all example $e_i \in D, i \in \{1, ..., n\}$ **do**
    $weakness(e_i) := 0$
  end for
  for all continuous attribute $a_j, j \in \{1, ..., m_1\}$ **do**
    $D_j := QuickSort(D_j, a_j)$ *in*   *incr.*   *order*
    $D_j = ReSort(D_j)$
    for all example $e_i \in E_j, i \in \{1, ..., n\}$ **do**
      if e_i is not border **then**
        $weakness(e_i) := weakness(e_i) + 1$
      end if
      NLC $(a_j) :=$ NumberLabelChanges(D,j)
    end for
  end for
  for all discrete attribute $a_j, j \in \{1, ..., m_2\}$ **do**
    for all value $v_i^j \in a_j$ **do**
      $V := \{e | value(e, a_j) = v_i^j\}$
      Let $\bar{e}$ be an example such that $weakness(\bar{e}) = min_{i \in V}\{weakness(e)\}$
      for all $e_i \in V$ except $\bar{e}$ **do**
        $weakness(e_i) := weakness(e_i) + 1$
      end for
    end for
    NLC $(a_j) :=$ NumberLabelChanges(D,j)
  end for
  for all example $e_i \in D, i \in \{1, ..., n\}$ **do**
    if $weakness(e_i) \geq m \cdot \lambda$ **then**
      remove e_i from D
    end if
  end for
  NLC Attribute Ranking
  Select the k first attributes

Algorithm 2 NLC - Number Label of Ghanges

INPUT D, m: data base, number of attributes
OUTPUT nlc: number of label changes
  if att(u[j],i) $\in$ subsequence of the same value **then**
    $nlc := nlc +$ Change Same Value
  else
    if label(u[j]) $<>$ lastLabel **then**
      $nlc := nlc + 1$
    end if
  end if

Mapping Ontologies in an Air Pollution Monitoring and Control Agent-Based System

Mihaela Oprea

University Petroleum-Gas of Ploiesti, Department of Informatics
Bd. Bucuresti Nr. 39, Ploiesti, 100680, Romania

Abstract. The solution of multi-agent system could be applied for air pollution monitoring and control systems modelling in the context of extending the area of web based applications to environmental systems. As the intelligent agents that compose such a multiagent system need to communicate between them and also with external agents they must share parts of their ontologies or they must identify the correspondent common terms. In this paper, we focus on the topic of ontology mapping in such a multi-agent system.

1 Introduction

The last decade has registered a strong and more concentrated challenge on the improvement of our environment quality under the international research framework of durable and sustainable development of the environment. The main concern of this challenge is to assure a healthy environment that allows the protection of ecosystems, and human health. One of the key aspects of this challenge is the air pollution control in urban regions [1]. The problems raised by air pollution includes the greenhouse effect, acid rain, smog, and the "holes" in the ozone layer. Therefore, in order to solve the current environmental problems more efficient tools have to be developed. Artificial intelligence provides several techniques and technologies that can solve efficiently the environmental problems. As one of the main concerns of making decisions in environmental protection management is the real-time reaction (in both directions, from the environment and to the environment), a multiagent system (MAS) technology could be applied in the near future as a realistic solution. In this context, we are developing such a system for air pollution monitoring and control in urban regions in a first step as a simulation. In this paper, we focus on the topic of ontology mapping, as the intelligent agents that compose the MAS need to communicate with other agents (e.g. meteorological agents) that are external to the system, and thus, do not know all the terms included in their ontologies.

2 *MAS_AirPollution* - An Air Pollution Analysis and Control System

An air pollution monitoring and control system modelled as a MAS could be a feasible solution in case the web-based applications are more extended to the

N. Lavrač, L. Todorovski, and K.P. Jantke (Eds.): DS 2006, LNAI 4265, pp. 342–346, 2006.

environmental systems. In such a solution the agents could monitor different sites and inform the supervisor agents about the problems that occur (e.g. the exceedance of the maximum allowed concentration for a specific air pollutant), so that a real-time decision could be taken and applied. Figure 1 shows the architecture of the system *MAS_AirPollution*, which is under development as a simulation. The goal of the system is to monitor and control air pollution in urban regions.

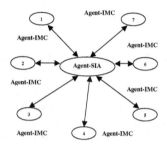

Fig. 1. The architecture of *MAS_AirPollution* system

The system is composed of a set of intelligent monitoring and control agents (Agent_IMC, specific to each site of the monitoring network) and a supervisor intelligent agent (Agent_SIA) in a star like architecture. We have developed an ontology specific to air pollution and control, `AIR_POLLUTION_Onto` [2], that is shared by all agents. In the cases when external agents (e.g. agents of other systems, used to exchange information related to the state of weather or air pollution), which have different ontologies, interact with the agents of *MAS_AirPollution*, it is used an ontology mapping mechanism that will be described in the next section.

3 Ontology Mapping

Suppose we have two ontologies O_1 and O_2. A mapping between the two ontologies is a partial function *map* that finds the maximal number of potential mapping pairs (t_1, t_2), where t_1 and t_2 are terms from the ontologies O_1 and O_2, respectively.

$$map(O_1, O_2) = \{(t_1, t_2) \mid t_1 \in O_1, t_2 \in O_2\}$$

The role of a mapping function is to find for each term (concept, relation, instance) from one ontology the corespondent term from the other ontology, i.e. the term that has the same intended meaning. The discrepancies (i.e. different terms) between two ontologies could appear due to different semantic structures (structural conflicts), different names for the same type of information or the same name for different types of information (type conflicts), and different representations of the same data (data conflicts).

Several ontology mapping methods and tools have been reported in the literature in the last years. A recent review of them is presented in [3]. The general remark is that there is no fully-automated ontology mapping tool. Such a tool is still difficult to be developed, and thus, all existing mapping tools need human validation. Two main classes of methods could be identified: (1) lexicon-based methods ([4]), and (2) structural methods ([3], [5]). The two types of ontology mapping methods, lexicon-based and structural could be combined in hybrid methods. In broker/matchmaker-based systems such as RETSINA/LARKS ([6], [7]), two types of matchings are used, context matching, and syntactical matching. WordNet [8], the most large, general purpose, machine readable and public available thesaurus, is an important tool that could be used when doing synset matching (see e.g. [3]), i.e. exploring the semantic meaning of the words by searching for synonims.

4 *OntoMap* - The Ontology Mapping Mechanism

In order to solve the heterogeneity problem of the ontologies we have designed a mechanism (implemented in the architecture of each agent) that tries to solve the discrepancies during the agents communication process. The ontology mapping method is based on the generalized Levenshtein distance [9], on the synset matching, done through WordNet and through AIR_POLLUTION_Onto (which memorize also the synonyms of the concepts), and on a simple structural mapping mechanism that uses the relations with the direct neighbour concepts, this being the novelty of our proposed method. Figure 2 shows the sketch of the ontology mapping algorithm.

```
Algorithm OntoMap (O1, O2, matched)
    for * each discrepancy (difference) of a term w1 in O1 do {
        if compound_word (w1) then
            generate_word_constituents(w1, Lw1)
        else Lw1={w1}
        string_matching(Lw1, O2, d_Levenshtein)
        if d_Levenshtein <= T_sm then return matched = 1;
        else
        {   synset_matching(Lw1, O2, WordNet, similarity1)
            if similarity1 >= T_synset then
                return matched = 1;
            else
            {   do structural_mapping(Lw1, O2, similarity2)
                if similarity2 <= T_structural then  return matched = 1;  }   }
        return matched = 0; }

    where T_sm is the preset threshold for string matching,
    T_synset is the preset threshold for the synset matching,
    and T_structural is the preset threshold for the structural matching.
```

Fig. 2. A sketch of the ontology mapping algorithm

When different terms appear during agents conversation, then the agent starts the *OntoMap* mechanism. First, if the unknown word is a compound word it will generate its constituents. Example of compound word is *Speed_of_the_Wind*. In general, a compound word is composed by words that are linked with a hyphen, an underscore, a linking word (e.g. *of*, *the*, *a*, *in*) or the words start with a capital letter. After that, it is started a string matching procedure that will try to find similar words that exist in the ontology *O2*. The string matching procedure uses the generalized Levenshtein distance as a measure of similarity between two words. It will return the most similar word. If the word is not recognized it is started the procedure of *synset matching* that will try to find synonyms of the word by using WordNet and the ontology of the system (that memorize also the synonyms of the terms used: concepts, relations, properties) in a similar way as do LOM [4]. If the similarity value is greater than a preset threshold then the word is recognized, otherwise it is applied a *structural mapping* by taking into account the relations betweeen the term and its direct neighbours in the ontology tree. The structural similarity computes the distance between the neighbourhood description vectors of the two terms. If the similarity value is less than a preset treshold, the word is recognized, otherwise only the intervention of a human could solve the discrepancy. Figure 3 shows a sequence from two ontologies, *O1* and *O2*, mapped by Onto_Map.

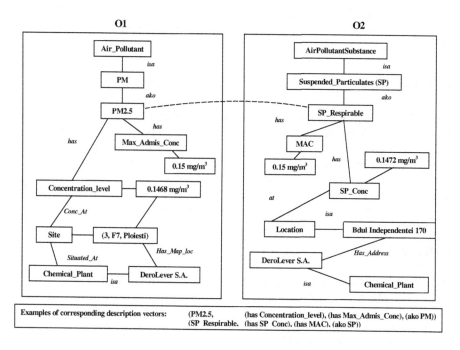

Fig. 3. Parts of the ontologies O1 and O2

5 Experimental Results

We have experimented the ontology mapping mechanism that was included in the multiagent system *MAS_AirPollution* for different ontologies used by the agents, either generic and specific to the task of air pollution monitoring and control, and meteorology forecasting, and/or with different hierarchical structures. The quality of the ontology mapping mechanism was measured by using two metrics: the precision and the recall: *precision* = N_1 / N_2; *recall* = N_1 / N_3, where N_1 is the number of found mappings that are correct, N_2 is the number of found mappings, N_3 is the number of existent mappings. Table 1 summarizes the results obtained so far, compared with LOM [4]. We have used three ontologies with different hierarchical structures: O_1, O_2 and O_3, with comparable dimensions (i.e. about 200 words). The experimental results are good, but they could be improved for example, by extending the structural analysis to more neighbours of a term, not only to the direct linked ones. Also, other similarity measures could be used as given in [10] and [11].

Table 1. Experimental results

Ontologies	Onto_Map		LOM	
	Precision	Recall	Precision	Recall
O_1, O_2	0.91	0.87	0.79	0.65
O_3, O_2	0.96	0.94	0.82	0.74
O_3, O_1	0.98	0.95	0.93	0.87

6   Conclusion

Depending on the application domain and the complexity of the ontologies, two classes of ontology mappings methods could be applied: lexicon-based and/or structural. We have presented a case study of ontology mapping in an air pollution monitoring and control multiagent system that is under development as a simulation. The experimental results showed a good behavior of the ontology mapping mechanism that combines a lexicon-based method with a simple structural analysis of the direct neighbours of the concepts and their relations.

References

[1]   Moussiopoulos, N. (Ed): Air Quality in Cities, Springer, Berlin (2003).
[2]   Oprea, M.: A case study of knowledge modelling in an air pollution control decision support system, AiCommunications, IOS Press, 18(4), (2005), 293-303.
[3]   Kalfoglou, Y., Schorlemmer, M.: Ontology mapping: the state of the art, The Knowledge Engineering Review, 18(1), (2003), 1-31.
[4]   Li, J.: LOM: A Lexicon-based Ontology Mapping Tool, Teknowledge Corporation, research report, 2004.
[5]   Noy, N.F., Musen, M.A.: The PROMPT suite: interactive tools for ontology merging and mapping, International Journal of Human-Computer Studies, 59, (2003), 983-1024.
[6]   Sycara, K.: In-context information management through adaptive collaboration of intelligent agents, in M. Klusch (ed.), Intelligent Information Agents, Springer (1999).
[7]   Sycara, K., Klusch, M., Widoff, S., Lu, J.: LARKS: Dynamic Matchmaking Among Heterogeneous Software Agents in Cyberspace, Journal of Autonomous Agents and Multi-Agent Systems, Kluwer, March 2001.
[8]   Fellbaum, C.: WordNet, An Electronic Lexical Database, MIT Press, (1998).
[9]   Levenshtein, I.V.: Binary codes capable of correcting deletions, insertions, and reversals, Cybernetics and Control Theory, (1966).
[10] Maedche, A., Staab, S.: Measuring Similarity between Ontologies, Proc. of EKAW-2002, Madrid, Spain, vol. 2473 of LNCS/LNAI, Springer (2002).
[11] Su, X., Hakkarainen, S., Brasethvik, T.: Semantic enrichment for improving systems interoperability, Proc. of the ACM Symposium on Applied Computing, (2004), 1634-1641.

Information Theory and Classification Error in Probabilistic Classifiers

Aritz Pérez, Pedro Larrañaga, and Iñaki Inza

Intelligent Systems Group, Computer Science and A.I Dept.
University of The Basque Country, Spain
{aritz, ccplamup, inza}@si.ehu.es

Abstract. This work shows, using bivariate continuous artificial domains, the relation that seems to exist between some measures based on the information theory and the expected classification error.

The relations that seem to be found in this work could be applied to the improvement of the classifiers which assign *a posteriori* probabilities to each class value. They also could be used in other tasks related to the supervised classification such as feature subset selection or discretization.

1 Introduction

Supervised classification is an outstanding task in data mining and pattern recognition. It lies in selecting a class label $c \in \{1, ..., r\}$ of the class variable C given an instantiation $\boldsymbol{x} = (x_1, ..., x_n)$ of the predictor variables $\boldsymbol{X} = (X_1, ..., X_n)$. This work is centered on classifiers which assign a probability to each class c given an instantiation $\boldsymbol{x}$, when the predictors are continuous $\boldsymbol{x} \in \mathbb{R}^n$. In order to measure the inaccuracy of a probabilistic classifier M the expected error of classification ϵ_M, is commonly used. ϵ_M is defined as follows:

$$\epsilon_M = \sum_{c=1}^{r} \int p(c) f(\boldsymbol{x}|c)(1 - p_M(c|\boldsymbol{x})) d\boldsymbol{x}$$

where $p(\cdot)$ and $f(\cdot)$, are respectively, the underlying true probability distribution and density function which model the domain. $p_M(c|\boldsymbol{x})$ is the class distribution conditioned to the predictors modelled by the classifier M.

The classifier which minimizes the classification error is the Bayes classifier and its error is known as the Bayes error. The Bayes classifier uses the underlying *multivariate* true distribution of the class conditioned to the predictors $p_{mul}(c|\boldsymbol{x}) = p(c|\boldsymbol{x}) \propto p(c) f(\boldsymbol{x}|c)$. In this work we call this error multivariate error ϵ_{mul}. On the other hand, we call the error of the classifier which uses each underlying *univariate* true distribution $p_{uni}(c|\boldsymbol{x}) \propto p(c) \prod_{i=1}^{n} p(x_i|c)$ univariate error ϵ_{uni}. Finally, the difference between the univariate and multivariate error is called difference error $\epsilon_{dif} = \epsilon_{uni} - \epsilon_{mul}$. This can be seen as the gain in the error when $p_M(c|\boldsymbol{x})$ is modelled in a multivariate way instead of a univariate one.

The use of the *information theory* (IT) [1] applied to the supervised classification is very extended. For example, the IT is used for the discretization of

N. Lavrač, L. Todorovski, and K.P. Jantke (Eds.): DS 2006, LNAI 4265, pp. 347–351, 2006.

continuous variables [2], variable subset selection [3] or the induction of classifiers [4]. However, there are few works that relate the IT to the classification
error.

The main concepts of the IT are entropy, conditioned entropy and mutual
information for discrete variables, and their differential versions for continuous
one. The definitions are analogous for continuous and discrete variables, changing
integrals by sums.

The *entropy* of a discrete variable C with values $c \in \{1...r\}$ is defined as:

$$H(C) = -\sum_{c=1}^{r} p(c)log_2 p(c)$$

where $p(c)$ is the probability distribution of C. The entropy of C can be interpreted as a measure of its uncertainty or disorder. Given $p(c)$, the error related to
the classifier $p_M(c|\boldsymbol{x}) = p(c)$ monotonically increases with $H(C)$. The differential
entropy of a continuous variable X is defined analogously:

$$h(X) = -\int f(x)log_2 f(x)dx$$

where $f(x)$ is the density function of X. It also can be interpreted in terms of
uncertainty or disorder. Besides, it can be considered a generalization of the variance. Given two continuous random variables X and Y, the differential entropy
of X when the density function of Y is known is defined as:

$$h(X|Y) = -\int \int f(x,y)log_2 f(x|y)dxdy$$

This can be interpreted as the remaining uncertainty or disorder of X when the
density of the variable Y, $f(y)$, is known.

The differential *mutual information* between X and Y is defined as:

$$I(X;Y) = \int \int f(x,y)log_2 \frac{f(x,y)}{f(x)f(y)}dxdy$$

We can rewrite the definition of mutual information as $I(X;Y) = h(X) -
h(X|Y)$. This can be interpreted as the amount of uncertainty or disorder shared
by X and Y. Generalizing, the mutual information common to the three random
variables X, Y and C can be defined as $I(X;Y;C) = I(X;Y) - I(X;Y|C)$ [1].

The discrete and continuous versions of these measures share a large extent
of their properties [1]. Therefore, we hope that most results of this work can be
generalizable to discrete and mixed domains. The relation between the different
IT based measures can be represented using a Venn diagram [1] (see Figure 1).

In the remainder we try to answer the following questions: given the *a pos-
teriori* probability $p_M(c|(x,y))$, what kind of relation exists between the uncer-
tainty that surrounds the class variable and the classification errors ϵ_{mul} and
ϵ_{uni}? Intuitively, the error should increase with the increase in the uncertainty
associated to $p_M(c|(x,y))$. When is more advisable to use $p_{mul}(c|(x,y))$ instead
of $p_{uni}(c|(x,y))$ for classification? It can be hoped that a relevant measure is the
difference between the IT based measures related to ϵ_{uni} and ϵ_{mul}.

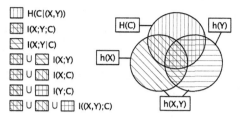

Fig. 1. Relation between different IT based. Each region specifies a part of the uncertainty that surrounds the variables.

2   Finding the Relationship Between the Expected Error and the Mutual Information

This section tries to find monotonically increasing/decreasing relations between different IT based measures and the errors ϵ_{mul}, ϵ_{uni} and ϵ_{dif}, when the predictors are continuous. For this purpose, we randomly generate a set of continuous bivariate artificial domains for computing some IT based measures and the errors introduced.

2.1   Random Generator of Domains

The random generator of domains (RGD) function employed in the experimentation generates, in a random way, a set of artificial bivariate domains (see Figure 2). Then it computes in each domain some measures based on the IT and the exposed errors ϵ_{mul}, ϵ_{uni} and ϵ_{dif}. Each generated domain is specified by the *a priori* distribution of the class $p(c)$ and the density joint function of the predictors (X, Y) given each class value $f(x, y|c)$. In order to model the densities $f(x, y|c)$, we have decided to use the Gaussian kernel based density functions [5] because of its flexibity modeling different density shapes. The used kernel based functions depend on the number of kernels to be used m, on the coordinates of each of those kernels $\{k_1, ..., k_m\}$, being $k_i = (x_i, y_i)$, and on a unique smooth parameter $h^2 = m^{-\frac{1}{6}}$.

2.2   Experiments

In order to study the relation between IT based measures and introduced errors, two sorts of experiments have been performed. The first experiment generates artificial domains with the same distribution $p(c)$ (RGD function with *constant* = true). Therefore, the entropy $H(C)$ is constant across them. Any real-world domain, with n predictors, can be seen as a set of $\binom{n}{2}$ bivariate domains fulfilling the constraint that $H(C)$ is constant. In this sense, we hope that the conclusions of this study can be used in real-world domains. In order to study the robustness of the IT based measures, the second experiment study the

Inputs
  the maximum number of kernels (m_{max}), number of classes (r), number of domains to be generated (d), a boolean which specifies if the distribution $p(c)$ is constant for every domain $constant = \{true, false\}$, and the range of the variables X and Y ($ranX$ and $ranY$).
Outputs
  The errors ϵ_{mul} and ϵ_{uni} and some IT based measures for each domain generated.
Algorithm
  If($constant$) **then** randomly generate $p(c), c \in \{1, ..., r\}$.
  Repeat d **times**:
     If($\overline{constant}$) **then** randomly generate $p(c), c \in \{1, ..., r\}$.
     Set a number of kernel components $m_c \in \{1, ..., m_{max}\}$ for each density $f(x, y|c)$.
     Specify the coordinates $(x_1, y_1), ..., (x_{m_c}, y_{m_c})$ inside the domain $ranX \times ranY$ for each kernel
         component of each density $f(x, y|c)$.
     Compute some IT based measures and the errors ϵ_{mul}, ϵ_{uni} and ϵ_{dif} for the current domain.
  End repeat
  Plot ϵ_{mul}, ϵ_{uni} and ϵ_{dif} versus IT based measures.

Fig. 2. The pseudocode of the RGD function

relation between IT based measures and the error across artificial domains with different randomly generated distributions $p(c)$ (RGD with $constant = $ false). Therefore, the entropy $H(C)$ varies through the artificial domains generated.

Figure 3 has been obtained using the RGD function with parameters $m_{max} = 40$, $r = 4$, $constant = \{true, false\}$, $d = 10000$, $ranX = [0, 15]$ and $ranY = [0, 15]$. The experiment has been performed with different parameter values obtaining similar conclusions.

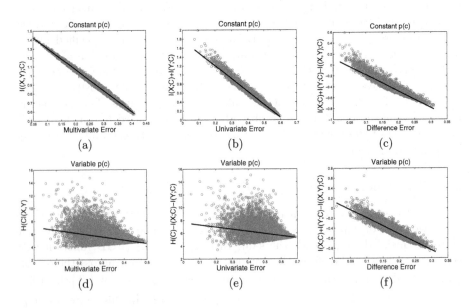

Fig. 3. Plot Errors ϵ_{mul}, ϵ_{uni} and ϵ_{dif} versus IT based measures

Constant $p(c)$. It seems that $I((X,Y);C)$ and $I(X;C)+I(Y;C)$ are inversely proportional to ϵ_{mul} and ϵ_{uni} respectively (see Figures 3(a) and 3(b)). Besides, $I(X;Y;C) = I(X;C)+I(Y;C)-I((X,Y);C)$ seems to be inversely proportional to ϵ_{dif} (see Figure 3(c)). Therefore, given a pair of continuous predictors, as lower $I(X;Y;C)$ becomes, the more advisable it is to use a classifier which models $p_M(c|\boldsymbol{x})$ in a multivariate way.

Variable $p(c)$. There does not seem to be any clear linear relation between the errors ϵ_{mul} and ϵ_{uni} and the IT based measures studied. Examples of the most relevant measures for the errors ϵ_{mul} and ϵ_{uni} are shown at Figures 3(d) and 3(e) respectively. On the other hand, $I(X;Y;C)$ seems to be again inversely proportional to ϵ_{dif} (see Figure 3(f)).

3   Conclusions

Setting any distribution of the class $p(c)$, in bivariate continuous domains, $I((X,Y);C)$ seems to be inversely proportional to the error ϵ_{mul} associated to the multivariate model $p_{mul}(c|x,y) \propto p(c)f(x,y|c)$. Besides $I(X;C) + I(Y;C)$ seems to be inversely proportional to ϵ_{uni}, the error of the classifier $p_{uni}(c|x,y) \propto p(c)f(x|c)f(y|c)$, which model the distribution in a univariate way. On the other hand, when the distribution $p(c)$ varies, there does not seem to be any clear relation between the introduced errors and the information theory based measures studied. Anyway, as $I(X;Y;C)$ decreases, it seems to be more advisable to use classifiers which model $p_M(c|x,y)$ in a multivariate way.

We are extending the work to discrete and mixed domains. We think that the relations that seem to exist between the measures based on the information theory and the classification error could be used for the improvement of the classifiers which assign *a posteriori* probabilities $p_M(c|\boldsymbol{x})$ to each class $c \in \{1,...,r\}$. They also could be used in other tasks related to the supervised classification such as feature subset selection or discretization, among others.

References

1. Cover, T.M., Thomas, J.A.: Elements of Information Theory. John Wiley and Sons (1991)
2. Fayyad, U., Irani, K.: Multi-interval discretization of continuous-valued attributes for classification learning. In: Proceedings of the 13th International Conference on Artificial Intelligence. (1993) 1022–1027
3. Hall, M.A., Smith, L.A.: Feature subset selection: A correlation based filter approach. In: Proceeding of the Fourth International Conference on Neural Information Processing and Intelligent Information Systems. (1997) 855–858
4. Pérez, A., Larrañaga, P., Inza, I.: Supervised classification with conditional Gaussian networks: Increasing the structure complexity from naive Bayes. International Journal of Approximate Reasoning. (2006) In press.
5. Silverman, B.: Density Estimation for Statistics and Data Analysis. Chapman and Hall: London (1986)

Checking Scientific Assumptions by Modeling

Joseph Phillips[1], Ronald Edwards[2], and Raghuveer Kumarakrishnan[1]

[1] DePaul University, School of Computer Science, Telecommunications and Information Systems
243 S. Wabash Ave., Chicago, IL 60604
jphillips@cs.depaul.edu, krv4u@yahoo.com
http://facweb.cs.depaul.edu/jphillips
[2] DePaul University, Department of Biological Sciences
2325 N. Clifton Ave., Chicago, IL 60614
redwards@depaul.edu
http://condor.depaul.edu/~biology/edwards

Abstract. We describe extensions to the science querying system Scilog to enable it to make efficient simulators. Given a scientific model's details and assumptions Scilog can create a C++ simulator to check assumption consistency. We used extended Scilog to test claims that Intelligent Design makes about Evolution and found them to be at odds with basic Biology.

1 Introduction

Computational Scientific Discovery (CSD) differs from Machine Learning (ML) in at least one crucial aspect: *bias*. Both ML and CSD prefer models that are small and accurate. CSD, however, also values *consistency with existing scientific literature*. This literature may be thought of as *bias in declarative form*.

Given that we must have this declarative bias, what else may we do with it? We may temporarily flip the relationship between bias and model. Instead of using the bias to constrain model search, we may start with a good model and double-check our bias.

The main contribution of this paper is the description of a procedural extension to Scilog that is rich enough to test specific cultural assumptions (biases). We used it to explore the differences in belief between Evolutionary Biologists and believers in Intelligent Design. Intelligent Design states that living things have more complexity than could have arisen by chance, so some kind of rational designer (*e.g.* a god, a space alien, *etc.*) must have helped. (It differs from Creationism which states that it definitely was God who created life.) This required quantifying the effect of a mechanism other than maximal selection or mutation for increasing allelic frequencies, which could lead to speciation.

Langley *et al* [3] created a system that fits time-varying data given abstract processes with generic equations. Our work generalizes this prior work by defining a specific modeling language and by allowing specific cultural assumptions to be stated.

N. Lavrač, L. Todorovski, and K.P. Jantke (Eds.): DS 2006, LNAI 4265, pp. 352–357, 2006.

2 Assumptions and Their Application

Our models have five components: three empirical ones (Theory, Laws and Data) and two non-empirical ones (Metaphysics and Analytics). (By "empirical" we mean "open to revision based upon observation".) Theory holds high level empirical knowledge (*e.g.* Newton's Laws of Motion plus Gravitation), Data holds low level measurements (*e.g.* Tycho Brahe's observations) and Laws holds knowledge that is supported by the Theory but closer to Data (*e.g.* Kepler's Laws of Planetary Motion). Metaphysics contains cultural assumptions made by a scientist (*e.g.* the ontology of domain objects, definitions of dimensions and units, and *a priori* limitations that a scientist may place on a model). Analytics gives information on how to transform problems without adding empirical content, like the knowledge needed to change co-ordinate systems.

Our component architecture is consistent with our previous work [4] and is supported by models of science held by the modern philosophers. The notion of "Metaphysics" (even that word) comes straight from the philosophy of science. We follow Goodman [1] and Lakatos [2] by admitting that cultural assumptions impose constraints on observations and the other aspects of empirical content. We place them in the Metaphysics component where they are centralized and potentially open to critique. The Analytics component comes from the Logical Empiricist idea of tautology. Finally, we follow Thagard [6] by splitting our empirical knowledge into Theory, Law and Data. Our conceptualization of "Law" may not be universally shared. "Law" is structurally similar to "Theory" but is supported by (perhaps provable from) Theory, Metaphysics and Analytics.

Storing assumptions gives our system (1) limited ability to automatically check the consistency of the Theory, Laws and Data with the Metaphysics, (2) the ability to constrain the search for revisions to Theory, Laws and Data using metaphysical knowledge, and, (3) the usage of the Theory, Laws and Data to define a simulation to check the suitability of the Metaphysics (which is the subject of this paper). Also having five components gives us the ability to support several queries. Metaphysics is always checked first to enforce conformity with basic assumptions. Analytics is always checked last to recast a query after all other components have failed.

This is how our system creates a simulator. First, the user loads one particular model, including its metaphysics. When the user chooses to investigate how well the metaphysics fits with the rest of the model, the system serially considers all process classes given in the metaphysics. For each one, if it specifies at least one constraint and has a procedure that calls another process class, then the system loops through the called process class and its subclasses to consider if simulation can be made.

A simulation can be made when a metaphysical process class (P_M) makes claims about an empirical process class (P_E) and certain conditions are met. All process classes recursively called by P_E must have procedures and are incorporated into the simulator. When a procedure needs to compute an attribute the system does the following:

1. Sees if the value is the same for all objects that it could describe. If it is, then the system fills it in with a constant.
2. Sees if an equation or decision tree associated with the current process class can answer that query. If so, then that assertion is recursively incorporated.
3. Sees if some other equation or decision tree can answer the query.
4. Sees if domain information exists for that attribute. If so, then it is assigned a random value from that domain at the beginning of the simulation.

If all of the steps fail, then it gives up that approach to making a simulator.

After the simulator is made it revisits the preconditions of P_M. Preconditions concerning the random values are incorporated into the simulator's random value choosing operation.

3 Experimental Domain: Population Genetics

We used Scilog (Phillips [5]) to build a population simulator that used genetic knowledge. The simulator kept track of the frequencies of alleles in a population, where an allele is a form of a gene. This simulator was used to test whether a series of improbable events is necessary to increase a non-maximally favored allele's frequency (as Intelligent Design states) or whether reasonably probable events can still increase a non-maximally favored allelic frequency (as implied by Evolution).

Both models have the following knowledge:

1. **Logistic growth.** When the population in a fixed area has a size **N** that is small relative to how many organisms that area can support (the area's *carrying capacity*, or **K**) then the population grows exponentially. However, as it grows it levels off to approach **K** asymptotically. This is given in the differential equation **dN = r * N * (1 – N/K)** where **r** is a growth rate that is characteristic to the species and environment.
2. **Mendelian genetics.** We assume that genes will have two alleles: a dominant (*e.g.* **A**) and a recessive (*e.g.* **a**). If the organism has the pairing **AA** or **Aa** then it exhibits the dominant phenotype. Having **aa** will make it exhibit the recessive phenotype.
3. **The remaining Hardy-Weinberg conditions.** We assumed that there is no mutation (as assumed by I.D.), no immigration or emigration (*i.e.* geographic isolation), and only same-generation random mating. These assumptions are standard in population dynamics modeling. The last is necessary to avoid defining a specific mating function that accounts for differing mating opportunities, fertility, brood sizes, age of sexual maturity, longevity, and breeding season's timing and duration.

Though simple, these assumptions are powerful. They make few claims about "real-world" genetics and behavior and therefore can encompass a wide variety of cases. (The most unrealistic simplification may be that a gene uniquely "controls" a trait, but developmental genes may influence hundreds of others.) Another advantage of using relatively basic biology is that these ideas are not challenged by either Evolutionists or believers in Intelligent Design. This knowledge goes into the Theory component.

We then defined a "Best of a Bad Circumstance" model to see if an intermediate choice can still increase, even though it is not the best. The idea is that there are two genes: **gene_a** and **gene_b**. If an organism has at least one copy of **gene_a**'s dominant allele **A** then it has maximal fitness. If, however, it has two copies of the recessive allele **a** then it has one of two possible fitnesses. The "Best of a Bad Circumstance" fitness (which in general is worse than the maximal fitness) is available if it has at least one copy of **gene_b**'s dominant allele **B**. If, however, it has genotype **aabb** then it has the worst.

Intelligent Design states that "The three conditions (geographic isolation, new trait creation, and superior fitness) are all simultaneously necessary for new allelic frequency increase, and this is improbable." **B** always encodes for a strategy that is never better, and most often worse, than **A**'s. The test we wish to conduct is "*Does B's frequency ever increase higher than due to fitness alone?*" This is currently inexpressible, so instead we gave Intelligent Design's Metaphysical component the process constraint "*Does B's frequency increase?*" We did the more detailed analysis by SAS afterward.

We translated our "Best of a Bad Circumstance" model to a Scilog program with five process classes. *Population_over_time* runs *create_1st_generation* and then *create_subsequent_generation* 100 times. *Create_1st_generation* calls *randomly_create_organism* the number of times given by the initial generation size. *Randomly_create_organism* uses stochastic decision trees to initialize alleles for **gene_a** and **gene_b** according to the free parameter initial probabilities of **A** and **B**. A third decision tree then computes the organism's fitness. *Create_subsequent_generation* uses Logistic Growth to compute how many organisms to create. For each it randomly chooses parents (weighted by their fitness) and calls *randomly_mate_organisms*. *Randomly_mate_organisms* uses stochastic decision trees to define alleles for **gene_a** and **gene_b** according to parents and Mendelian genetics. A third computes its fitness.

We encoded the Intelligent Design assumption set in its Metaphysics as a process class *creationist_over_time* which calls *population_over_time*. It is also given the preconditions that the initial generation is not larger than the carrying capacity, and that *fitness(aabb)* is not greater then *fitness(aaB?)*. Finally, it has the constraint that the fraction of **B**'s in the initial generation is expected to be greater than or equal to the fraction of **B**'s in the final generation. The "null hypothesis" (that these conditions *can* increase **B** allelic freq.) is the metaphysics for Evolution.

We gave the following free parameters these ranges: initial generation size and carrying capacity: x where x = floor(10^y + 0.5), y∈{1.0, 1.0625, 1.125, ... 3.0}; growth rate, p(**A**) in init. population and fitness(**aabb**): {0, 0.1, 0.2, ... 1.0}; # generations: {100}; p(B) in init. population and fitness(**aaB?**): {0.1, 0.2, 0.3, ... 1.0}; fitness(**A???**): {1.0}.

Finally, we told our system to create a simulation program. The simulator that randomly choose free parameters from their legal domains (re-choosing all parameters if the initial generation size was greater than the carrying capacity or if *fitness(aabb)* was greater than *fitness(aaB?)*) and did the simulation. It was called 100,000 times by a wrapper program that re-seeded the simulator's random number generator with the wrapper's index.

The frequency of **A** always increased, while the frequency of **B** increased in 57,038 of the 100,000 trials. Although this contradicts the Intelligent Design's assertion that **B** should never increase, its proponents could say that this is just selection. However, a more detailed look at when **B** increases shows that selection is only one effect. Principal component analysis by SAS gives us the results in Table 1:

Table 1. Principal Component Analysis

	Eigenvalue	Proportion		Greatest abs vect	2nd greatest abs vect
V1	1.47065	29.41%	r	v3: 0.999433	v2: -0.029435
V2	1.16894	23.38%	K	v5: -0.707107	v1: 0.706781
V3	0.999898	20.00%	n0	v5: 0.707053	v1: 0.706742
V4	0.830874	16.62%	fit(aaB)- fit(aabb)	v4: 0.706908	v2: -0.706071
V5	0.529637	10.59%	inc. B	v4: 0.707044	v2: 0.706863

All variance was accounted for by five eigenvectors:

1. Increase in carrying capacity with the initial generation's size (29.41%)
2. Increase in the frequency of **B** without selection for B (23.38%)
3. Variation in birth rate (20.00%)
4. Increase in the frequency of **B** with an increase in selection for **B** (16.62%)
5. Increase in the initial generation's size without carrying capacity (10.59%)

This shows two independent trends concerning **B**. First, Vector 4 (17% of the variance) accounts for selection. **B** increases as *fitness(aaB?)* becomes greater than *fitness(aabb)*. This is expected by Intelligent Design. Second, however, is Vector 2 (with 23% of the variance) which shows an increase in **B** *without* a high selection. This is the *Founder Effect:* that random variations in the initial generation influence later generations. This effect leads **B** to become disproportionately frequent in *some* smaller environments (*e.g.* small islands). Apparently below a certain size the favorable effects of selection are enhanced by a tendency for the Founder Effect to reinforce the frequency and potential fixation of **B**. **B** was never selectively favored nor artificially increased by mutation. The only required (and reasonable) condition was the geographic isolation of a diverse subpopulation.

4 Conclusion

This work has neither "proved" nor "disproved" Intelligent Design because the vague statement "X, Y and Z make A too unlikely" is inherently difficult to prove or disprove. Instead, we have shown Intelligent Design is *inconsistent* with basic biological theory and attribute distributions that we think reasonable more than 50% of the time.

In this work we have (1) described how allelic frequency can change, (2) extended a system to do simulations, (3) built a simulator with a general algorithm that used specific domain knowledge of process classes and their relationships, and (4) given scientists an opportunity to test some of their assumptions and those of others.

We gratefully acknowledge DePaul's University Research Council for support.

References

1. Goodman, N. (1966) The New Riddle of Induction, in: *J. of Philosophy* 63: 281-331.
2. Lakatos, I. (1970) Falsification and the methodology of scientific research programmes. In Lakatos, I. and Musgrave, A. (ed.) *Criticism and the growth of knowledge*. Cambridge University Press: Cambridge.
3. Langley, Pat, Sanchez, Javier, Todorovski, Ljupco, Dzeroski, Saso (2002) "Inducing process models from continuous data" in *Proc. of 19th Int'l Conf. of Machine Learning*, Morgan Kaufmann, San Francisco, p 347-354.
4. Phillips, J. (2001) Towards a method of searching a diverse theory space for scientific discovery. *Discovery Sci.*, Springer-Verlag, Berlin, p 304-322.
5. Phillips, J. (2003) Scilog: A language for scientific processes and scales. *Discovery Sci.*, Springer-Verlag, Berlin, p 442-451.
6. Thagard, P. (1988) Computational Philosophy of Science, MIT, Cambridge MA.

Incremental Algorithm Driven by Error Margins *

Gonzalo Ramos-Jiménez, José del Campo-Ávila, and Rafael Morales-Bueno

Departamento de Lenguajes y Ciencias de la Computacion
E.T.S. Ingenieria Informatica. Universidad de Malaga
Malaga, 29071, Spain
{ramos, jcampo, morales}@lcc.uma.es

Abstract. Incremental learning is a good approach for classification when data-sets are too large or when new examples can arrive at any time. Forgetting these examples while keeping only the relevant information lets us reduce memory requirements. The algorithm presented in this paper, called IADEM, has been developed using these approaches and other concepts such as Chernoff and Hoeffding bounds. The most relevant features of this new algorithm are: its capability to deal with datasets of any size for inducing accurate trees and its capacity to keep updated the estimation error of the tree that is being induced. This estimation of the error is fundamental to satisfy the user requirements about the desired error in the tree and to detect noise in the datasets.

1   Introduction

Machine learning systems are currently required to deal with large datasets. Moreover, data streams have recently become a new challenge for data mining because of certain features [1] of which infinite data flow is the most notable. Trying to extract knowledge from such numbers of examples can become an inaccessible problem for traditional algorithms such as C4.5 [2] due to memory requirements. Probabilistic representation provides the opportunity of exploring data and keeping the most relevant information from seen examples [3]. Therefore, required memory does not have to depend on the number of examples in the dataset, but on the structure and associated statistics.

The concept "incremental learning" has been used rather loosely but we consider that there two main features: its capability to incorporate new examples into the knowledge base [4] and its capability to evolve this knowledge base from a very basic concept to another more complex one [2].

Chernoff [5] and Hoeffding [6] bounds have recently been used in machine learning area [7, 8, 9] in conjunction with probabilistic representation. They have been used to provide statistical evidence in favour of a particular split test, to ensure a minimum number of scans through a sequence database, etc.

The paper is organized as follows. In Section 2 we introduce IADEM, explaining its main contributions and how they are used. The experiments that we have made and the results are presented in Section 3. Finally, in Section 4, we summarize our conclusions and suggest future lines of research.

---

* This work has been partially supported by the FPI program and the MOISES-TA project, number TIN2005-08832-C03-01, of the MEC, Spain.

N. Lavrač, L. Todorovski, and K.P. Jantke (Eds.): DS 2006, LNAI 4265, pp. 358–362, 2006.

2   IADEM

IADEM is an incremental algorithm with no example memory which knowledge base is represented using decision trees. IADEM receives examples to learn from, but once they are processed they are forgotten. No example is saved, only the relevant information for IADEM is kept using counters stored in the decision tree. Thus, the memory requirements of this algorithm only depend on the size of the decision tree and its associated counters. This way of processing data lets us deal with any kind of dataset or data stream independently of size. The examples used by the algorithm are sampled with replacement from the dataset or they can be taken directly from the data stream. It is important that the source of data does not have concept drift because IADEM is not ready to afford it yet.

In this algorithm, a very important point are the user's requirements about maximum level of error and confidence. Thus, the user must set two arguments for the algorithm: the maximum desired error for the tree that is going to be induced (ε) and its confidence $(1 - \delta)$. IADEM calculates some statistics using the counters stored in the decision tree. In addition to the estimated values for that statistics, the error margins for those estimated values are calculated too. Using the statistics and their error margins, IADEM maintains, at every moment, an estimation of the superior and inferior bounds of the error in the tree. That estimation is fundamental for the operation of this algorithm:

- *Stop* condition: the superior bound of the error ($\sup(error)$) is used to stop the algorithm when the user's requirements are satisfied ($\sup(error) \leq \varepsilon$). Thus, the number of examples needed to stop the algorithm does not depend on the size of the dataset. If the dataset is too small, it will continue sampling with replacement until the maximum error in the tree comes below the desired level. On the other hand, it is possible that the algorithm uses less examples than the number of examples in the dataset.
- *Expansion* condition: the inferior bound of the error ($\inf(error)$) is used to limit the number of expansions. If $\inf(error)$ is lower than ε, that means, in the best case, that the error in the tree can be below the desired error, so we can continue sampling and trying to reduce $\sup(error)$. Thus, the algorithm could find a solution without any additional expansion. Therefore, the first condition for considering to do an expansion is the proximity of $\inf(error)$ to ε.
- *Noise* condition: the different values of $\sup(error)$ when expansions occur are used to detect noise. When IADEM expands a leaf node, $\inf(error)$ and $\sup(error)$ usually decrease. But, when the error in the tree is close to the noise in the dataset, the behaviour changes and it is usual that $\sup(error)$ increase after an expansion. We consider a maximum level of increases of $\sup(error)$ in the last expansions, and the algorithm stops if this level is exceeded.

IADEM uses two kinds of nodes in the border of the tree: the real and the virtual nodes. The real ones are the leaves of the tree and they constitute the real border of the tree. Every real node (or leaf) has as many virtual nodes as attributes are unused in the branch that ends in that leaf. The set of virtual nodes that corresponds to a real node represents all the possible expansions for that leaf and they register all the information that will be needed to do an expansion. The information stored in the real nodes is the following: the total number of examples sampled since the node exists as virtual ($t \in \mathbb{N}$), the number of those that match with the branch that reaches that node ($m \in \mathbb{N}$) and the number of those examples depending on the class label ($m_k \in \mathbb{N}$ where $k \in \{1 \dots z\}$ and z is the number of classes). On the other hand, the information stored in the virtual nodes is: the total number of examples sampled since the virtual node exists ($t' \in \mathbb{N}$), the number of those that match with the branch that reaches that virtual node depending on the value for the corresponding virtual node attribute ($m'_v \in \mathbb{N}$ where $v \in \{1 \dots r\}$ and r is the number of values of the attribute) and the number of those examples depending on the class label ($m'_{v,k} \in \mathbb{N}$).

Considering these counters and the number of examples that have been processed, and using the Chernoff and Hoeffding bounds, IADEM provides estimated values and error margins for different calculated elements: the probability of one example reaching a real node ($w = m/t$); and the probability of an example being classified as k class in a real node ($p_k = m_k/m$) or in a virtual node ($p'_{v,k} = m'_{v,k}/m'$). To calculate the error margins for them we use the following general expression:

$$\varepsilon\_margin(x) = \min \left\{ \sqrt{\frac{3 \cdot a}{b} \ln(2/\delta)}, \sqrt{\frac{1}{2 \cdot b} \ln(2/\delta)}, 1 \right\} \tag{1}$$

and the values for a and b depend on the calculated element (x) which error margin is being considered. Thus, for $x = w$ we have $a = m/t$ and $b = t$; for $x = p_k$ we have $a = m_k/m$ and $b = m$; and for $x = p'_{v,k}$ we have: $a = m'_{v,k}/m'_v$ and $b = m'_v$. We use both bounds because none of them gives always the minimum value. It depends on the input (a and b) and using equation 1 we select the minimum value.

With those elements and their error margins we can calculate the error produced in every leaf node of the tree. Using those errors we can estimate the superior and inferior bound of the error in the tree. Those calculated elements are also used in the expansion process. To do an expansion we need to identify which is the node that will be expanded ($worst\_leaf$). That leaf will be the one that contributes the most error to the tree. Once we have selected the leaf node, we find the best attribute to do the expansion using the information in the virtual nodes.

---

1.   INITIALIZE
2.   **while** ($\neg Stop \wedge \neg Noise$) **do:**
2.1.   SAMPLE_AND_RECALCULATE
2.2.   **if** ($\neg Stop \wedge Expansion \wedge Is\_Expansible(worst\_leaf)$) **then:**
2.2.1.   EXPAND_TREE

Fig. 1. IADEM algorithm

3   Experiments and Results

In this section, we will focus on synthetic datasets using the LED dataset from *UCI* [10] and synthetic datasets randomly generated using decision trees. The main aim for using them is controlling the size and noise. IADEM has been compared with other well-known algorihtms: C4.5 [2], ITI [11] and VFDT [7] using the default configuration for all the algorithms. The values presented were obtained from a 10-fold cross validation, and the average and standard deviation are given. A Wilcoxon signed rank test has been conducted and significant differences are shown ($\oplus$ for better and $\ominus$ for worse).

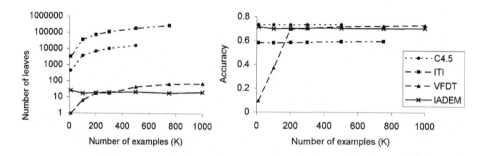

Fig. 2. LED dataset. Accuracy and size of the decision trees depending on the size of the dataset.

Experiments with LED dataset have the level of noise fixed to 10%. Considering the results (see Figure 2) we can see that VFDT and IADEM induce decision trees which size is very stable, while the trees induced by C4.5 and ITI are bigger when the number of examples in the dataset increase (and they go out of memory). The accuracy reached by IADEM is clearly better than the accuracy reached by ITI and it is very close to the one reached by C4.5 or VFDT.

Table 1. Results for the synthetic datasets randomly generated using decision trees

Dataset	Algorithm	Leaves	Accuracy	Time
Syn_1	VFDT	449.00 ± 0.00 $\ominus$	99.55 ± 0.03 $\ominus$	**453**
398 leaves	ITI	6461.70 ± 250.73 $\ominus$	99.66 ± 0.03 $\ominus$	704
0% noise	IADEM	**424.00** ± 33.57	**100.00** ± 0.00	12184
Syn_2	VFDT	1357.20 ± 356.10 $\ominus$	73.23 ± 5.99 $\ominus$	**1188**
663 leaves	ITI	114052.00 ± 257.09 $\ominus$	79.03 ± 0.13	8168
15% noise	IADEM	**742.90** ± 84.87	**80.07** ± 3.23	22544
Syn_3	VFDT	659.60 ± 121.33 $\ominus$	92.07 ± 1.12 $\ominus$	1627
200 leaves	ITI	5101.00 ± 345.73 $\ominus$	99.40 ± 0.07 $\ominus$	**523**
2% minority class	IADEM	**261.0** ± 0.00	**100.00** ± 0.00	4105

We have generated three kinds of synthetic datasets with one million examples: deterministic, non-deterministic and imbalanced datasets. In Table 1, we show the results achieved by the algorithms and we can see the following: IADEM usually induces decision trees which accuracy is significantly better than other ones. Their size are also significantly smaller than the ones produced by ITI and VFDT.

4   Conclusion

This paper introduces IADEM, an incremental classifier for learning from increasingly common high-volume datasets and data streams. Using probabilistic representation and Chernoff and Hoeffding bounds as the core of the algorithm, we have been able to design IADEM and provide it with some useful features:

- Independence from the size of the data source. Memory requirements only depend on the knowledge structure that is being induced.
- Information about the estimated error for the decision tree at every moment. This estimation lets IADEM stop when the user's requirements (ε and δ) are satisfied.
- Detection of noise in the dataset, and consequently stop of execution.

Out aim of improving IADEM involves some issues. We are working to improve the detection of noise in order to get accuracies nearer to the maximum ones. We also expect to include the capability of working with continuous attributes. The improvement of the prediction method incorporating functional leaves and learning in the presence of concept drift are another promising points.

Acknowledgement. The authors wish to thank Alejandra Cabaña, for her assistance and advice about the statistical aspects of the experimental section. The authors also want to thank João Gama for his advice about the experimental section.

References

1. Wang, H., Fan, W., Yu, P.S., Han, J.: Mining concept-drifting data streams using ensemble classifiers. In: Proc. 9th ACM SIGKDD Int. Conf. on Knowledge Discovery and Data Mining, ACM Press (2003) 226–235
2. Quinlan, J.R.: C4.5: Programs for Machine Learning. Morgan Kaufmann (1993)
3. Fisher, D.H., Schlimmer, J.C.: Models of incremental concept learning: A coupled research proposal. Technical Report CS-88-05, Vanderbilt University (1998)
4. Schlimmer, J.C., Fisher, D.H.: A case study of incremental concept induction. In: Proc. 5th Nat. Conf. on Artificial Intelligence, Philadelphia, Morgan Kaufmann (1986) 496–501
5. Chernoff, H.: A measure of asymptotic efficiency for tests of a hypothesis based on the sums of observations. Annals of Mathematical Statistics **23** (1952) 493–507
6. Hoeffding, W.: Probability inequalities for sums of bounded random variables. Journal of the American Statistical Association **58** (1963) 13–30
7. Domingos, P., Hulten, G.: Mining high-speed data streams. In: Proc. of the 6th ACM SIGKDD Int. Conf. on Knowledge Discovery and Data Mining, ACM Press (2000) 71–80
8. Yang, J., Wang, W., Yu, P.S., Han, J.: Mining long sequential patterns in a noisy environment. In: Proc. ACM SIGMOD Int. Conf. on Management of Data, ACM Press (2002) 406–417
9. Gama, J., Rocha, R., Medas, P.: Accurate decision trees for mining high-speed data streams. In: Proc. 9th ACM SIGKDD Int. Conf. on Knowledge Discovery and Data Mining, ACM Press (2003) 523–528
10. Blake, C., Merz, C.J.: UCI repository of machine learning databases. University of California, Department of Information and Computer Science (2000)
11. Utgoff, P.E., Berkman, N.C., Clouse, J.A.: Decision tree induction based on efficient tree restructuring. Machine Learning **29**(1) (1997) 5–44

Feature Construction and δ-Free Sets in 0/1 Samples

Nazha Selmaoui[1], Claire Leschi[2], Dominique Gay[1],
and Jean-François Boulicaut[2]

[1] ERIM, University of New Caledonia
{selmaoui, gay}@univ-nc.nc
[2] INSA Lyon, LIRIS CNRS UMR 5205
{claire.leschi, jean-francois.boulicaut}@insa-lyon.fr

Abstract. Given the recent breakthrough in constraint-based mining of local patterns, we decided to investigate its impact on feature construction for classification tasks. We discuss preliminary results concerning the use of the so-called δ-free sets. Our guess is that their minimality might help to collect important features. Once these sets are computed, we propose to select the essential ones w.r.t. class separation and generalization as new features. Our experiments have given encouraging results.

1 Introduction

We would like to support difficult classification tasks (from, e.g., large noisy data) by designing well-founded processes for building new features and then using available techniques. This is challenging and our thesis is that the recent breakthrough in constraint-based mining of local patterns might provide some results. Considering the case of 0/1 data whose some attributes denote class values[1], many efficient techniques are now available for computing complete collections of patterns which satisfy user-defined constraints (e.g., minimal frequency, freeness, closeness). Our goal is not only to consider such patterns as features but also to be able to predict (part of) the classification behavior based on these pattern properties. In this paper, we discuss preliminary results concerning the so-called frequent δ-free sets in 0/1 samples. When $\delta = 0$, these sets have been studied as minimal generators for the popular (frequent) closed sets. Otherwise ($\delta > 0$), they provide a "near equivalence" perspective and they have been studied as an approximate condensed representation for frequent sets [1]. Furthermore, the minimality of δ-free sets has been exploited for class characterization (see, e.g., [2]) and non redundant association rule mining (see, e.g., [3]). Our guess is that this minimality, in the spirit of the MDL principle, might help to collect relevant features. This is suggested in [4] as a future direction of work, and we provide some results in that direction. Section 2 introduces δ-freeness and our feature construction process. Section 3 reports about classification tasks on both UCI data sets [5] and a real-world medical data set. Section 4 concludes.

---

[1] It is trivial to derive Boolean data from categorical data and discretization operators can be used to transform continuous attributes into Boolean ones.

N. Lavrač, L. Todorovski, and K.P. Jantke (Eds.): DS 2006, LNAI 4265, pp. 363–367, 2006.
© Springer-Verlag Berlin Heidelberg 2006

2   Feature Construction by Using δ-Free Sets

Given a potentially large labeled 0/1 data set, our feature construction process consists in three main steps: (1) mining frequent δ-free sets associated to their δ-closure [1, 6] and select those whose δ-closure includes a class attribute; (2) further select the essential patterns w.r.t. some interestingness criteria; (3) encode the original samples in the new representation space defined by these descriptors.

Let $r = (T, I)$ a 0/1 data set where T is a set of objects and I a set of Boolean attributes. An itemset A is subset of I and we recall the definition of useful evaluation functions on itemsets. The frequency of A in r is defined as $freq(A, r) = |support(A, r)|$ where $support(A, r) = \{t \in T/A \in t\}$. Let γ be an integer, A is called γ-frequent if $freq(A, r) \geq \gamma$. The closure of A in r denoted $closure(A, r)$ is the largest superset of A with the same frequency. An itemset A is closed if $closure(A, r) = A$. Since [7], it is useful to formalize this by means of the same-closure equivalence relation. Two itemsets A and B are said equivalent in r ($A \sim_f B$) if $closure(A, r) = closure(B, r)$. Indeed, we have the following properties :

(i) $A \sim_f B \equiv freq(A, r) = freq(B, r)$;
(ii) Each equivalence class contains exactly one maximal (w.r.t. set inclusion) itemset which is a closed set, and it might contain several minimal (w.r.t. set inclusion) sets which are called 0-free sets or free sets for short.
(iii) If A and B are in the same equivalence class and $A \subseteq V \subseteq B$ then V is in the same equivalent class.

Definition 1 (δ-free itemsets and δ-closures). *Let δ be an integer. A is a δ-free itemset if $\forall S \subset A, |freq(S, r) - freq(A, r)| > \delta$. The δ-closure of an itemset A is defined as $closure_\delta(A) = \{X \in I/freq(A, r) - freq(A \cup \{X\}, r) \leq \delta\}$.*

The intuition is that the δ-closure of a set A is the superset X of A such that every added attribute is almost always true for the objects which satisfy the properties from A: at most δ false values are enabled. The computation of every frequent δ-free set (i.e. sets which are both frequent and δ-free) can be performed efficiently [6]. Given threshold values for γ (frequency) and δ (freeness), our implementation outputs each δ-free frequent itemset and its associated δ-closure. Notice that when $\delta = 0$, we collect all the free frequent itemsets and their corresponding closure, i.e., we compute a closed set based on the closure of one minimal generator. Since we are interested in classification, we also assume that some of the attributes denote the class values.

Interestingness measures are needed to select the new features among the δ-frees. Our first measure is based on homogeneity and concentration (HC) [8]. It has been proposed in a clustering framework where formal concepts (i.e., closed sets) were considered as possible bi-clusters and had to be as homogeneous as possible while involving "enough" objects. Our second measure is the well-known information gain ratio (GI). Selecting features among the frequent δ-free itemsets is crucially needed since 0/1 data might contain a huge number patterns which are relevant neither for class discrimination nor for generalization.

At first, the homogeneity and concentration measures were used to respectively maximize the intra-cluster similarity and to limit the overlapping of objects between clusters. Homogeneity is defined as:

$$Homogeneity(A, r) = \frac{|support(A, r)| \times |A|}{divergence(A) + (|support(A, r)| \times |A|)}$$

where $divergence(A) = \sum_{t, A \in t} |t - A|$. If an itemset is pure then its divergence is equal to 0, and its homogeneity is equal to 1. This measure enables to keep the itemsets having many attributes shared by many objects. The concentration is defined as:

$$Concentration(A, r) = \frac{1}{|support(A, r)|} \times \sum_{X \in t, \forall t, A \in t} \frac{1}{|X|}$$

Then, the interestingness measure of an itemset is defined as the average of its homogeneity and its concentration. The more the interestingness is close to 1, the more an itemset is considered as essential for classification purposes.

The filtering of new descriptors is performed in three steps. Once frequent δ-free sets (say A) and their δ-closures (say X) have been extracted, we first retain only the sets X which include a class attribute. The associated minimal generators, i.e., the frequent δ-free sets whose δ-closures involve a class attribute, are selected as potentially interesting new features. We further focus on their supporting sets of objects among the classes. Then, we retain only the patterns that are very frequent in one class and merely infrequent in the other ones. We formulate this condition through the function Gr_Rate defined as follows (C_i is a class attribute):

$$Gr\_Rate(A) = \frac{|support_{C_i}(A)|}{\sum_{j \neq i} |support_{C_j}(A)|}$$

The selected patterns are those for which Gr_Rate is greater than a user-defined threshold. That selection criterion is clearly related to the nice concept of emerging pattern [9]. Finally, we use one of the two interestingness measures cited above to further reduce the number of new descriptors while preserving class separation.

We can now use the selected itemsets as features for encoding a new representation for the original data. In our experiments, we computed the value taken by each new attribute $NewAttr_A$ for a given object t as follows:

$$NewAttr_A(t) = \frac{|A \cap t|}{|A|}$$

It quantifies the number of occurrences of the object in the corresponding itemset. The normalization aims to avoid "unbalanced" values due to occurrences in large versus small itemsets. As a result, we obtain a numerical database with a number of features derived from the selected frequent (δ-free) itemsets.

3   Experimental Results

We have tested our approach on a real-world medical data set meningitis[2] and two data sets from the UCI Repository (Vote and Cars [5]). We report the best accuracy provided by three classical classification algorithms: Naive Bayes (NB), J48 (i.e., C4.5), and Multilayer Perceptron (NN) available within the popular WEKA platform [10]. The best accuracy results on the original data are 92.36% with J48 and 99.54% with NN for Cars; 97.26% for Vote; 94% for meningitis. We present the results in Table 1 for different values of γ and δ thresholds. For Cars, we give the rate for J48 and NN to show that the classifier built from J48 is improved. We improve the NN classifier when $minfreq = 2$ and $\delta = 2$ (99.71%). In general, the results are quite encouraging. Using the Information Gain Ratio instead of homogeneity and concentration has given better results. It should be explained by the fact that the later measures have been targeted towards closed sets and not δ-free ones.

Table 1. Accuracy for Vote, Cars, and Meningitis

Datasets	$minfreq,\delta$	#Patterns	Measure	#Selected	Accuracy
Cars	2%,2	450	GI	21	J48=99.25, NN=99.25
			HC	36	J48=97.39, NN=99.71
	4%,4	167	GI	18	J48=98.49, NN=99.31
			HC	16	J48=86.40, NN=86.98
Meningitis	5%,2	60045	GI	7	98.79
			HC	14	93.92
	10%,2	31098	GI	6	97.88
			HC	8	92.71
Vote	5%,3	19859	GI	9	98.40
			HC	15	95.17
	10%,3	10564	GI	10	97.01
			HC	8	96.10

4   Conclusion

We have investigated the use of δ-free sets for feature construction in classification problems. In that context, the main issue was the selection of good patterns w.r.t. class separation and generalization purposes. We specified a filtering process based on both their local and global properties w.r.t. data. Using the new features on several classification tasks with different algorithms has given rise

---

[2] meningitis concerns children hospitalized for acute bacterial or viral meningitis (329 samples described by 60 Boolean attributes.

to quite encouraging results. Our next step will be to use other types of local patterns (e.g., closed or almost-closed ones, emerging patterns) and provide a fair comparison of their interest for feature construction.

Acknowledgments. This research has been carried out while N. Selmaoui and D. Gay were visiting INSA Lyon. This work is partly funded by EU contract IQ FP6-516169 (FET arm of IST). The authors wish to thank P. Francois and B. Crémilleux who provided meningitis, and J. Besson for his technical support.

References

1. Boulicaut, J.F., Bykowski, A., Rigotti, C.: Approximation of frequency queris by means of free-sets. In: Proceedings the 4th European Conference on Principles and practice of Knowledge Discovery in Databases (PKDD). (2000) 75–85
2. Boulicaut, J.F., Crémilleux, B.: Simplest rules characterizing classes generated by delta-free sets. In: 22nd SGAI International Conference on Knowledge Based Systems and Applied Artificial Intelligence, ES'02. (2002) 33–46
3. Becquet, C., Blachon, S., Jeudy, B., Boulicaut, J.F., Gandrillon, O.: Strong association rule mining for large gene expression data analysis: a case study on human SAGE data. Genome Biology **12** (2002)
4. Li, J., Li, H., Wong, L., Pei, J., Dong, G.: Minimum description length principle : generators are preferable to closed patterns. In: Proceedings 21st National Conference on Artificial Intelligence, Menlo Park, California, The AAAI Press (2006) To appear.
5. Newman, D., Hettich, S., Blake, C., Merz, C.: UCI repository of machine learning databases (1998)
6. Boulicaut, J.F., Bykowski, A., Rigotti, C.: Free-sets : A condensed representation of boolean data for the approximation of frequency queries. Data Mining Knowledge Discovery **7** (2003) 5–22
7. Bastide, Y., Taouil, R., Pasquier, N., Stumme, G., Lakhal, L.: Mining frequent patterns with counting inference. SIGKDD Explorations **2** (2000) 66–75
8. Durand, N., Crémilleux, B.: Ecclat : a new approach of clusters discovery in categorical data. In: 22nd SGAI International Conference on Knowledge Based Systems and Applied Artificial Intelligence, ES'02. (2002) 177–190
9. Dong, G., Li, J.: Efficient mining of emerging patterns: discovering trends and differences. In: Proceedings of the fifth ACM SIGKDD International Conference on Knowledge Discovery and Data Mining, New York, NY, USA, ACM Press (1999) 43–52
10. Witten, I.H., Frank, E.: Data Mining : Practical machine learning tools and techniques (2nd edition). Morgan Kaufmann Publishers Inc., San Francisco, USA (2005)

Visual Knowledge Discovery in Paleoclimatology with Parallel Coordinates*

Roberto Therón

Departamento de Informática y Automática,
Universidad de Salamanca, Salamanca, 37008, Spain
theron@usal.es

Abstract. Paleoclimatology requires the analysis of paleo time-series, obtained from a number of independent techniques. Analytical reasoning techniques that combine the judgment of paleoceanographers with automated reasoning techniques are needed to gain deep insights about complex climatic phenomena. This paper presents an interactive visual analysis method based in Parallel Coordinates that enables the discovery of unexpected relationships and supports the reconstruction of climatic conditions of the past.

1 Introduction

The need to foresee abrupt climatic changes is an urgent challenge for the society and paleoclimatology research. Computers have played a key role in our understanding of the climatic dynamics. Nowadays, the improvement of data acquisition methods offer us the opportunity to gain the needed depth of information to diagnose and prevent any natural disaster. By means of an analysis of such data, paleoceanographers are expected to assess (understand the past) and forecast (estimate the future). Although massive amounts of data are available, the development of new tools and new methodologies is necessary to help the expert extract the relevant information.

Some of the data needed to understand paleoclimate are time-series of specific attributes related to the oceans. Thus, one problem scientists must face is how to know environmental parameters, such as sea surface temperature (SST), at each given past moment. One approach for the quantitative reconstruction of environmental conditions of the past is based in a nearest neighbor prediction.

Visualization provides insight through images. These visual representations combined with interaction techniques that take advantage of the human's visual reasoning allow experts to see, explore, and understand large amounts of information at once. Thus, this paper presents an interactive visual analysis method

---

* This work was supported by the MCyT of Spain under Integrated Action (Spain-France) HF2004-0277 and by the Junta de Castilla y León under project SA042/02. The author would like to acknowledge Ana Mayordomo García for her assistance in the implementation of this work.

N. Lavrač, L. Todorovski, and K.P. Jantke (Eds.): DS 2006, LNAI 4265, pp. 368–372, 2006.

that, through the combination of techniques coming from statistics, information theory, and information visualization, enables the discovery of unexpected relationships and supports the reconstruction of climatic conditions of the past.

2   Visualization and Interaction: Parallel Coordinates

The use of proper interactive visual representations can foster knowledge discovery. Using the computed nearest neighbors data, it is possible to automatically find patterns in information, and represent such information in ways that are meant to be revealing to the analyst. On the other hand, by interacting with these representations, using their expert knowledge, it is possible to refine and organize the information more appropriately. This way, it is possible, not only to reconstruct paleoenvironmental features, but to visualize what information is being used to estimate them, and help the paleoclimatologists to decide upon using particular data or not, according to their field experience.

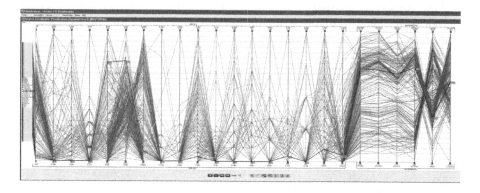

Fig. 1. Reconstruction visually driven by Parallel Coordinates

Scatter plots, maps and animations are common methods for geovisualization that have a long history in cartography and information visualization. Parallel Coordinates Plots (PCP) [3] are also a common method of information visualization, used for the representation of multidimensional data, and is an emerging practice in geovisualization [5][6]. The use of PCP (see figure 1) as part of the interactive visual analysis of multidimensional data can provide paleoclimatic knowledge discovery. Furthermore, instead of just use PCPs as another way of visualizing the data, they are used as a highly interactive tool that permits both gaining insight about the paleodata and visually reconstruct the paleoenvironmental features.

Having faunal census estimates of one or more fossil samples (the core) and one or more sets of faunal data from modern samples (sites) with the related environmental features (the database), the k-nearest neighbor method is used to predict environmental variables for each sample down the core (i.e, going back in time). The output of this process is depicted using an interactive PCP to facilitate the exploration of relationships among attributes and provide a mechanism for extracting patterns. In the example of figure 1 each site of the database is drawn as a polyline passing through parallel axis, which represent the species, and the environmental variables that we want to reconstruct (last four axis on the right). The polyline corresponding to a particular sample (20 cm of depth in this example) in the core is represented as a yellow polyline. Note that, since the core only have the species data and we want to reconstruct the environmental variable for each sample, there are no yellow segments in the environmental axis. Each polyline of the database is color coded using the k-nearest nighbor results: the more red the polyline is, the more dissimilar is to that particular sample of the core.

This static picture is already showing many things that were hidden in k-nearest nighbor results. For instance, it can be observed that the most similar sites for sample 20, are clustered in the low temperatures (four first axis in the environmental zone at right hand side). This means that the core site at sample 20 was cold.

However, several interaction techniques [4] have been integrated with this PCP to allow brushing [1], linking, animation, focus + context, etc., for exploratory analysis and knowledge discovery purposes.

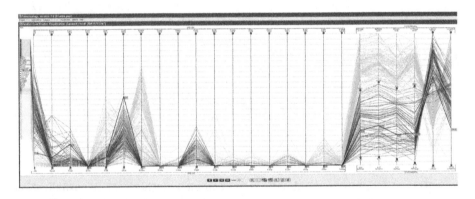

Fig. 2. Analytical reasoning and reconstruction by means of interaction

PCPs can be enhaced by means of dynamic queries in the form of axis filtering [2][7]. The range of an attribute can be specified by moving the handles at the top and bottom of the axis range sliders. To prevent users from losing global context during dynamic filtering, all the polylines are maintained on the background. Users can see the position of a polyline. Figure 2 shows a reconstruction already computed. After filtering the sites (the current ranges are shown between the handles) that were too dissimilar (kept in soft blue on the background), the expert decided to reconstruct the SSTs for sample 20. Note that now the yellow segments of the polyline for that sample also occupy the environmental axis. As expected, the values are an average of the values of the blue polylines. On the left hand side, only the interesting site labels are highlited. In the snapshot, the expert is comparing a site (thicker black polyline) with the reconstructed core sample.

Also note that in figure 2, all axis have the same scale (a percentage, as opposed to figure 1) in order to compare the relative abundances of the species, and help discover relationships between species and climatic features. Furthermore, any axis can be dragged and dropped, so the order of the axis is changed. This way the shape of the polyline also changes, helping to reveal hidden patterns and making analysis easier.

On the contrary to other approaches such as [6], time is not represented on an axis. In paleoclimatology time is most relevant, so the PCP is animated in order to visualize the evolution of different species trough geological time, and their relationships, both among themselves and between some species and environmental variables.

Location plays a key role in paleoceanography analysis. One possibility is to represent the latitude and/or longitude values for each database site. This can be interesting in order to highlight how temperature varies with time for the same latitude, for example. On the other hand, the benefits of using interactive maps are well known.

A common coordination technique is brushing and linking [1], where users can select objects in one view and the corresponding objects in all the other views are also automatically selected. This technique is the natural approach for the problem at hand. This way, all the benefits explained above can be put together in order to provide the paleoceanographers with the best interactive visual tool to discover knowledge and support decisions about climatic reconstructions.

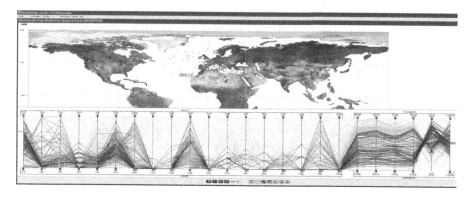

Fig. 3. Multiple linked views in PaleoAnalogs

In figure 3 the brushing and multiple linked views approach can be seen. As in [2], three modes of brushing and linking interaction that are coordinated among all the views described in the previous sections are possible:

- probing: this mode is used to view more details about an object (e.g. site labels and dissimilarity values) and to get an understanding of the relationships between the different views. Probing is a transient operation. Moving the mouse pointer over an object, highlights the object (e.g, a polyline) and as the mouse pointer is moved away, the highlighting disappears.

- selecting: this mode is used to mark objects that are of short-term interest, in order to further examine or perform operations on them (e.g see the values on every axis of a selected polyline). Clicking on an object selects it and marks it. If a selected object is filtered, then it becomes deselected.
- painting: this mode is used to mark objects that are of long-term interest, in order to use them as references for comparisons (e.g compare two polylines of two sites). Objects remain painted until they are reset explicitly.

3   Conclusions and Future Work

This work is an example of how interactive analysis can help knowledge discovery in the paleoclimatology field. We have shown how a black-box machine learning technique can be greatly improved so the reconstructions of paleoenvironmental conditions can be more accurate. This is accomplished by fostering a user driven reconstruction procedure where the expert get more insight from the data and can decide on the validity of potentials reconstructions. Finally, we can add that more complex interactive analysis can be designed that will help to gain a deeper knowledge about the climatic evolution of a given area.

References

1. Richard A. Becker and William S. Cleveland. Brushing scatterplots. *Technometrics*, 29(2):127–142, 1987.
2. D. Brodbeck and L. Girardin. Design study: using multiple coordinated views to analyze geo-referenced high-dimensional datasets. In *Proceedings. of the International Conference on Coordinated and Multiple Views in Exploratory Visualization*, pages 104–111, 2003.
3. Alfred Inselberg. The plane with parallel coordinates. *The Visual Computer*, 1:69–91, 1985.
4. Daniel A. Keim. Information visualization and visual data mining. *IEEE Transactions on Visualization and Computer Graphics*, 8(1):1–8, 2002.
5. E.L. Koua and M.-J. Kraak. A usability framework for the design and evaluation of an exploratory geovisualization environment. In *Proceedings. Eighth International Conference on Information Visualisation*, pages 153–158, 2004.
6. A. C. Robinson, J. Chen, E. J. Meyer, and A. M. MacEachren. Combining usability techniques to design geovisualization tools for epidemiology. *Cartography and Geographic Information Science,*, 32:243–255, 2005.
7. Jinwook Seo and Ben Shneiderman. Interactively exploring hierarchical clustering results. *IEEE Computer*, 35(7):80–86, 2002.

A Novel Framework for Discovering Robust Cluster Results

Hye-Sung Yoon[1], Sang-Ho Lee[1], Sung-Bum Cho[2], and Ju Han Kim[2]

[1] Ewha Womans University, Department of Computer Science and Engineering,
Seoul 120-750, Korea
comet@ewhain.net, shlee@ewha.ac.kr
[2] Seoul National University Biomedical Informatics (SNUBI), Seoul National
University College of Medicine, Seoul 110-799, Korea
csb1749@snu.ac.kr, juhan@snu.ac.kr

Abstract. We propose a novel method, called heterogeneous clustering
ensemble (HCE), to generate robust clustering results that combine multiple partitions (clusters) derived from various clustering algorithms. The
proposed method combines partitions of various clustering algorithms by
means of newly-proposed the selection and the crossover operation of the
genetic algorithm (GA) during the evolutionary process.

1 Introduction

Data mining techniques have been used extensively as approaches to uncover
interesting patterns within large databases. Of these, genomic researchers are
willing to apply clustering algorithms to gain a better genetic understanding
of and more biological information from bio-data, because there is insufficient
prior knowledge of most bio-data. However, no single algorithm has emerged
as the method of choice within the bio-data analysis community, because most
of the proposed clustering algorithms largely are heuristically motivated, and
the issues of determining the correct number of clusters and choosing a good
clustering algorithm are not yet rigorously resolved [6].

Recent research shows that combining clustering merits often yields better
results, and clustering ensemble techniques have been applied successfully to
increase classification accuracy and stability in data mining [1][3][4]. Still, it remains difficult to say which clustering result is best, because the same algorithm
can lead to different results as a result of repetition and random initialization.
One of the major dilemmas associated with clustering ensembles is also how to
combine different clustering results. Therefore, a new mechanism to combine the
different numbers of cluster results is needed.

In this paper, we want to demonstrate a new heterogeneous clustering ensemble (HCE) method, based on a genetic algorithm (GA) that combines clustering
results from diverse clustering algorithms, thereby resulting in clustering ensemble improvement. We focus on optimizing the information provided by a
collection of different clustering results, by combining them into one final result,
using a variety of proposed methods.

N. Lavrač, L. Todorovski, and K.P. Jantke (Eds.): DS 2006, LNAI 4265, pp. 373–377, 2006.
© Springer-Verlag Berlin Heidelberg 2006

The paper is organized as follows. Section 2 explains the proposed hetero-geneous clustering ensemble method based on genetic algorithm. Section 3 de-scribes the experimental data and significant experimental results obtained by using the newly-proposed method. Finally, Section 4 presents our conclusions.

2   Algorithm

This section explains our proposed heterogeneous clustering ensemble (HCE) method, which functions via the modified application of basic GA operations.

2.1   Genetic Algorithm (GA) for the Better Selection Problem

The following modified genetic operators, Reproduction and Crossover, were applied in this paper.

Reproduction. Once a suitable chromosome is chosen for analysis, it is nec-essary to create an initial population to serve as the starting point for the GA. We applied different types of clustering algorithms to a dataset and constructed a paired non-empty subset with two clusters, among all clustering results of clustering algorithms. For example, one clustering algorithm generates three clusters (1, 2, 3) and the other also generates three clusters (A, B, C) using different parameters. These six clusters are created as an initial population and are comprised of 30 paired non-empty subsets (Figure 1). This natural repro-duction process uses the fitness function as a unique way to determine whether each chromosome will or will not survive. In this paper, we select a pair (subset) with the largest number of highly-overlapped elements among all paired subsets, which is the fitness function to select a pair for the next crossover operation.

For instance, suppose that bio-data with 10 elements, as shown in Figure 2, generate an initial population through the reproduction operation. If two subsets (1, 2) and (1, 3) are selected paired subsets from Figure 1, the first cluster (1, 2, 3) in {(1, 2, 3) (4, 5, 6) (7, 8, 9, 10)} is compared with the other clusters {(1, 2) (3, 4) (5, 6) (7, 8, 9, 10)}. That is, the first cluster (1, 2, 3) and the first cluster (1, 2) from the other cluster results have 2 values to the highly-overlapped than {(3, 4) (5, 6) (7, 8, 9, 10)} clusters, as shown in Figure 2. Moreover, the (4, 5, 6) cluster has a value of 2 to the highly-overlapped with the other cluster (5, 6) and the (7, 8, 9, 10) cluster has a representative value equal to 4, derived by comparing it to the other cluster (7, 8, 9, 10) of {(1, 2) (3, 4) (5, 6) (7, 8, 9, 10)}. This process adds the representative values of each cluster and selects a subset for the crossover operation, by comparing all population pairs. As shown as (A) and (B) in Figure 2, the subsets of (1, 2) and (1, 3) each have 17 and 15, and finally, the (1, 2) subset was selected with greater selection probability.

Crossover. The selected subset produces offspring from two parents, such that the offspring inherit as much meaningful parental information as possible. This operator process is based on [2] where the methodological ideas can be found. These procedures exchange the cluster traits from different cluster results and

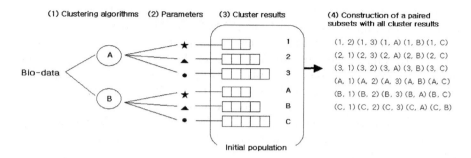

Fig. 1. Initial generation of the population

Elements 10 : 1, 2, 3, 4, 5, 6, 7, 8, 9, 10
(A) Subset (1, 2)
$[\{(1, 2, 3) \ (4, 5, 6) \ (7, 8, 9, 10)\}, \{(1, 2) (3, 4) (5, 6) (7, 8, 9, 10)\}]$
$\Rightarrow [\{(2, 2, 4), (2, 1, 2, 4)\}] \Rightarrow [8, 9] \Rightarrow \mathbf{17}$
(B) Subset (1, 3)
$[\{(1, 2, 3) (4, 5, 6) (7, 8, 9, 10)\}, \{(1, 2) (3, 6) (4, 5) (7, 8) (9, 10)\}]$
$\Rightarrow [\{(2, 2, 2), (2, 1, 2, 2, 2)\}] \Rightarrow [6, 9] \Rightarrow 15$

Fig. 2. Selection method for the evolutionary reproduction process

elements with highly-overlapped and meaningful information being inherited by
the offspring, until finally we achieve an optimal final cluster result.

2.2   Heterogeneous Clustering Ensemble (HCE)

Our proposed HCE method based on the GA operation is as follows.

Algorithm. Heterogeneous Clustering Ensemble (HCE)

Input :

(1) The data set of N data points $D = X_1, X_2,.., X_N$
(2) A set of different clustering algorithms, K_i
(3) The different clustering results, C_j
(4) The clustering result is $S= \{Sk_1c_j, Sk_2c_j,....., Sk_ic_j\}$
    - Sk_ic_j are different numbers of clustering results, C_j, of the ith algorithm

Output : The optimal cluster results on the dataset D

1. Run clustering algorithms K_i on D
2. Construct a paired non-empty subset, $SM^{(g)}$, from the cluster results, S
3. Iterate n until convergence:
    3.1 Compute the fitness function $F(t)$ and select two parents from $SM^{(g)}$
    3.2 Crossover two parents
    3.3 Replace $SM^{(g)}$ parent subsets by newly-created offspring

---

In the present experiment, we aim to identify associations between patients. Therefore, the input data of our algorithm is executed to a vector for each gene, based on patients (samples). The end result demonstrates similar patient clusters. The first stage presented in our algorithm applies different clustering algorithms to the input data. From that result, we construct, $SM^{(g)}$, a paired subset with only two elements from the cluster results, S, of different clustering algorithms. The third stage is the GA application stage within the HCE algorithm. We select two parents as a couple, with the largest number of highly-overlapped elements to fitness function $F(t)$ for crossover manipulation within the population $SM^{(g)}$.

3   Application

This section explains the experimental data and our experimental results.

3.1   The Datasets

We used the published CAMDA 2006 conference dataset. This dataset contains microarray, proteomics, single nucleotide polymorphisms (SNPs), and clinical data for chronic fatigue syndrome (CFS) [5]. In our experiments, both microarray and clinical, were used for application and verification [2].

3.2   Experimental Results

We applied the AVADIS analysis tool to generate cluster results from different clustering algorithms. We also compared the results that were generated using AVADIS to those generated by our proposed method [2]. Table 1 lists the comparisons between the four clusters created by our method, and the four clusters of three different clustering algorithms created by the parameter change. The applied different clustering algorithms - KM, HC, PCA and HCE - indicate k-Means, Hierarchical Clustering, Principal Component Analysis, and our proposed method, respectively.

For validity testing, we used the two categories of clinical datasets: clinical assessment data to determine whether CFS is a single or heterogeneous illness; and actual classified clinical data about the symptomatic groups. Thereafter, we compared the final four clusters that resulted by means of our proposed method with the four cluster results derived using the other clustering algorithms. As shown Table 1, we chose a representative symptomatic group from every cluster results of the two categories data. The most similar representative values between the two categories are written in **bold** typeface. As a result, we found that our HCE method mostly agrees with the clusters classified by the two categories of clinical data. Here, L/M and M/W are said to cluster in the same ratio as the number of patients classified symptomatically as least/moderate and moderate/worst.

Table 1. Cluster results comparison of three clustering algorithms and the HCE method

Algorithms	KM				HC				PCA				HCE			
Cluster results #	1	2	3	4	1	2	3	4	1	2	3	4	1	2	3	4
Category 1	L	M	L	L/M	L/M	L	M	L	M	L	M	M	L	M	M	L
Category 2	W	W	W	L	L	L	W	W	M	W	W	M	L	M	L/M	L

4 Conclusions

Experimental results using a real dataset prove that our method can search effectively for possible solutions and improve the effectiveness of clusters. Combining different clustering algorithms by considering bio-data characteristics and analysis of cluster results, also can overcome the instability inherent in clustering algorithm problems. And our proposed HCE method improves its performance as the number of iterations increase. Moreover, we need not remove elements for preprocessing, nor fix the number of clustering during the first application step, as required by existing clustering algorithms, because the GA is rapidly executed. Therefore, it can extract more reliable results than other clustering algorithms.

References

1. Greene, D., Tsymbal, A., Bolshakova, N., Cunningham, P.: Ensemble clustering in medical diagnostics. Proceedings of the 17th IEEE Symposium on Computer-Based Medical Systems, (2004) 576–581
2. Hye-Sung, Y., Sun-Young, A., Sang-Ho, L., Lee, Sung-Bum, C., Ju Han, K.: Heterogeneous clustering ensemble Method for combining different cluster results. Proceedings of Data Mining for Biomedical Applications, PAKDD Workshop BioDM, **LNBI 3916** (2006) 82–92
3. Jouve, P. E., Nicoloyannis, N.: A new method for combining partitions, applications for distributed clustering. Proceedings of the International Workshop on Parallel and Distributed Machine Learning and Data Mining, (2003)
4. Qiu, P., Wang, Z. J., Liu, K.J.: Ensemble dependence model for classification and prediction of cancer and normal gene expression data. Bioinformatics, **21** (2005) 3114–3121
5. Whistler, T., Unger, E. R., Nisenbaum, R., Vernon, S. D.: Integration of gene expression, clinical, and epidemiologic data to characterize Chronic Fatigue Syndrome. Journal of Translational Medicine, **1** (2003)
6. Xiaohua, H.: Integration of cluster ensemble and text summarization for gene. Proceedings of the 4th IEEE Symposium on Bioinformatics and Bioengineering, (2004) 251–258

Gene Selection for Classifying Microarray Data Using Grey Relation Analysis

Li-Juan Zhang[1] and Zhou-Jun Li[2]

[1] Computer School, National University of Defence Technology, Changsha, 410073, China
[2] School of Computer Science & Engineering, Beihang University, Beijing, 100083, China
nudtzlj@126.com, zhoujun.li@263.net

Abstract. Gene selection is a common task in microarray data classification. The most commonly used gene selection approaches are based on gene ranking, in which each gene is evaluated individually and assigned a discriminative score reflecting its correlation with the class according to certain criteria, genes are then ranked by their scores and top ranked ones are selected. Various discriminative scores have been proposed, including t-test, S2N,RelifF, Symmetrical Uncertainty and x^2-statistic. Among these methods, some require abundant data and require the data follow certain distribution, some require discrete data value. In this work, we propose a gene ranking method based on Grey Relational Analysis (GRA) in grey system theory, which requires less data, does not rely on data distribution and is more applicable to numerical data value. We experimentally compare our GRA method with several traditional methods, including Symmetrical Uncertainty, x^2-statistic and ReliefF. The results show that the performance of our method is comparable with other methods, especially it is much faster than other methods.

Keywords: gene selection, gene ranking, grey relational analysis.

1 Introduction

Recent advanced technologies in DNA microarray analysis are intensively applied in disease classification, especially for cancer classification. Different classification approaches have been used to analyze the microarray data [1],[2],[3]. However, classification based on microarray data is very different from previous classification problems in that the number of features (typically tens of thousands) greatly exceeds the number of instances (typically less than one hundred). It is thus important to first apply feature selection methods prior to classification. It has been shown that selecting a small set of informative genes can lead to improved classification accuracy [1]. The most commonly used gene selection approaches are based on gene ranking, in which each gene is evaluated individually and assigned a discriminative score reflecting its correlation with the class according to certain criteria, genes are then ranked by their scores and the top ranked ones are selected. Various discriminative scores have been proposed, including t-test [4], S2N ratio [2], ReliefF[5], symmetrical uncertainty[6] and etc. These methods are widely used in microarray data analysis and proved to be effective and efficient.

N. Lavrač, L. Todorovski, and K.P. Jantke (Eds.): DS 2006, LNAI 4265, pp. 378–382, 2006.
© Springer-Verlag Berlin Heidelberg 2006

However, some (such as t-test) of these methods are based on statistics, which requires plenty of data and requires the data follow certain distribution, some (such as symmetrical uncertainty) require discrete data value. Unfortunately, these requirements are apparently not satisfied by microarray data, in which the number of tissue samples is very limited, and the data value is numerical. Here we propose a method based on Grey Relational Analysis (GRA) from grey system theory [7], which requires less data, does not rely on data distribution and is more applicable to numerical data value.

The remainder of this paper is organized as follows. In section 2, we introduce the basic knowledge of GRA. In section 3, we describe our GR-GRA method. Section 4 contains the experimental evaluation. Section 5 concludes this work.

2 Grey Relational Analysis

Grey system theory was put forward by Deng [7] in 1982. It is specially designed for handling situations in which only limited data are available and has been becoming very popular in many areas such as image coding, pattern recognition, and etc.[8],[9]. GRA is one of the most important parts of grey system theory. GRA can measure the mutual relationships among factors of grey system. It is based on the level of similarity and variability among all factors to establish their relation. As compared to traditional methods, there are three advantages in GRA: (1) Large amount of data sample is not needed; (2) No specific statistical data distribution is required; and (3)easy construction and strong discriminable ability are the unique merit of GRA. For computation of GRA, three concepts (grey relational coefficient, grey relational grade and grey relational ordering) are vital. They are described as follows.

Grey Relational Coefficient

Let $X=\{X_i \mid i \in I\}$ be a space sequence where $I=\{1,2,...,n\}$. X_i is a factor of the system, and its value at the k-th item in the sequence is $x_i(k),k=1,2,...,m$. If we denote the reference sequence X_0 by $(X_0(1), X_0(2),..., X_0(m))$, the compared sequence X_i by $(X_i(1), X_i(2),..., X_i(m))$, then the grey relational coefficient between X_0 and X_i at the kth-item is defined to be:

$$\gamma_{0,i}(k) = \frac{\Delta_{min} + \beta\Delta_{max}}{\Delta_i(k) + \beta\Delta_{max}} \qquad (1)$$

Where $\Delta_i(k)=\mid X_0(k)-X_i(k)\mid$ is the absolute difference of the two comparing sequences X_0 and X_i. $\Delta_{max} = \max_{i \in I} \max_k \Delta_i(k)$ and $\Delta_{min} = \min_{i \in I} \min_k \Delta_i(k)$ are respectively the maximum and minimum values of the absolute differences of all comparing sequences, and $\beta \in [0,1]$ is a distinguishing coefficient, usually let it be 0.5, the purpose of which is to weaken the effect of Δ_{max} when it gets too big, and enlarges the difference significance of the relational coefficient. $\gamma_{0,i}(k)$ reflects the degree of closeness between the two comparing sequences at k. At Δ_{min}, $\gamma_{0,i}(k) = 1$, that is, the relational coefficient attains its largest value. While at Δ_{max}, $\gamma_{0,i}(k)$ attains the smallest value. Hence $0 < \gamma_{0,i}(k) \leq 1, \forall i$

Grey Relational Grade

The relational grade of two comparing sequence can be quantified by the mean value of their grey relational coefficients:

$$\gamma_{0,i} = \frac{1}{m} \sum_{k=1}^{m} \gamma_{0,i}(k) \tag{2}$$

Here, $\gamma_{0,i}$ is designated as the grey relational grade between X_0 and X_i

Grey Relational Ordering

In relational analysis, the practical meaning of the numerical values of grey relational grades between elements is not absolutely important, while the grey relational ordering between them yields more subtle information. If $\gamma_{0,i} > \gamma_{0,j}$, we say X_i to X_0 is better than X_j to X_0. If $\gamma_{0,i} < \gamma_{0,j}$, we say X_i to X_0 is worse than X_j to X_0. If $\gamma_{0,i} = \gamma_{0,j}$, we say X_i to X_0 is worth equally this X_j to X_0.

3   Gene Ranking Using Grey Relational Analysis

A microarray data set can be represented by a numerical-valued expression matrix $M = \{w_{ij} \mid 1 \leq i \leq n, 1 \leq j \leq m\}$, where w_{ij} is the measured expression level of gene i in sample j. Each row corresponds to one particular gene and each column to a sample. For classification of microarray data, each sample has a class label. Let $C = (c_1,...,c_m)$ denote the class, we look C as the reference sequence, gene $g_i(1 \leq i \leq n)$ as the compared sequence. Then the discriminative score of a gene g_i is no other than the grey relational grade between g_i and C, and it can be computed by formula 1 and 2. By the principle of Grey Relational Ordering, a gene g_i is better than g_j if and only if the grey relational grade between g_i and C is larger than that of g_j.

GRA requires the value for all gene expression sequences be non-dimension and equal-scale. Therefore, it is indispensable to preprocess the data before grey relational analysis can be performed. We here use *mean value processing* method from grey system theory: namely, we first compute the mean values of all the primitive gene expression sequences $g_1,...,g_n$. Then we use these mean values to divide values of the corresponding sequences to obtain a collection of new sequences.

Based on the methodology presented before, we develop an algorithm, named GR-GRA (Gene Ranking using Grey Relational Analysis). It can be described as follows:(1)Perform data preprocessing using *mean value processing* method.(2)Compute the grey relational grade of each gene with respect to the class attribute.(3)Order the genes in descending grey relational grade value. And select top k genes as the optimal gene subset. The value of k is predefined by the user.

4   Experiments and Results

In this section, we empirically evaluate the efficiency and effectiveness of our method on gene expression microarray data. We choose three gene ranking methods (ReliefF, χ^2-statistic and Symmetrical Uncertainty) in comparison with our GR-GRA. In addition, we also select two well-known classifiers, namely the decision tree learner C4.5 and the simple Bayesian classifier naïvebayes to demonstrate the advantages and disadvantages of different gene ranking methods.

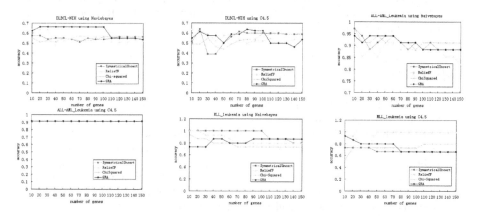

Fig. 1. Comparison of classification accuracy using different methods

Table 1. Comparison of average running time using different methods

	GRA		SymmetricalUnc		ReliefF		ChiSquared	
	C4.5	bayes	C4.5	bayes	C4.5	bayes	C4.5	bayes
DLBCL-NIH	0.74s	0.63s	2.64s	2.46s	40.34s	41.01s	2.89s	2.55s
ALL-AML	0.48s	0.22s	0.84s	0.86s	3.36s	3.30s	0.95s	0.90s
MLL	0.54s	0.53s	2.59s	2.55s	12.67s	12.70s	2.68s	2.64s

Our experiments are conducted using Weka's implementation of all these existing algorithms and GR-GRA is also implemented in Weka environment [11]. In our experiments, we used the following three published data sets: MLL_leukemia [12] (12582 genes, 72 samples in three classes), DLBCL-NIH [13] (7399 genes, 240samples in two classes) and ALL-AML_leukemia [2] (7129 genes, 72 samples in two classes).

For each data set, we compare the classification accuracy on testing data set and the average run time of GR-GRA with other gene ranking methods, including Symmetrical Uncertainty, χ^2-statistic and ReliefF, when the top 10,20, …,150 genes are selected. Figure 1 shows the classification accuracy of different methods, and Table 1 records the average running time of each method. It can be seen from Table 1 that GR-GRA is much faster than other methods. And Figure 1 tells us that the classification accuracy using GR-GRA is comparable with other methods.

5 Conclusions and Future Work

In this work, we propose a new gene ranking method based on GRA. Our method is simple and fast, requires less data, does not rely on data distribution and is more applicable to microarray data. We have experimentally compared our proposed method with other three methods, and the results suggest that GR-GRA is much faster than others and the accuracy of GR-GRA is comparable with other methods.

As a gene ranking method, GR-GRA does not take into account the interaction between genes. We are studying the redundancy among genes and trying to reduce it using GRA. In the near future, we will present an improved GR-GRA method, which will limit redundancy in resulting gene set.

References

1. Lu, Y. and Han, J. (2003): Cancer classification using gene expression data. Information Systems, 28(4), pp. 243-268.
2. T. R. Golub, D. K. Slonim, P. Tamayo, et al. Molecular classification of cancer: Class discovery and class prediction by gene expression monitoring. *Science*, 286(5439):531.537, 1999.
3. Cho, S.B. and Won, H.H. (2003): Machine learning in DNA microarray analysis for cancer classification. Proc. of the First Asia Pacific Bioinformatics Conference, Adelaide, Australia, 19, pp. 189-198, Australian Computer Society.
4. T.H. Bø and I. Jonassen. New feature subset selection procedures for classification of expression profiles. *Genome Biology*, 3(4):research0017.1–0017.11, 2002.
5. Kononenko, I. (1994). Estimating attributes: analysis and extensions of Relief. In De Raedt, L. and Bergadano, F., editors, *Machine Learning: ECML-94,* pages 171-182. Springer Verlag.
6. L. Yu and H. Liu. Redundancy based feature selection for microarray data. In *Proc. Of SIGKDD*, 2004.
7. J. Deng, " Control systems of grey systems", *Systems and Control Letter*, 1982, 5, pp. 288-294.
8. Cheng-Hsiung Hsieh, "Grey image hiding", *The Journal of Grey System*, 2000, 3, pp. 275-282.
9. Y.T. Hsu, "High noise vehicle plate recognition using grey system", *The Journal of Grey System*,1998, 10, pp. 193-208.
10. David K.W.Ng. Grey system and grey raltional model. ACM SIGICE, October 1994.
11. I. Witten and E. Frank. *Data Mining – Pracitcal Machine Learning Tools and Techniques with JAVAImplementations*. Morgan Kaufmann Publishers, 2000.
12. Scott A. Armstrong. Et al "MLL Translocations Specify A Distinct Gene Expression Profile that Distinguishes A Unique Leukemia". Nature Genetics,30: 41-47, January 2002.
13. Andreas Rosenwald, et al. The use of molecular profiling to predict survival after chemotherapy for diffuse large-B-cell lymphoma. The New England Journal of Medicine, 346(25): 1937-1947, June 2002.

Author Index

Lecture Notes in Artificial Intelligence (LNAI)

Vol. 4087: F. Schwenker, S. Marinai (Eds.), Artificial Neural Networks in Pattern Recognition. IX, 299 pages. 2006.

Vol. 4068: H. Schärfe, P. Hitzler, P. Øhrstrøm (Eds.), Conceptual Structures: Inspiration and Application. XI, 455 pages. 2006.

Vol. 4065: P. Perner (Ed.), Advances in Data Mining. XI, 592 pages. 2006.

Vol. 4062: G. Wang, J.F. Peters, A. Skowron, Y. Yao (Eds.), Rough Sets and Knowledge Technology. XX, 810 pages. 2006.

Vol. 4049: S. Parsons, N. Maudet, P. Moraitis, I. Rahwan (Eds.), Argumentation in Multi-Agent Systems. XIV, 313 pages. 2006.

Vol. 4048: L. Goble, J.-J.C.. Meyer (Eds.), Deontic Logic and Artificial Normative Systems. X, 273 pages. 2006.

Vol. 4045: D. Barker-Plummer, R. Cox, N. Swoboda (Eds.), Diagrammatic Representation and Inference. XII, 301 pages. 2006.

Vol. 4031: M. Ali, R. Dapoigny (Eds.), Advances in Applied Artificial Intelligence. XXIII, 1353 pages. 2006.

Vol. 4029: L. Rutkowski, R. Tadeusiewicz, L.A. Zadeh, J.M. Zurada (Eds.), Artificial Intelligence and Soft Computing – ICAISC 2006. XXI, 1235 pages. 2006.

Vol. 4027: H.L. Larsen, G. Pasi, D. Ortiz-Arroyo, T. Andreasen, H. Christiansen (Eds.), Flexible Query Answering Systems. XVIII, 714 pages. 2006.

Vol. 4021: E. André, L. Dybkjær, W. Minker, H. Neumann, M. Weber (Eds.), Perception and Interactive Technologies. XI, 217 pages. 2006.

Vol. 4020: A. Bredenfeld, A. Jacoff, I. Noda, Y. Takahashi (Eds.), RoboCup 2005: Robot Soccer World Cup IX. XVII, 727 pages. 2006.

Vol. 4013: L. Lamontagne, M. Marchand (Eds.), Advances in Artificial Intelligence. XIII, 564 pages. 2006.

Vol. 4012: T. Washio, A. Sakurai, K. Nakajima, H. Takeda, S. Tojo, M. Yokoo (Eds.), New Frontiers in Artificial Intelligence. XIII, 484 pages. 2006.

Vol. 4008: J.C. Augusto, C.D. Nugent (Eds.), Designing Smart Homes. XI, 183 pages. 2006.

Vol. 4005: G. Lugosi, H.U. Simon (Eds.), Learning Theory. XI, 656 pages. 2006.

Vol. 3978: B. Hnich, M. Carlsson, F. Fages, F. Rossi (Eds.), Recent Advances in Constraints. VIII, 179 pages. 2006.

Vol. 3963: O. Dikenelli, M.-P. Gleizes, A. Ricci (Eds.), Engineering Societies in the Agents World VI. XII, 303 pages. 2006.

Vol. 3960: R. Vieira, P. Quaresma, M.d.G.V. Nunes, N.J. Mamede, C. Oliveira, M.C. Dias (Eds.), Computational Processing of the Portuguese Language. XII, 274 pages. 2006.

Vol. 3955: G. Antoniou, G. Potamias, C. Spyropoulos, D. Plexousakis (Eds.), Advances in Artificial Intelligence. XVII, 611 pages. 2006.

Vol. 3949: F. A. Savacı (Ed.), Artificial Intelligence and Neural Networks. IX, 227 pages. 2006.

Vol. 3946: T.R. Roth-Berghofer, S. Schulz, D.B. Leake (Eds.), Modeling and Retrieval of Context. XI, 149 pages. 2006.

Vol. 3944: J. Quiñonero-Candela, I. Dagan, B. Magnini, F. d'Alché-Buc (Eds.), Machine Learning Challenges. XIII, 462 pages. 2006.

Vol. 3937: H. La Poutré, N.M. Sadeh, S. Janson (Eds.), Agent-Mediated Electronic Commerce. X, 227 pages. 2006.

Vol. 3930: D.S. Yeung, Z.-Q. Liu, X.-Z. Wang, H. Yan (Eds.), Advances in Machine Learning and Cybernetics. XXI, 1110 pages. 2006.

Vol. 3918: W.K. Ng, M. Kitsuregawa, J. Li, K. Chang (Eds.), Advances in Knowledge Discovery and Data Mining. XXIV, 879 pages. 2006.

Vol. 3913: O. Boissier, J. Padget, V. Dignum, G. Lindemann, E. Matson, S. Ossowski, J.S. Sichman, J. Vázquez-Salceda (Eds.), Coordination, Organizations, Institutions, and Norms in Multi-Agent Systems. XII, 259 pages. 2006.

Vol. 3910: S.A. Brueckner, G.D.M. Serugendo, D. Hales, F. Zambonelli (Eds.), Engineering Self-Organising Systems. XII, 245 pages. 2006.

Vol. 3904: M. Baldoni, U. Endriss, A. Omicini, P. Torroni (Eds.), Declarative Agent Languages and Technologies III. XII, 245 pages. 2006.

Vol. 3900: F. Toni, P. Torroni (Eds.), Computational Logic in Multi-Agent Systems. XVII, 427 pages. 2006.

Vol. 3899: S. Frintrop, VOCUS: A Visual Attention System for Object Detection and Goal-Directed Search. XIV, 216 pages. 2006.

Vol. 3898: K. Tuyls, P.J. 't Hoen, K. Verbeeck, S. Sen (Eds.), Learning and Adaption in Multi-Agent Systems. X, 217 pages. 2006.

Vol. 3891: J.S. Sichman, L. Antunes (Eds.), Multi-Agent-Based Simulation VI. X, 191 pages. 2006.

Vol. 3890: S.G. Thompson, R. Ghanea-Hercock (Eds.), Defence Applications of Multi-Agent Systems. XII, 141 pages. 2006.

Vol. 3885: V. Torra, Y. Narukawa, A. Valls, J. Domingo-Ferrer (Eds.), Modeling Decisions for Artificial Intelligence. XII, 374 pages. 2006.

Vol. 3881: S. Gibet, N. Courty, J.-F. Kamp (Eds.), Gesture in Human-Computer Interaction and Simulation. XIII, 344 pages. 2006.

Vol. 3874: R. Missaoui, J. Schmidt (Eds.), Formal Concept Analysis. X, 309 pages. 2006.

Vol. 3873: L. Maicher, J. Park (Eds.), Charting the Topic Maps Research and Applications Landscape. VIII, 281 pages. 2006.

Vol. 3864: Y. Cai, J. Abascal (Eds.), Ambient Intelligence in Everyday Life. XII, 323 pages. 2006.

Vol. 3863: M. Kohlhase (Ed.), Mathematical Knowledge Management. XI, 405 pages. 2006.

Vol. 3862: R.H. Bordini, M. Dastani, J. Dix, A.E.F. Seghrouchni (Eds.), Programming Multi-Agent Systems. XIV, 267 pages. 2006.